Stalin's Great Science
The Times and Adventures of Soviet Physicists

History of Modern Physical Sciences

Aims and Scope

The series will include a variety of books dealing with the development of physics, astronomy, chemistry and geology during the past two centuries (1800–2000). During this period there were many important discoveries and new theories in the physical sciences which radically changed our understanding of the natural world, at the same time stimulating technological advances and providing a model for the growth of scientific understanding in the biological and behavioral sciences.

While there is no shortage of popular or journalistic writing on these subjects, there is a need for more accurate and comprehensive treatments by professional historians of science who are qualified to discuss the substance of scientific research. The books in the series will include new historical monographs, editions and translations of original sources, and reprints of older (but still valuable) histories. Efforts to understand the worldwide growth and impact of physical science, not restricted to the traditional focus on Europe and the United States, will be encouraged. The books should be authoritative and readable, useful to scientists, graduate students and anyone else with a serious interest in the history, philosophy and social studies of science.

Series Board

Professor Igor Aleksander
Imperial College, London, UK

Professor Stephen G Brush
University of Maryland, USA

Professor Richard H Dalitz
Oxford University, UK

Professor Freeman J Dyson
Princeton University, USA

Professor Chris J Isham
Imperial College, London, UK

Professor Maurice Jacob
CERN, France

Professor Tom Kibble
Imperial College, London, UK

Dr Katherine Russell Sopka
Consultant and Historian of Science

Professor Roger H Stuewer
University of Minnesota – Twin Cities, USA

Dr Andrew C Warwick
Imperial College, London, UK

Professor John Archibald Wheeler
Princeton University, USA

Professor C N Yang
*State University of New York,
Stony Brook, USA*

Forthcoming Title

Critical Edition of the Peierls Correspondence
edited by Dr Sabine Lee, Birmingham University, UK

HISTORY OF MODERN PHYSICAL SCIENCES – VOL. 2

Stalin's Great Science

The Times and Adventures of Soviet Physicists

Alexei B. Kojevnikov
University of Georgia, Athens, USA
Institute for History of Science and Technology, Moscow, Russia

Imperial College Press

Published by

Imperial College Press
57 Shelton Street
Covent Garden
London WC2H 9HE

Distributed by

World Scientific Publishing Co. Pte. Ltd.
5 Toh Tuck Link, Singapore 596224
USA office: Suite 202, 1060 Main Street, River Edge, NJ 07661
UK office: 57 Shelton Street, Covent Garden, London WC2H 9HE

British Library Cataloguing-in-Publication Data
A catalogue record for this book is available from the British Library.

Cover illustration:
"Mass celebration on the occasion of the opening of Second Congress of Comintern, Uritsky square [in front of the Winter Palace – AK], Petrograd, 1921," painting by Boris Kustodiev. Two figures dressed in black jackets are easily identifiable physicists P. L. Kapitza (with the pipe) and N. N. Semenov. The artist had recently met two young and arrogant graduates of the Petrograd Polytechical Institute, who told him that one day they would become very famous. He agreed to paint their portraits, but recycled one of the sketches and included the same characters in a larger painting of an outdoor mass performance, typical of the revolutionary era. Eventually, both Semenov and Kapitza won Nobel Prizes, the former in chemistry for the discovery of chemical chain reactions, the latter in physics for discoveries and inventions related to liquid helium.

STALIN'S GREAT SCIENCE
The Times and Adventures of Soviet Physicists

Copyright © 2004 by Imperial College Press

All rights reserved. This book, or parts thereof, may not be reproduced in any form or by any means, electronic or mechanical, including photocopying, recording or any information storage and retrieval system now known or to be invented, without written permission from the Publisher.

For photocopying of material in this volume, please pay a copying fee through the Copyright Clearance Center, Inc., 222 Rosewood Drive, Danvers, MA 01923, USA. In this case permission to photocopy is not required from the publisher.

ISBN   1-86094-419-1
ISBN   1-86094-420-5 (pbk)

Printed in Singapore by World Scientific Printers (S) Pte Ltd

To the memory of my teacher,
Soviet physicist Igor Yurievich Kobzarev (1932–1991)

Contents

Introduction		xi
Acknowledgments		xvii
Abbreviations		xxi
Note on Transliterations		xxiii
1.	**The Great War and the Invention of Soviet Science**	1
	Science, Industry, and Military in the Late Russian Empire	2
	The War Crisis	5
	The Idea of Research Institutes	12
	The Study of Natural Productive Forces	17
	The Network of State-Sponsored Research and Development	19
2.	**Socialist, or Big, Science**	23
	Revolution and the War's Legacy	24
	From Optical Glass to the Optical Institute	32
	From the Optical Institute to an Optical Industry: Modernization, Soviet Style	37
	The Transition to Big Science	43
3.	**Freedom, Collectivism, and Electrons**	47
	Freedom as Problem	48
	The Origin of the Collectivist Metaphor	52
	Liberated Holes	56
	The Collectivized Electron and the Bloch Electron	59

	The Phonon and Quantum Individuality	64
	Shared Excitation	69
4.	**Lev Landau's *Wanderjahre*, or Theoretical Physics in the Context of Cultural Revolution**	73
	Education of a Soviet Scientist	74
	Rockefeller Philanthropists and Bolshevik Russia	80
	A Quantum Rebel	85
	Theoretics and the Cultural Revolution	88
	The Kharkov School	92
5.	**Scientist under Stalin's Patronage: The Case of Piotr Kapitza**	99
	Expatriate	100
	Prisoner	106
	Missionary	109
	Client	114
	Minister	120
6.	**"To Catch Up and To Surpass…"**	126
	Before Fission	127
	As if it were not about the Bomb	131
	Atomic Secrets	135
	Strategic Choices	141
	Socialist Management at Work	146
	The Bomb and the Fallout	152
7.	**President of Stalin's Academy: The Mask and Responsibility of Sergei Vavilov**	158
	Early Career	160
	Rise with and within the Academy	165
	The Call	170
	Political Profile	175
	Political Writings and the Ideological Image of Soviet Science	179
8.	**Games of Soviet Democracy**	186
	The Campaign of Ideological Discussions in Sciences, 1947–1952	188
	Exercises on the Philosophical Front	191
	Games of Intra-Party Democracy	197
	Opening Pandora's Box	203
	Resolving the Controversy and Defining Consensus	207
	Soviet Ideology and Science	214

Contents

9.	**Modernist Science, Ideological Passions**	217
	Debates about the New Physics	219
	Professors vs. Academicians	226
	Anatomy of a Discussion	229
	Paradigm Shift, Soviet Style	235
	Changing Guards	240
10.	**Collective Excitations**	245
	Electrons Free and Trapped	248
	Arrested Electrons and the Polaron	251
	The Exciton and the Collectivist Alternative to Band Theory	254
	The Roton and Collective Excitations	259
	The Plasmon and the Collective Movement	265
	The Prose of Collectivism	272
11.	**Dialogues about Knowledge and Power in Totalitarian Political Culture**	276
	The Trinity of Higher Knowledge	277
	Elements of Soviet Metaphysics of Knowledge and Power	279
	The Bolshevik Pact with the Specialists, and Its Failure	283
	The Stalinist Pact with the Intelligentsia	287
	Post-War Negotiations about Power	291
	The Post-Stalin Settlement	295

Conclusion	301
Bibliography	307
Name Index	343
Subject Index	349

Introduction

In the midst of World War II, concerned over the fate of thousands of displaced scholars and academic refugees from Central Europe, his native Austria annexed to Hitler's Third Reich, the philosopher Karl Popper (1902–1994) argued from his temporary refuge in New Zealand that science for its normal functioning required political democracy. From observing the commonality that both social institutions are grounded in freedom of critical debate, Popper inferred a causal relationship: "ultimately, [scientific] progress depends very largely on political ... democracy," while, on the other hand, "political power, when it is used to suppress free criticism, or when it fails to protect it, can impair the functioning of [social] institutions, on which all progress, scientific, technological, and political ultimately depends" (Popper 1950, 404; 1957, 155).

With or without reference to Popper, the thesis that science and democracy require and support each other subsequently became one of the main postulates of post-war liberalism, taken as so obvious that justification or debate seemed rather superfluous. Although many must have noticed, few if any stated explicitly that the generalization based on the experience of Nazi-ruled Europe did not work in the case of the Soviet Union. There, precisely in Popper's times, science and technology were making spectacular advances under a political regime that was strongly suppressing critical discourse in politics, and sometimes also in science. The worst decades of Stalin's dictatorial rule were also the period of arguably the greatest progress achieved by science and technology on Russian soil since the time of Peter the Great. That such an impressive cultural achievement took place in the context of a repressive political system, severe ideological orthodoxy, and an almost complete ban on international contact can be called the main paradox of Soviet science.

Coming to terms with this obvious fact has presented a huge problem, however, since during the Cold War neither of the powerful ideological adversaries wanted to acknowledge the paradox; rather, for different reasons, they preferred to consider it nonexistent. While vehemently opposed on many other issues, both Soviet and American officials claimed emphatically that science and democracy were natural allies. Communists had been promoting this thesis since long before Popper, and their attachment to it lasted without wavering from the very beginning to the very end of Soviet power. By communist lights, Soviet policies were as solidly scientific and democratic as science itself, and the paradox simply did not appear to them because the obvious successes of Soviet science merely confirmed in their eyes the superiority of "Soviet democracy."

Anticommunist ideologues denounced the Soviet system as the very opposite of democracy and, by implication, also accused it of being harmful to science. To maintain the latter claim and to see science as suffering rather than prospering at the hands of communists, they had to turn their attention away from the achievements to the weaknesses and failures of Soviet science. The most spectacular example of such a blunder—the Lysenko episode, or the ban on genetics research between 1948 and 1964—alone probably received more coverage in the Western literature than all the rest of Soviet scientific developments combined. This proverbial case so perfectly suited propaganda purposes and was invoked so often that it has acquired symbolic status as the first, and sometimes the only, event that many a general reader and professional historian in the West considered worth knowing from the history of Soviet science. Thousands of references to this selectively chosen example established the standard trope of connecting the failures and problems of Soviet science and technology to the pernicious influences of politics and ideology, while refusing to see the very same forces at work in the cases of achievements and triumphs. One could thus save the existing stereotype and avoid acknowledging the paradox by adopting a vision that was blind in one eye.

The present work chooses, instead, to set aside the stereotype in order to develop a less impaired view of the history of Soviet science in its social, political, and ideological contexts. The book grew out of a series of historical investigations undertaken over the last ten years, some of which were originally published as separate articles. In the course of these studies, I had to gradually abandon a fair amount of wishful thinking that stood in the way of serious understanding of the historical phenomenon of Soviet science. The same, at least to some degree, will also be asked of the reader of this book who wants to follow the argument, which may be easy for some and difficult for others, or perhaps easy and difficult at the same time.

Easy, because some beliefs such as Popper's thesis have in the meantime lost much of their former status as universally valid claims, and historians are no longer constrained by them when writing about the development of science in many other societies and time periods, for example, 17th century French absolutism, 19th century British imperialism or 20th century American militarism. It has indeed been shown that from the beginnings of modern science on, freedom of scientific discourse could be maintained precisely by keeping it separate from the much more dangerous discourse on political, religious, and moral issues (Shapin and Schaffer 1985). Difficult, because the Soviet case is still often judged by special criteria and remains sensitive. Many of the Cold War ideological stereotypes were not laid to rest with the end of the Cold War itself—in fact, they have survived particularly well in the field of Soviet history and continue to underlie the bulk of popular and professional literature in the field. The ubiquity of such inherited prejudices may, for example, make the title of this book appear to some readers as deliberately ironic, when in fact it has to be understood quite literally, as plain talk.

I will be exploring in this book how and why a truly great science developed in Russia and the Soviet Union between the mid 1910s and the mid 1950s—a period of utter social and political turmoil, characterized by the disastrous First World War, the Revolution and its social upheaval, an even more devastating Civil War, the crisis of collectivization and industrialization, the paranoid Stalinist purges, the epic Great Patriotic War against Hitler's Germany, the euphoria of victory, and the subsequent plunge into the Cold War. During much of this period, Stalin's rule not only dominated the politics, but also left its heavy mark on many other aspects of the country's life, including science. In fact, writing about science in the Stalinist era is practically impossible without taking into account its social and political environment, which was neither democratic nor liberal, nor economically prosperous, and definitely unsafe.

To offer a formulaic, one-phrase general answer to the question why science advanced so spectacularly under such circumstances would be to engage once again in delusion. Instead, one needs detailed historical investigation, which here will follow the life trajectories of several important scientists through the turbulence of Soviet history, and try to understand and interpret their work and actions in the context of the politics, ideology, and culture of their time. These stories will throw a light on what opportunities and restrictions existed for science in its specific Soviet setting, and what concrete results, successes, and failures followed. The majority of the case studies are taken from the field of physics, arguably the most developed and in many respects the model science in the Soviet Union. Taking this as a paradigmatic example, one can analyze and explain many of the general characteristics of Soviet science; but when necessary, evidence from other sciences, in particular chemistry, geology, biology, and linguistics, will also be considered.

The first two chapters investigate the origin of a new institutional system of research and development that emerged in Soviet Russia shortly after the Revolution. The roots of that phenomenon can be traced back to the period of World War I, the responses of Vladimir Vernadsky and other Russian academics to the war crisis and to the lack of connections between university research and industry. Their visions of a new national system of science materialized during the subsequent Civil War, thanks to the revolutionary momentum of radical changes and to the establishment of a special pact between the Bolshevik party and research-oriented scientists. The resulting Soviet system of research institutes marked the beginning of 20th century big science with its characteristic nexus between advanced research, government, and the military.

Chapters 3 and 4 describe how the radical scientific breakthroughs of the early 20th century, which included Einstein's relativity theory and quantum mechanics and overturned the existing conceptual order in science, played out in the context of revolutionary Russia of the 1920s. The perceived affinity between the two revolutions, the social and the scientific, abetted the enthusiastic reception of modernist science by many Soviet academics, including Yakov Frenkel and Lev Landau. The cultural revolution of the late 1920s further advanced younger radicals in science and allowed them to institutionalize the new discipline of theoretical physics. The language of the Russian revolution even influenced the way some Soviet theorists defined research problems in the new physics and formulated solutions to them.

The next two chapters explore how science, state patronage, and technological modernization were linked together in the mature Stalinist society. The problem is first analyzed from the perspective of an individual

Introduction

scientist, Piotr Kapitza, who was the most successful at designing a way of establishing close, client-patron relationships with Stalin and other leading politicians, and of promoting his scientific discoveries and industrial inventions. The same question is then investigated in a case where the Stalinist government organized and managed a scientific and technological project of top state priority, namely the production of nuclear weapons. The existing Soviet tradition of big science undertakings proved itself in the process to be perfectly suitable for replicating the American Manhattan Project.

Chapters 7 through 9 discuss the relationship between science and ideology under Stalin, often mistakenly characterized as one of simple conflict and opposition. The actual patterns were almost as complex as in the historical relationship between science and religion—a subject on which simplistic interpretations were also once dominant, but ultimately failed to withstand closer historical investigations. By far the main strategy in the Soviet case was that of a mutual rhetorical adaptation, as illustrated by the case study of Sergei Vavilov, who presided over the Academy of Sciences during its major post-war expansion, was primarily responsible for formulating the public image of Soviet science, and mobilized Marxist dialectics in the service of modern science. Conflicts in and about particular scientific theories, however, could often assume ideological forms, as happened most dramatically in the late 1940s, when an initiative by politicians to stimulate criticism in science coincided with intensified competition among scholars for newly available resources and prestige. A new reconstruction of the sequence of events leading to the infamous 1948 ban on Mendelian genetics will be offered that will also help explain why similar ideological discussions in other fields, in particular linguistics and physics, produced entirely different results.

Chapter 10 explicates how some of the most fundamental concepts in contemporary science—quasiparticles and other collectivist models in condensed matter—had their roots in the socialist worldview and in the collectivist philosophy of freedom. This can be seen as an example of how Soviet ideology contributed constructively to a distinctively socialist approach in science, which has subsequently become universally accepted and has had a major impact on the general development of science in the past century. The final chapter of the book summarizes the history of the reciprocal relationship between power and knowledge in the Soviet Union. Science was not only influenced by, but constituted one of the central elements of the Soviet polity, being an important creator as well as a creation of Soviet civilization. By following this approach one can begin to understand how a society that started with a pact between a revolutionary ruling party and scientific experts ended up in their mutual—and mutually self-destructive—alienation.

As a field of research, the history of Soviet science and technology is still in an early stage, and the overall state of knowledge remains rather fragmentary. Some pieces of the picture are already known in great detail, but many important aspects, trends and episodes remain as yet virtually untouched by historians. Confident as I may be in the conclusions reached in the chapters that follow, I am also aware that additions, modifications, and corrections will be forthcoming in subsequent research. I am hopeful nevertheless that the current book will make a contribution to the emerging new, post-Cold War history of Soviet science, and of the Soviet Union in general.

Acknowledgments

A great many colleagues and friends provided input, encouraging and critical comments, and advice during the research and writing of individual papers that have finally come together in this book. I am grateful to all of them and beg their pardon that only a few are named below, mainly those who have particularly influenced my development as a historian and have been my teachers in the profession.

In Moscow during the 1980s, Igor Yurievich Kobzarev, physicist and polymath, proposed that I write my doctoral dissertation on the history of quantum electrodynamics. His teaching style, consisting of requests to hear an interesting new story about every three months, and initial disagreement with any historical reconstruction I could come up with, was perfect for me. In return, he traded many intriguing stories from the lives of Soviet physicists, thus introducing me to this exciting era in the history of science, the story of which at the time could neither be published nor researched without serious limitations in the Soviet Union. Meanwhile, Vladimir Pavlovich Vizgin taught me some of the most important professional skills of a historian of science, and Igor Serafimovich Alekseev provided a model of philosophical rigor.

At the height of Gorbachev's *perestroika*, around 1990, when official censorship ceased to present an obstacle to inquiries into Soviet history, several freshly minted Ph.D.'s declared themselves representatives of a new field, the social history of Soviet science. They burst through the opening door with all the guts, naiveté, and excitement of that thrilling era. Non-stop collective discussions and brainstorming sessions within this closely-knit group, which at the time included Daniel Alexandrov, Olga Elina, Nikolai Krementsov, Kirill Rossiianov, and Irina Sirotkina, provided me with the most intensive mutual learning experience I have ever had the opportunity to enjoy, one for which I will remain forever nostalgic (Kojevnikov 2002a). Meanwhile, in the works of Loren Graham, David Joravsky and Paul Josephson I came across the first examples of how similar topics were discussed among historians of science in the United States.

In Germany, Helmut Rechenberg kindly suggested that I apply for a Humboldt fellowship, which helped me overcome the Soviet taboo against self-promotion. In the United States, Paul Forman, John L. Heilbron, and Sheila Fitzpatrick have always provided intellectual inspiration and moral encouragement during my years of postdoctoral travels, while Daniel Kevles, Spencer Weart, and Stephen Brush in addition offered institutional support and helped me to understand the culture of American academe. Cathryn Carson was a most attentive and critical reader of many of the papers that eventually became parts of this book. James Barlament and Katherine Livingston read the whole manuscript as it approached its final form and corrected many mistakes in it. I am also particularly indebted to Jessica Wang for being such a wonderful, patient, and supportive partner and a critical colleague at the same time. Without all the aforementioned colleagues, this study would have never been completed.

Some chapters of the book rely, fully or in part, on the following earlier publications:

"The Great War, the Russian Civil War, and the Invention of Big Science," *Science in Context* (2002) 15: 239–275.
"Freedom, Collectivism, and Quasiparticles: Social Metaphors in Quantum Physics," *Historical Studies in the Physical and Biological Sciences* (1999) 29 (2): 295–331.
"Lev Landau: Physicist and Revolutionary," *Physics World* (2002) 15 (6): 35–39.
"Filantropiia Rokfellera i sovetskaia nauka," *Voprosy Istorii Estestvoznaniia i Tekhniki* (1993) no. 2: 80–111.
"Piotr Kapitza and Stalin's Government: A Study in Moral Choice," *Historical Studies in the Physical and Biological Sciences* (1991) 22 (1): 131–164.
"President of Stalin's Academy: The Mask and Responsibility of Sergei Vavilov," *Isis* (1996) 87: 18–50.

"Rituals of Stalinist Culture at Work: Science and the Games of Intraparty Democracy circa 1948," *The Russian Review: An American Quarterly Devoted to Russia's Past and Present* (1998) 57: 25–52.

"Dialogues about Knowledge and Power in Totalitarian Political Culture," *Historical Studies in the Physical and Biological Sciences* (1999) 30 (1): 227–247.

"Dialoge über Macht und Wissen" in *Im Dschungel der Macht: Intellektuelle Professionen unter Stalin und Hitler*, edited by Dietrich Beyrau (Göttingen: Vanderhoeck & Ruprecht, 2000), 45–64.

I am grateful to Blackwell Publishers, Cambridge University Press, the Institute of Physics Publishing Ltd., the University of California Press, the University of Chicago Press, and Vanderhoeck & Ruprecht for permission to reproduce previously published work in this volume.

Over the years, parts of the research that contributed to this study have been supported financially by the following institutions: the Soviet (and later Russian) Academy of Sciences, the Alexander von Humboldt Stiftung, the Max-Planck-Institut für Wissenschaftsgeschichte, the American Institute of Physics, the Volkswagen-Stiftung; the Rockefeller Archive Center, and the National Science Foundation (grant SES-9911008).

Abbreviations

I would like to thank the following archives and their staff for kind assistance:

ARAN—Arkhiv Rossiiskoi Akademii Nauk (Archive of the Russian Academy of Sciences), Moscow

ARAN SPb—Arkhiv Rossiiskoi Akademii Nauk, Sankt-Peterburgskoe Otdelenie (Archive of the Russian Academy of Sciences), St. Petersburg

AHQP—Archive for the History of Quantum Physics, Berkeley

AIP—Niels Bohr Library, American Institute of Physics, College Park, MD

BCL—Birkbeck College Library, University of London

CCA—Rossiiskii Gosudarstvennyi Arkhiv Noveishei Istorii (RGANI); formerly Tsentr Khraneniia Sovremennoi Dokumentatsii (TsKhSD); originally Arkhiv TsK KPSS, or the Central Committee Archive, Moscow

CPA—Rossiiskii Gosudarstvennyi Arkhiv Sotsial'no-Politicheskoi Istorii (RGASPI); formerly Rossiiskii Tsentr Khraneniia i Izucheniia Dokumentov Noveishei Istorii (RTsKhIDNI); originally Tsentral'nyi Partiinyi Arkhiv (TsPA), or Communist Party Archive, Moscow

GARF—Gosudarstvennyi Arkhiv Rossiiskoi Federatsii (Russian State Archive), Moscow

IFP— Institut Fizicheskikh Problem im. P. L. Kapitsy (Archive of the Kapitza Institute of Physical Problems), Russian Academy of Sciences, Moscow

IPPI—Institut Problem Peredachi Informatsii (Institute for Information Transmission Problems), Russian Academy of Sciences, Moscow
NBA—Niels Bohr Archive, Copenhagen
RAC—Rockefeller Archive Center, Pocantico Hills, NY
RGAKFD—Rossiiskii Gosudarstvennyi Arkhiv Kinofotodokumentov (Russian State Archive of Cinema and Photographic Documents), Krasnogorsk
VIA—Voenno-Istoricheskii Arkhiv (Archive of Military History), Moscow

Note on Transliterations

For the most part, I followed the transliteration rules of the Library of Congress, except in the case of a few personal names. Since Russian authors did not obey uniform rules of transliteration when publishing in foreign languages, the bibliography may include different versions of the same Russian name, e.g. Alexandrov and Aleksandrov, Bogolubov and Bogoliubov, J. Frenkel and Ya. Frenkel, Fock and Fok, Joffe and Ioffe, Iwanenko and Ivanenko, Kapitza and Kapitsa, Kojevnikov and Kozhevnikov, Mandelstam and Mandel'shtam, Rossianov and Rossiianov.

Chapter 1

The Great War and the Invention of Soviet Science

During the early years of the 20th century, the prevailing motivation among Russian academic researchers was to catch up with their European colleagues in contributing to the world's body of knowledge in 'pure' sciences, all the while demonstrating sometimes benign but mostly arrogant neglect of practical, 'applied' research. Similar attitudes were common in other European countries as well, but Russian scholars took the ideology of pure science much more seriously and literally.[1] After all, industry offered practically no career opportunities for them, and the only available jobs for scientists in Russia—with very few exceptions—existed at universities and other teaching institutions. The drawbacks of this situation and the degree of disengagement between science, industrial, and in particular military production became obvious with the outbreak of World War I.

The war crisis produced a major shift in scientists' attitudes towards research and its goals. Even university-based scholars started searching for practical and military applications of their knowledge and establishing links with industry. Relatively little could be accomplished in the course of the war itself to compensate for the previous almost total absence of such links, yet the technological inadequacy exposed in the time of a major national crisis stimulated plans and proposals by Russian

---

[1] On British scientists and their ideology in that period, see (Edgerton 1996).

academics for serious changes in the goals and the infrastructure of the country's scientific effort. Their drafts envisioned the recognition of science as a separate profession from teaching, the creation of a network of research institutes, and a turn towards practical, applied research linked to the military and industrial needs of the nation. Some of the key ideas of those proposals, eventually realized, defined the main characteristic features of the post-revolutionary Soviet system of science.

1. Science, Industry, and Military in the Late Russian Empire

"I don't think it is dishonorable for a Russian professor of chemistry to work in the applied direction," Vladimir Markovnikov (1837–1904) of Moscow University defended himself in 1901. Despite the common recognition that chemistry was industrially and economically the most important scientific discipline of the late 19th century, Russian academic chemists took pride in working on "pure" topics. Even though Markovnikov's investigations of the chemical composition of petroleum from Caucasus oil fields had led him to discover important new classes of organic substances, they deviated from the accepted norm and needed a special apology (Solov'ev 1985, 310).

Markovnikov had been drawn into studies of Russian oil by chance twenty years earlier through an invitation to review the state of the country's chemical industry. As no direct industrial statistics were available, he analyzed the existing data on foreign trade and concluded that national industry was capable of producing only the most primitive of chemicals while almost all products that required special chemical knowledge or expertise were imported from abroad. The rapidly developing Moscow textile industry in particular depended on imports not only of synthetic dyes but even of soda from Germany. It was the "fate of all nations who are culturally less developed than others" to be economically disadvantaged when new artificial products are developed in more advanced countries, warned Markovnikov in his 1879 public lecture "Modern Chemistry and Russian Chemical Industry." Having concluded that "one can hardly expect from Russian industry any stimulus for the development of chemistry in Russia," he argued in favor of protectionist tariffs to encourage the production of more sophisticated chemicals (Markovnikov [1879] 1955, 646, 666). After the lecture Markovnikov was approached by the entrepreneur V. I. Ragozin, who offered financial support for a research project on petroleum from Baku.

In contrast with the sorry state of the chemical industry, academic research in chemistry—as well as in mathematics and physiology—achieved a very high level of development in late Imperial Russia. By the end of the century, most universities and engineering schools equipped advanced chemical laboratories for their professors' research (Lomonosovsky 1901). In Markovnikov's judgment, Russian chemists were "sometimes ahead of others," even though their studies had a "predominantly theoretical character." His observation resembled typical British complaints of the time: the Russian chemist Nikolai Zinin discovered how to synthesize anilin, which was then used by the British chemist William Perkin to synthesize the first artificial dyestuff, but it was Germany that took all the profits from monopolizing the world's industrial application of these discoveries. Markovnikov complained further that Russian industry suffered from a severe shortage of qualified technicians and at the same time had no jobs to offer to chemists with university diplomas (Markovnikov [1879] 1955, 642, 648).

Even Markovnikov, however, paid tribute to the common cult of his Russian and German peers by insisting that university instruction in chemistry should deviate in no way from the strict norms of pure science, lest its level be compromised. He waited for industry to become sophisticated enough to engage in useful interaction with academic science rather than looked for the latter to step down from its pedestal (Markovnikov [1879] 1955, 672–675). Twenty years later, around 1900, the debate was still alive in Russia over whether the country should industrialize or remain agricultural, even as industrialization was already developing apace with foreign capital and imported technologies, but without a single advanced industrial laboratory or any significant contact with academic science. Markovnikov's focus on applied topics remained, as before, a controversial exception to the dominant attitude among his colleagues. Russian academic chemists enjoyed a great international reputation, and petroleum exports were rising rapidly, but the chemical industry accounted for a pitiful 2.1 percent of the country's total industrial production (Kovalevsky 1900, 242).

Another distinct exception to the prevailing norm was Dmitry Mendeleev (1834–1907), Markovnikov's colleague from St. Petersburg, famous for his invention of the periodic table of chemical elements. A very unconventional academic, Mendeleev ventured far outside his special field of inorganic chemistry, publishing and advising government officials and private individuals on various matters of industrial production, metrology, technology, economics, foreign trade—with regard to which he also favored and helped introduce protectionist tariffs—and politics. Even more unusual for a university professor, Mendeleev also did some important research for the military. Ivan Chel'tsov, chemistry professor at the School of Mine Officers in St. Petersburg, asked him to help the Russian Navy discover the secret of the latest French invention, smokeless gunpowder. Hoping to learn some secrets from the allies, Mendeleev undertook a special trip to France and Britain, reporting afterwards that the French did not want to share much. The British were more open, but their powder was no good. What little he had learned enabled Mendeleev to start experiments in his university laboratory and eventually to offer the Navy his own version of smokeless gunpowder, together with a piece of advice:

> The safe and timely achievement of the goal of providing the Russian Navy with appropriate types of smokeless gunpowder is possible only with the help of an independent scientific and practical study of the problem in Russia, whereby all details would have to be developed by us ourselves, and the appropriate temporary secrecy maintained at the level with which similar activities are pursued in England, France, and Germany (Dmitriev 1996, 137).

The Navy agreed to establish a Scientific-Technical Laboratory in 1891 with which Mendeleev was affiliated until 1895, yet his efforts proved abortive. Mendeleev's version of smokeless gunpowder was tested in 1893, but the rivalry between the Navy and the War ministries impeded its full-scale industrial production (Gordin 2001, ch. 7). By the time of Mendeleev's death in 1907, the project had been abandoned for at least three years, for which Russia paid dearly during the Great War, buying Mendeleev-invented gunpowder from the United States. As for the War Ministry, it did not have any research facility until December 1914—several months into the full-scale European war—when it established its own Central Scientific-Technical Laboratory (Nauka 1920, 117–118).

The situation with applied research at military schools had not advanced significantly further. As a young officer studying at the Grand Duke Mikhail Artillery Academy in St. Petersburg, Vladimir Ipatieff (1867–1952) nourished aspirations to become a chemist. Among his Academy professors was one famous scholar—metallurgist Dmitry Chernov—who had made fundamental discoveries in the fields of steel production and gun manufacturing, but teaching and occasional consulting rather than original research occupied the rest of the faculty. As there was no chemical laboratory to speak of, Ipatieff assembled a small private laboratory in his apartment. Upon graduation in 1892, he became the Academy's instructor in chemistry and—in order to produce the dissertation necessary for the promotion to professorial rank—attended a laboratory at St. Petersburg University. Ipatieff's request for funding at the Artillery Academy was initially turned down by a superior, who

> explained in a soft, insinuating voice, with which he was used to overruling his opponents, why the chemical laboratory never received more than half of its allotted money.... He said there was no reason why the Academy should appropriate money for a laboratory which had not produced a single scientific investigation in ten years and whose only published dissertation did not describe a single experiment (Ipatieff 1946, 59, 102–109).

Only after Ipatieff had defended his dissertation and been promoted in 1899 to become the Academy's first Professor of Chemistry and Explosives did he manage to organize the chemical laboratory properly. There he soon discovered a new class of catalytic organic reactions occurring under high temperatures with iron as a catalyst, which brought him recognition and fame in the academic world and opened the way for his subsequent groundbreaking studies of chemical catalysis. Even though he taught at a military school and rose to the rank of general by 1910, Ipatieff regarded his laboratory research as "pure organic chemistry," made little if any effort to put his discoveries to use, either military or civilian, and blamed industry for the lack of interest:

> Unfortunately, the Russian chemical industry was too immature to use the scientific discoveries even then available. I still did not bother to take out patents, and once told one of my friends that I was a scientist and wanted complete freedom in my work, which I would not have if I had to be concerned with patents. Had I been a German chemist, I should probably have been infected by the same patent disease as were others. The German chemical industry made full use of my data at no cost to itself (Ipatieff 1946, 174, 178).

Ipatieff recalled that his attitude towards industrial applications started changing in 1913, after a German engineer took out a patent on his discoveries. A major shift in his understanding of what it was to be a scientist, however, occurred as a result of the Great War.

2. The War Crisis

Those who watched Russian industry perform during the first months of the Great European War could not escape the conclusion that the country's degree of economic, industrial and scientific dependence upon

Germany was intolerable, bordering on colonial. This was not surprising in the fields of high technology, such as machines and chemicals, where less than half of the needed products were manufactured in Russia (Grinevetsky 1919, 33). But even in industries that could have relied entirely on native materials and supplies some essential parts had to be imported. When the border with Germany closed in August 1914, chaos ensued in the Russian industry, which was unable to find or quickly produce substitutes for previously imported goods. Shocked by this degree of dependence, some observers even suspected a pre-war German conspiracy behind the arrangement (Novorussky 1915). Although diplomats had been expecting a showdown with Central powers for years, plans for the wartime mobilization of industry had not been prepared. The crisis and shortages at the start of the war made many in Russia—as in other countries—envy the German war-oriented management of industries and demand centralized and rational planning of the economy several years before this principle would be declared 'socialist' by the revolutionary Bolshevik government.[2]

Foreign investors dominated the Russian civilian industry, with the usual consequence that it relied on imported technologies and know-how rather than on independent research and expertise. Military industry and major munitions factories were owned by the state, yet even in this traditional field of governmental concern, the prevailing strategy had been buying and copying foreign innovations. As one economic historian has concluded, while some of these factories "yielded to no one in the quality of their product and stood the test of international comparison and competition... there is no sign that the state sector was the locus of technical innovation or innovation in management style" (Gatrell 1994, 258). The Putilov, Obukhov, and Okhta military factories established modest laboratory facilities during the war, but they were used primarily for routine control of production (Bastrakova 1973, 45). General Aleksei Manikovsky, who during the war was responsible for the supplies of the Russian army, could only complain that "Germany had supplied the entire world, including Russia, with tools of war, and we had paid our money for the development of expensive German military industry" (Manikovsky 1920, 237).

Once the war broke out, the military followed its traditional instincts and turned to allies and neutrals with requests for supplies and technology. Hastily Manikovsky tried to place orders for military equipment in

---

[2] See the analysis and comparison of mobilization-style economies in Germany, Russia, and other belligerent countries during the Great War in (Bukshpan 1929).

Japan and the United States, an effort that resulted in huge expenditures and limited satisfaction. A year later he came to consider those purchases a mistake and to think that much better results could have been achieved had the resources been directed to the development of native industrial production from the very beginning of the war: "After having spent more than 300 million rubles on foreign automobiles, we now [November 1915] came to the decision to develop our own manufacturing" (Manikovsky 1920, 248). Military officials also gradually recognized the need to develop or greatly expand the production of aircraft, chemicals, radios, optical devices, and other war-related products requiring cutting-edge scientific knowledge and expertise.

The extreme shortage of shells was considered the major cause of the Russian Army's difficulties during the first year of the war. To deal with the problem, the Army's Chief Artillery Administration established a commission for the procurement of munitions. The commission started its work by arranging for major purchases of toluen and crude benzol abroad, primarily in the United States, rather than developing their manufacture in Russia. Ipatieff, who was appointed a member of the commission, recalled that in the early period of the war "[t]he mood in general was one of pessimism with a lack of confidence in our own forces and a feeling of inferiority in the face of German technology" (Ipatieff 1946, 196). That initial decision was regretted and was reversed in 1915 when the commission ordered the construction of the first state benzol plant in the Donets coal basin. The plant started producing in September 1915, and its inauguration was followed immediately by the construction of some twenty more state and private factories and by further work on the production of other war-needed chemicals: benzene, toluen, trotyl, and xylene.[3]

On 26 January 1915 professor-chemist and General Grigory Zabudsky, the commander of the newly created Central Scientific-Technical Laboratory of the army, called a meeting to discuss the technical component of the war. Among other topics, the meeting briefly considered the use of "suffocating and intoxicating gases in shells," which the majority of officers opposed arguing that "such methods can be regarded inhuman and have not been previously used by the Russian army." Still, Zabudsky

---

[3] On the 'shell crisis' and attempts to purchase materials abroad see (Stone 1975, ch. 7). On wartime production see (Ipatieff 1946) and "Materialy o nalichii proizvodstva i rasshirenii proizvodstva vzryvchatykh veshchestv i promezhutochnykh produktov, 1915" Voenno-istoricheskii arkhiv, Moscow (hereafter VIA) 507-2-37.

did not completely rule out possible future uses of poison gases "in case of the enemy's gross abuse of such methods." He ordered the laboratory's department of powder and explosives to conduct research on appropriate substances in order to be ready "in case of an emergency, to start production."[4] The emergency was not long in coming. In late May 1915, one month after the first massive attack with poison gas on the Western front, the Germans used chemical weapons also in the East, at Rawka near Warsaw, causing the Russian army about 9,000 casualties, more than 1,000 of which were fatalities. Only then did the Chief Artillery Administration organize a special Commission on Poison Gases for the production of liquid chlorine, phosgene and other gases for use in shells (Haber 1986, 36–39; Ipatieff 1946, 197–215).

The commissions on explosives and on poison gases merged in 1916 to form the War Chemical Committee, which later added three more departments: Incendiaries and Flame Throwers, Gas Masks, and Acids. The first tests of shells filled with chlorine took place in June 1915. Industrial production of chlorine started in early 1916 and of phosgene later that year. The committee decided to abstain from using cyanide-containing substances unless the Germans used them first, but went ahead with producing chloropicrin—tear gas—and a few other chemicals.[5] Driven by the huge military demand, Russian industrial production swelled during the war years. Employment in the chemical industry rose between 1913 and 1917 from 33,000 to 117,000 workers (Gatrell 1986, 185; Strumilin 1935). The crisis with the production of explosives and shells was resolved, at least partly, by 1916 (Barsukov 1938, 351). Hundreds of tons of poisonous chemicals were also produced, but, according to the available statistics, by April 1917 most of them remained at storage and production sites with only the miniscule amount of 138 pud (just over 2 tons) of liquid chlorine actually delivered to the front.[6] It thus does not seem likely that the

---

[4] Zabudsky to the Secretary of the General Staff M. A. Beliaev, 29 January 1915. (VIA. 507-3-192, pp. 1–2).
[5] "Pervye ispytaniia snariada s khlorom, 11-12.6.1915," (VIA, 507-3-192, pp. 43–44); "Khod rabot komissii po zagotovke udushlivykh sredstv, 1915–1916" (VIA, 507-3-1, p. 26); "Doklady o khode rabot komissii po zagotovleniiu udushaiushchikh sredstv, 1915–1917," (VIA, 507-3-2).
[6] For statistical data on Russian production of war-related chemicals in 1916–1917 see (Bukshpan 1929, 362–366). For data on the production and delivery of suffocating substances see "Perepiska i doneseniia ob izgotovlenii i perevozkakh udushaiushchikh sredstv, 1915–1917," (VIA, 504-16-20, p. 590).

already decaying Russian army had a chance to use any substantial amount of chemical weaponry.

The war changed much for the Russian academic community as well. It broke scientific communication and contact with colleagues from other belligerent nations, resulting in virtual scientific isolation that lasted about six years, till the end of the Civil War in 1920. No other problem of that period—not even enormous economic and political hardships—caused so

V. N. Ipatieff in 1916, then a Lieutenant General serving on the War Chemical Committee of the Russian Imperial Army. The war reoriented Ipatieff's work in chemistry from research on academic topics to the organization of industrial production of munitions and other military supplies. Later Ipatieff became a chief organizer of the Soviet chemical industry and military research and occupied responsible posts in the revolutionary Bolshevik government. After 1930 he lived and worked in the United States.
[*Courtesy*: Northwestern University Archives.]

many complaints among Russian scientists, but none also contributed so much to the development of their identity as a national community. Whereas before the war most Russian research was published in foreign and foreign-language periodicals, the war years saw the establishment of national scientific societies in those fields which were still lacking them in Russia and an upsurge in the number of Russian-language academic journals (Aleksandrov 1996).

Perhaps even more important, the war crisis led to a major shift in attitudes towards research and its goals. While Ipatieff and his military peers at the War Chemical Committee were building and mobilizing the chemical industry, civilian chemists started searching for ways to make their contributions to the war effort.[7] Aleksei Chichibabin, chemistry professor at Moscow Higher Technological School, published newspaper appeals to chemists inviting them to join research on medicaments, and to industrialists, arguing that in order to achieve economic independence from Germany "[t]he Russian chemical industry, from the very beginning, must find its basis in Russian science ... and take care of the establishment of most favorable conditions for the quickest and widest development of Russian chemical science" (Chichibabin 1914; 1915). Chichibabin put his laboratory to work on alkaloids for the needs of the pharmaceutical industry and started developing methods for the production of opium, codeine, morphine, aspirin, and other medicaments whose importation had stopped during the war. About 30 volunteers—chemists and chemistry students—joined him in this effort, and in March 1916 the Council of Ministers approved the establishment of an experimental pharmaceutical factory adjacent to the Moscow Higher Technological School for the production of war medicaments (Evteeva 1958, 332–334).

Probably the single most important contribution by a Russian academic scientist to the war effort came from Ipatieff's scientific rival, Nikolai Zelinsky of Moscow University. In 1915 Zelinsky started working on so-called "passive chemical warfare," or protection against poison gases. By the fall of that year he proposed using activated charcoal and developed appropriate chemical methods for the required activation (Zelinsky and Sadikov 1918; 1941). Engineer Eduard Kummant designed a special rubber mask with a container for charcoal, and the manufacturing of the Kummant–Zelinsky mask started in 1916 despite bureaucratic delays and rivalry with other inventors. By the end of the war, Russia had produced

---

[7] On Russian chemistry and the war, see (Brooks 1997).

Russian soldiers during World War I wearing Kummant–Zelinsky gas masks.
[*Source*: *World War 1914–1918: A Pictured History*, edited by J. A. Hammerton, vol. 1. (1934).]

some 15 million gas masks of several different types (Ipatieff 1946, 218–235; Nametkin 1954, 11).[8]

However important this particular invention was, it only started paying off in 1917 when the country's determination and willingness to wage the war were already collapsing. Overall, the degree of involvement of Russian scientists in the war hardly matched that of their German, British, and French peers.[9] Institutionally and as a community, Russian science came to the situation of national emergency unprepared. With no pre-existing working relationship with either military or civilian industry, even the available scientific expertise and potential could not be used effectively. Connections had to be established in the course of the war itself, which took time and started delivering modest results towards the end of the war. In contrast, the Russian scientists' response to the inadequacies

---

[8] See also "Opisaniia i chertezhi izobretenii protivogazov i priborov dlia bor'by s otravliaiushchimi veshchestvami, 1915–1917" (VIA, 507-5-72) and "Svedeniia o ezhednevnom proizvodstve protivogazov, 1917" (VIA, 504-16-179).

[9] For comparisons with other national developments and scientific research in the war, see (Dewey 1988; Haber 1986; Hardach 1992; Hartcup 1988; MacLeod 1993, 1998; Trebilcock 1993).

revealed by the war and their public outcry for major reforms were anything but modest.

3. The Idea of Research Institutes

In an Empire pregnant with revolution, many monarchists and conservatives were looking forward to progressive change, while among those who under normal conditions would be called moderates radicalism was in high fashion. Even many representatives of the noble and wealthy classes developed distaste for half measures and piecemeal, compromise solutions, preferring, at least in posture, revolution to reform. The spectrum of proposals for social change favored by Russians reflected many common international trends of the early 20th century, but tended to be more radical in demands, more uncompromising in tone, and more urgent in time schedule. Ideas for reform in science proposed by Russian scientists display very similar characteristics. For example, the public value of research versus teaching was rising in all major scientific powers, but in late Imperial Russia this tendency took the form of a demand that scientists should be liberated from teaching obligations altogether and recognized as a separate profession with their own specialized institutes for research. After several twists and turns during the turbulent second decade of the century, this idea materialized in revolutionary Russia and eventually became the single most characteristic feature of the Soviet system of science.

The adage that "the success of science (and technology) is impossible without emancipating the modern scientist from his obligations as a teacher" is due to Kliment Timiriazev (1843–1920), the famous plant physiologist from Moscow University, popularizer of Darwinism and a radical democrat by political convictions. Timiriazev came to this conclusion in 1911 in response to two important events of that year, the infamous Kasso affair in Moscow and the founding of the Kaiser Wilhelm Gesellschaft in Berlin. A few months earlier, the governing body of Moscow University, its Academic Council, was caught in a conflict between radical students and the police. A student meeting on the university campus in memory of the recently deceased Count Lev Tolstoy was viewed as political, and therefore illegal. After all, Tolstoy was not only the country's greatest novelist, but also a dissident religious thinker officially excommunicated from the Russian Orthodox Church. The police entry to campus in order to prevent students from meeting was also illegal, as it violated the principle of university self-governance. In protest of this violation, the rector, Aleksandr Manuilov, and two other high elected officials of the University, Mikhail Menzbir and Piotr Minakov, resigned from their administrative posts. To reprimand them, the minister of Enlightenment, Lev Kasso, not only accepted their resignations as administrators, but also fired them from their professorial positions. This abuse of power triggered a wave of solidarity resignations among other members of the Academic Council. In total, about a quarter of the faculty—more than 100 professors and privat-docents—resigned, a rather bold act since only a few of them could have reasonable hopes of obtaining positions elsewhere, outside the system of state schools.[10]

Never before—except perhaps in the devastating fire during Napoleon's occupation of 1812—had Moscow University experienced such a damaging blow, and the public outcry against government "obscurantism" ran high, especially in the Moscow press. Since Timiriazev had already reached retirement age, his personal resignation was largely an act of symbolic protest, but as a public figure he was one the principled and most vocal critics of the regime. Other timely news arrived from the foreign press, allowing Timiriazev to discuss the treatment of science in the favorite Russian genre of political discourse by juxtaposing and contrasting Russia to some mythical, undifferentiated "West." Typical for "us," in his narrative, was the pogrom of Moscow University faculty by Kasso and state bureaucrats, while characteristic for "them" was the opening ceremony of the Kaiser Wilhelm Gesellschaft in Berlin with its projected dozen research institutes. Although one might question the value of the genre of festive speeches as a source of real information, Timiriazev extracted from Emil Fischer's opening address the conclusion he wanted: that Germans held science in such high regard as to establish separate institutes for researchers "without teaching obligations" (Timiriazev [1911] 1963).

Timiriazev was not only an influential scientist but also a true democrat. His unreserved belief in science was matched only by his unreserved belief in democracy and, furthermore, by the insistence that the two had to go hand in hand. Such views in the early 20th century—when Germany led the world in many sciences—were somewhat counterfactual, but Timiriazev was not discouraged by this inconsistency. He was confident

---

[10] See the description of the Kasso affair in P. N. Lebedev's letter to F. A. H. Krüger, February 1911 (Lebedev 1990, 358–359). For a general study of the conflicts at Russian universities in the early 20th century, see (Kassow 1989).

K. A. Timiriazev late in his life, around 1920. The plant physiologist Timiriazev was the major authority on, and proponent and defender of, Darwinian evolutionary theory in Russia, where his role was similar to that of T. H. Huxley, "Darwin's bulldog," in England. In 1911 Timiriazev put forth the idea of organizing scientific research in special institutes outside the universities, which later became the dominant Soviet trend. Shortly before his death in 1920, he publicly endorsed the Bolshevik regime, thus helping to forge the pact between research-oriented scientists and the Soviet government.
[*Courtesy*: K. A. Timiriazev Museum, Moscow.]

not only that science under Anglo-Saxon democracies must be much better off than in imperial Germany, but also that it had surely advanced further in the progressive trend of liberating science from teaching. He had heard something about the Carnegie Institution of Washington and about the "endowment of research" in Britain, and he misinterpreted these examples as proof for his claim (Timiriazev [1911] 1963, 58).

In accordance with the rules of the genre of writings about "Russia and the West," Timiriazev depicted his country as backward for still having "all its science concentrated in universities" while "the entire civilized world" had recognized the highest value of research itself. The massive resignation of Moscow professors had proved to him that scientists could not be free as long as they remained in their teaching positions as state employees. The Kasso affair indeed destroyed the international pride of Moscow science, Piotr Lebedev's laboratory in the cellar of the Moscow University's Physics Institute.

Lebedev (1866–1912), a son of a Moscow merchant, defied his father's wishes by choosing an academic career. After receiving a Ph.D. in Germany, he returned to Moscow and defended another dissertation for a Russian doctoral degree, which was necessary for a professorial career at a Russian university. In 1903 the Moscow University opened a new specially designed building for its Physics Institute, and the recently appointed Extraordinary Professor Lebedev started creating there a German-style research school. He had already won an international fame for a series of very delicate experiments in 1899–1900 in which he succeeded in measuring the mechanical pressure produced by light, thus experimentally confirming the last remaining great prediction of Maxwell's electromagnetic theory. Lebedev regarded advanced research and the training of research students as his primary job obligation, and in this he was a perfect example of the so-called "research imperative" taking hold of Russia by the turn of the century. The Russian universities followed in this respect the most advanced system of research-oriented German universities, and this was also where Lebedev drew his main professional inspirations. The Physics Institute at Moscow University was designed according to the best German models as a three-story building with a lecture hall, apartments for professors, experimental *Praktikum* for students, and laboratories for research in the cellar. Like his German teacher August Kundt, Lebedev started gathering a following among advanced students and instructing them in performing cutting-edge research. By 1911, there were about two dozen of them at various stages of work in his laboratory, and a few were already completing dissertations and looking around for jobs. Though Lebedev was not very active politically, he felt obliged to resign in solidarity with other professors as a result of the Kasso affair. He and most of his students left the university in 1911, abandoning a well-equipped laboratory and the possibility of continuing a productive research program (Lebedev 1990).

Timiriazev through newspapers appealed to Moscow merchants to save "their" Lebedev and help create "safe havens for scientific research" in the form of research institutes independent of state universities. He remained fully convinced that his proposal to separate science and university education was following the general trend of more developed countries rather than embarking on an original path of institutional development (Timiriazev [1911] 1963, 58, 65). His call found a warm reception among many other scientists and publicists, but the idea developed further in two distinctively different forms. Timiriazev and other Moscow authors argued for the establishment of non-governmental facilities for research and looked towards private philanthropy for support. Like Lebedev, many Moscow professors came from the merchant estate themselves or had personal ties to families of major local merchants and industrialists, who by the early 20th century had developed a taste for cultural philanthropy (Buryshkin 1991). The initiative of professors who resigned resulted in the founding of the (awkwardly named) Moscow Society for Scientific Institute in 1912 with the purpose of raising private funds and donations for the construction of non-governmental research institutes (Zernov 1912). Four such institutes—in physics, chemistry, biology, and social sciences—were planned, and two were actually built despite the on-going war. After Lebedev's premature death of heart disease in 1912, his student Piotr Lazarev (1878–1942) continued pushing forward the construction of the Physical Institute and eventually became its director, while the Biological Institute came to be directed by Nikolai Kol'tsov. Both institutes opened in 1917 on the eve of the revolution, and they later became the nuclei of the much larger Soviet institutions.

In contrast to Moscow academics, their colleagues in St. Petersburg had much closer relations—personal and otherwise—to the state bureaucracy and typically looked to the government as a source of patronage. The Imperial Academy of Sciences in St. Petersburg used the occasion of Mikhail Lomonosov's bicentennial in 1911—marked by the festive official celebrations of the proclaimed founding father of national Russian science—to lobby for the establishment of a large Lomonosov Institute for research in three fields: physics, chemistry, and mineralogy. The proposal received His Majesty's approval but was later postponed because of the Great War and never materialized (Bastrakova 1999; Imperatorskaia 1917, 102–106). The idea of state research institutes, however, did not die. Another major spokesman for Russian science, Vladimir Vernadsky (1863–1945) of the Academy of Sciences, modified it according to the new situation during the war and took first steps towards its practical realization, with huge consequences for the post-revolutionary period.

4. The Study of Natural Productive Forces

World War I cultural propaganda centered around the theme of the holy struggle between civilization, culture, and barbarity. Russia's traditional dependence on Europe, particularly Germany, made it harder for Russian authors to use the language of militant cultural nationalism that permeated the writings of French, German, and British war ideologues. Russian educated elites could easily formulate their opposition to the "Teutonic race" in nationalistic, monarchist, religious, or moral terms, but not in terms of cultural superiority. Instead they pictured Russia's war as a war for cultural and economic independence against the cultural imperialism of Germans. Once the war broke out, Russian writers who had previously decried the country's backwardness in inflated terms started calling its lack of culture only "illusory" and claiming that "the victory over Germany is necessary in the name of [European] culture" (Brenchkevich 1915, 32; Grimm 1915, 14–15; Trubetskoi 1915).

Compared to religious philosophers, scientists in Russia had even fewer reasons to claim nationalistic cultural superiority. Unlike their Western colleagues, they typically produced rather moderate, almost internationalist statements and complained most strongly about the damage done by the war to international exchange and communication in science. The pacifist Timiriazev was on the left side of the political spectrum and even during the war continued to proclaim science as a universal, international, and rational activity, denouncing all its military applications (Timiriazev [1915] 1963). Vernadsky, geologist and geochemist, was much closer to the political center and also much younger. He was one of the leaders of the constitutional democrats, the political party of liberal opposition in late Imperial Russia that favored the establishment of a constitutional monarchy. It was nicknamed "the party of professors," since its Central Committee consisted mostly of established academics. Vernadsky had also resigned from Moscow University during the Kasso affair of 1911; he was later elected to the Imperial Academy of Sciences and moved to St. Petersburg. As the major wartime public spokesman for science, he represented the Academy rather than universities (Mochalov 1982; Bailes 1990).

The war had a great impact on Vernadsky's scientific and social views. He felt that the tremendous movement of human masses during the mobilization resembled the power of geological forces, which prompted his investigation of global geological effects of human activities, a long line of inquiry that established his later reputation as one of the founding fathers of ecological thought. In historical terms, Vernadsky predicted that the on-going war—like the decades of European wars after 1789—would mark the transition to a new historical era, in particular with regard to the role and the importance of science "because of its real applications to the interests of defense, despite moral reservations" (Vernadsky [1916a] 1922, 54). Like most Russian scientists, Vernadsky lamented the interruption of scientific contacts between belligerent nations, mistakenly believing that scientific research in other European countries continued on its prewar scale:

> As we know, we are continuing our scientific work with the same speed. Our work is developing and improving now, and it also had not been interrupted or slowed down in the years of our other national disturbances—either during the war with Japan or in the years of the revolution [of 1905] (Vernadsky [1915a] 1922, 135).

Whether Vernadsky understood it or not, his statement reflected the fact that Russian science was much less integrated into the total war effort than science in Britain, Germany, or France. He acknowledged, at least, that it was ill-prepared for the new tasks, and he expected major changes in the immediate postwar period: "[Although] the development of science will not stop the war, ... regardless of the outcome of the war, both winners and losers will have to direct their thought towards further development of scientific applications to the military and navy affairs" (Vernadsky [1915a] 1922, 131–132).

According to Vernadsky, the most important task facing postwar Russian science would not be the competition with other nations in pure science but the study of Russia's own natural resources and productive forces:

> Russian society has suddenly realized its economic dependence on Germany, which is intolerable for a healthy country and for an alive strong nation. ... [This dependence] has transcended the limits of necessary, unavoidable and profitable exchange of products of nature, labor, and thought between two neighboring nations. It has developed into an exploitation of one country by the other. ... One of the consequences—and also one of the causes—of Russia's economic dependence on Germany is the extraordinary insufficiency of our knowledge about the natural productive forces with which Nature and History had granted Russia (Vernadsky [1915b] 1922, 5).

By his count, only 31 out of 61 economically useful chemical elements were mined and produced in Russia. Even aluminum had to be imported, since deposits of bauxite had not yet been explored. Vernadsky believed that practically any useful mineral could be found in the country's enormous territory and referred to this work as "necessary for national security," since Russia had to catch up with other nations in this regard (Vernadsky [1916a] 1922, 65). Following his proposal, in February 1915 the Imperial Academy of Sciences abandoned its century-long tradition of concentrating on pure science and established a Commission for the Study of Natural Productive Forces of Russia (KEPS) (Kol'tsov 1999, 14–15). The task of KEPS, according to Vernadsky, encompassed the study of all kinds of national resources and called for collaboration and mobilization of geologists, mineralogists, zoologists, botanists, chemists, physicists, and even social scientists, following the example of the wartime "mobilization of various engineers who work on the basis of exact sciences, physicians, bacteriologists, and ... chemists." Vernadsky was aware that these plans could be realized fully only after the end of the war, but he insisted that preparations had to start right away. He did not worry too much about wartime expenditures on the project, which would bear fruit only in the long run because "one can establish all the necessary research institutes at the expense of just one super-dreadnought" (Vernadsky [1916a] 1922, 54–55, 68).

With regard to these mobilization plans, science and scientific manpower began to be considered among the country's most important resources. The Academy in Petrograd together with the editorial board of the Moscow scientific periodical *Priroda* established a joint commission to prepare the first national census of academic populations and institutions. In early 1917, questionnaires were distributed to all known scholarly institutions in Russia, and the answers were collected by the year's end from both capitals—Petrograd and Moscow—and from the majority of provinces. Because of the revolutionary unrest, the results of the survey could not be published as planned in early 1918. With financial help from the Bolshevik's Commissariat of Enlightenment, the two volumes were published a couple of years later, containing unprecedented demographic information on science and scientists in Petrograd and Moscow almost exactly on the eve of the November 1917 Bolshevik coup (Nauka 1920–1922).

5. The Network of State-Sponsored Research and Development

Vernadsky did not need to wait for the results of the census to declare in 1915 that the war had revealed that the existing scientific infrastructure was totally insufficient for the proposed grand project and that cardinal changes were needed as badly there as in Russia's political system:

> After the war of 1914–1915 we will have to make known and accountable the natural productive forces of our country, i.e. first of all to find means for broad scientific investigations of Russia's nature and for the establishment of a network of well-equipped research laboratories, museums and institutions.... This is no less necessary than the need for an improvement in the conditions of our civil and political life, which is so acutely perceived by the entire country (Vernadsky [1915a] 1922, 140).

Itself an assembly of research laboratories by statute, the Academy of Sciences in terms of its financing was one of the institutions of the royal court and thus naturally expected state patronage. The support it already had, however, was certainly inadequate for the proposed goal of KEPS, but Vernadsky insisted that the work had to start immediately without waiting for the war's end. As the first step, he proposed the preparation of a series of detailed summaries of the available knowledge on Russia's energy, minerals, ores, plants, animals, and chemical factories. He specifically listed some minerals that were in great demand and had to be searched for without delay. With an initial modest contribution from the Academy's budget, KEPS commissioned a number of scientists to write such reviews, which were published as the series *Materials for the Study of the Natural Productive Forces of Russia*.

Reviewing the results of the first year, Vernadsky formulated an ambitious plan of further reform directed towards "a new organization of scientific work" on the national level. His scheme consisted of (1) a national congress of scientists for the discussion of the study of productive forces, (2) coordination of scientific work for the sake of planned research, and (3) creation of new research institutions: museums, laboratories and institutes. Vernadsky specifically elaborated on the last point, arguing that the entire national network of specialized research institutes of applied, theoretical, and mixed nature was needed as a matter of state priority and that KEPS should draft plans for them. He was convinced that "the higher schools alone cannot satisfy the growing needs of scientific research" and therefore of the "impossibility and the disadvantage of the permanent linking of all scientific research work to the institutions of higher education" (Vernadsky [1916b] 1922, 29).

In his 1916 report to KEPS, Vernadsky mentioned plans for an institute on clay and aluminum, and for an experimental station on Kara-Bogaz Bay on the Caspian Sea (a deposit of raw salts). In the following year,

KEPS drafted proposals for a half-dozen more research institutes and laboratories of applied aims (Platinum, Physico-Chemical Analysis, Hydrology, Alloys and Metallography, Petroleum), while learned societies proposed several more (Metallurgy, Pharmaceuticals, Chemical Reagents, Coal) (Bastrakova 1973, 46–49). The idea of separate research institutes thus appeared again, but this time in the form of a much broader and comprehensive network and within the context of the important practical task facing the nation.

Many years later, some Soviet historians referred to Vernadsky's proposal of 1916 as a prophetic anticipation of the Soviet system of planned research with the Academy as its highest administrative body. "Anticipation" is probably too strong and simultaneously too weak a term. The two developments were separated by two decades of revolutionary social changes and are clearly not identical. At the same time, there is not just similarity but a real causal progression linking them together. Vernadsky's wartime proposal represented an important early stage in the process of reform that eventually resulted in the Soviet system of scientific research and development. This connection has been somewhat obscured by the persistent tendency among historians to see the revolution of 1917 and the Bolsheviks as the origin of all new and important developments of the Soviet era. The preceding period of World War I has thus been overshadowed in perceived importance by the subsequent great revolution and has received undeservedly little attention in the history of Russia, in comparison to that of other European countries. In summarizing the results achieved by the Russian Empire, historians typically chose 1913, the last year of peace, with the new society seen as starting in 1917 and the intermediate war presented as disintegration leading to a disgraceful end rather than as marking the emergence of something new and important.

Recently, several important historical studies have started challenging this stereotype, so that a new image of the Great War is gradually emerging, that of a beginning at least as much as an end.[11] More precisely, one can see that the new Soviet society, in many of its essential features, was born during a period of permanent war—1914 to 1921—and retained for at

---

[11] A similar approach has also been developed in (Holquist 1997; Hoffmann and Holquist, forthcoming). With regard to science, Nathan Brooks (1997, 360) has already noted that "for chemists, the period from World War I through the early years of the Soviet regime was one of relative continuity, not discontinuity." Several recent books indicate the rise of historical interest towards the role of World War I in Russian history (Pisarev and Mal'kov 1994; Mal'kov 1998; Smirnov 1999).

least several decades some of the acquired birthmarks. With respect to the image, institutions, and practice of Russian science, it is possible to say that the new Soviet system was invented even before the revolution of 1917.

To be sure, not all of the mature system's features could already be seen emerging at that early stage. For example, the Academy of Sciences was not yet as dominant among academic institutions as it would eventually become. Although some aspirations for administrative grandeur were present all along and were reflected in Vernadsky's 1916 proposal, the Academy would not be able to achieve this goal until some twenty years later in the mature Stalinist society (Vucinich 1984). For the immediate turbulent decade of wars and revolutions, however, the significance of the Academy and KEPS would be less in overseeing and administering than in part encouraging, in part coordinating, but most of all reflecting the general trend. Similar processes were developing largely independently at various locations and institutions throughout the Empire, as formerly "pure" academics were turning towards economically and militarily important work, establishing nuclei of new research institutions, and preparing blueprints and proposals for an expanded postwar activity along these lines. It was specifically these wartime proposals and activities, which—after 1917—provided the foundation for the emerging special relationship between science and the new revolutionary government.

Chapter 2

Socialist, or Big, Science

The revolution and its impetus towards radical social change helped advance the reform drafted by Russian scientists during the Great War and allowed it to materialize faster and more thoroughly than comparable proposals discussed by scientists in other countries. Many of the new scientific projects appealed to the revolutionary mentality and modernist ideology of the new Bolshevik government and were quickly adopted and adapted by it. The Bolsheviks endorsed particularly strongly the establishment of research institutes, which won them allies among research-oriented scientists while also helping the Bolshevik struggle for political control over universities. During the Civil War, amid economic destruction and hardships, radical institutional changes in the social infrastructure of science were pushed forward with astonishing ease. As the war and the attendant isolation of Russian science ended in 1921, the foundation of a novel government-sponsored system of research and development was already in place.

For contemporaries, this new system represented the socialist way of organizing scientific research. Yet it also contained the key features of the later, international phenomenon known as "Big Science." The new scientific institutes were organized and funded by the government, they tended to exist independently of, or at least separately from, universities and other institutions of higher education, and their workers received salaries for doing research rather than teaching. These institutes were significantly larger than the typical university-based institute or laboratory, with a tendency towards centralization or even monopolization of a particular research field or problem. They were usually organized around large interdisciplinary projects that required the collective work of engineers and scientists from different disciplines and that combined cutting-edge fundamental research with the simultaneous development and production of some sophisticated novel technology or the pursuit of some other major practical goal, military or civilian or both. This characteristic common pattern of a "symbiosis, ... a fusion of 'pure' science, technology, and engineering" (Pestre and Krige 1992, 93) allows one to conclude that, despite their different names, 'socialist' and 'big' science were parts of the same international trend that was central for the development of science during much of the 20th century.

1. Revolution and the War's Legacy

Timiriazev's pacifism contributed to his rapprochement with the Bolsheviks, the only major political party in Russia that opposed the imperialist war on principle. Ill health prevented the ailing patriarch of Russian science from participating personally in the organizational reform of scientific institutions, but his public position helped forge the pact between research-oriented academics and the new Soviet regime. Timiriazev became the first famous scientist to express open support for the Bolshevik-led government when it was still struggling to consolidate its power. Grateful Bolsheviks helped Timiriazev with the publication of his volume of collected public essays and speeches, *Science and Democracy*, which came out in 1920 and received a particularly warm welcome from Lenin (Timiriazev [1920] 1963). Timiriazev's 1911 article on the liberation of science from teaching was reprinted in the volume and thus appeared to have received the highest political approval.

Yet even without waiting for Lenin's indirect endorsement, scientists and Soviet officials quickly agreed that a research institute was the best and most progressive way of organizing science. To Nikolai Gorbunov (1892–1937)—then probably the most influential patron of science in Bolshevik circles—the concept looked so genuinely novel and revolutionary that he saw to it that a phrase about linking research to industrial production and establishing the "entire network of new scientific applied institutes, laboratories, experimental stations, and testing facilities" was inserted into the new Communist Party Program adopted at the 1919

N. P. Gorbunov, secretary to the chairman of the Soviet government, V. I. Lenin, in his Kremlin office in the 1920s. During the first Soviet decade, Gorbunov's technical expertise in chemical engineering was a rare commodity among leading Bolsheviks. He was largely responsible for many decisions and initiatives by the revolutionary government in building up Soviet institutions and activities in the field of science and technology. Gorbunov fell out of political favor after Stalin's power grab in 1928, but in 1935 he was elected to the Academy of Sciences for his role in geographical expeditions to the mountains of Central Asia, and he served as the Academy's Permanent Secretary. Gorbunov died in the Stalinist purges, like many communist officials of the early era.
[*Source*: (Rossiianov 2002, 288), with the author's permission.]

Party Congress (Bastrakova, Ostrovitianov *et al.* 1968, 91; Gorbunov 1986, 14–15). A young chemical engineering graduate from Petrograd Technological Institute, Gorbunov was one of the few important Bolsheviks who possessed technical expertise, and as such he played a crucial role in designing early Soviet policies with regard to science and technology.[1] Sergei Ol'denburg (1863–1934), the permanent secretary of

---

[1] Gorbunov's influence depended not so much on his relatively low formal rank in the party hierarchy as on his post as the secretary of the Soviet government, which provided him direct access to all important decision makers. For more on his role in the organization of Soviet science, see (Gorbunov 1986).

the Russian Academy of Sciences, summarized the consensus a few years later in the following formula: "while the 18th century was, for science, the century of academies, the 19th century became the century of universities, and the 20th century is starting to become the century of research institutes" (Ol'denburg 1927, 89).

The Bolsheviks bought into the idea even more readily, as it helped them to win over scientists as collaborators and simultaneously to sort out their ambivalent relationship with professors. While enthusiastically pro-science at the most extreme—for both ideological and pragmatic reasons—the Bolsheviks often perceived the most common real-life carriers of scientific knowledge, college professors, as "counter-revolutionary." Many university councils, indeed, openly protested the ousting of the Provisional government by the Soviets in November 1917, and, even more important, faculty representatives effectively rejected the proposal of a radically egalitarian, democratic reform of higher education drafted by the Commissariat of Enlightenment in spring 1918 (Smirnova 1979; 1984). Subsequently Soviet officials treated defenders of academic autonomy with suspicion and hostility, but they generally gave a much warmer welcome to the same individuals when they appeared in the role of scientists and experts rather than as faculty members. Even the leader of the liberal university opposition, biologist Mikhail Novikov, collaborated successfully with the economic commissariat as head of the Moscow commission of scientific experts. At the same time, in his other role as the rector of Moscow University, Novikov organized strikes of university professors and struggled bitterly against the Commissariat of Enlightenment (Novikov 1930). He ended up being exiled abroad in 1922 together with some 160 representatives of the academic opposition, but many more scholars found that the newly organized research institutes offered them a much more comfortable and freer shelter than the politically and administratively embattled universities. The resulting gradual flow of scientists from teaching to research positions made it easier for the Bolsheviks to start replacing old bourgeois professors with politically and ideologically more loyal teachers of students.

For scientists and scientific experts for the government, the criteria of political loyalty were generally much lower than for university teachers. General Ipatieff was a monarchist and did not welcome even the February revolution, thinking that "a constitutional monarchy would have served Russia's needs best." He had no belief in the Bolsheviks' utopian goals, although he gave them credit for one thing, that "having gained control of the nation, they made the only sensible move possible in concluding peace immediately" (Ipatieff 1946, 246, 263). When the Bolsheviks usurped political power in the capital, Ipatieff persuaded his subordinates at the War Chemical Committee that "army men had no right to stop their work in war time" and that "the government in power should be obeyed." He took a similar line at the meeting of the Physico-Mathematical Division of the Academy of Sciences:

> Control of the state belong[s] to the group capable of setting up a strong government.... In the period [of general dissatisfaction] the intelligentsia...should not try to oppose the new government. Our personal feelings toward the Bolshevik regime were ours alone and no one could force us to express them. The autocracy of the Tsarist regime had dissatisfied many of us; yet we had continued to do our duty (Ipatieff 1946, 259–260).

With such an understanding of Russian national interests and of his own duty, Ipatieff did not hesitate to start an immediate active collaboration with the new regime in his professional role as a military scientist. The Bolsheviks for their part entrusted the monarchist with high-level posts within the Soviet government. In May 1920 Ipatieff was appointed to direct the former Central Scientific-Technical Laboratory of the War Ministry, soon to be renamed the State Scientific-Technical Institute (GONTI). Subsequently, as a member of the governing collegium of the economic commissariat VSNKh, he was effectively in charge of the nation's chemical industry and research (Ipatieff 1946, 285–330).

As a liberal, and also a deputy Minister of Enlightenment in the last version of the Provisional Government, Vernadsky was much less inclined towards cooperating with the Soviet regime. Together with several other officials of the deposed government, he signed an appeal to the Russian public protesting the Bolshevik coup. After the Bolsheviks ordered the arrest of signatories and the closing of the newspapers that had published the document, Vernadsky decided to escape temporarily from Petrograd, taking a leave from the Academy of Sciences and departing for his family home in a Ukrainian province (Vernadsky 1994, 223–224). The permanent secretary of the Academy and the former minister in the Provisional government, Ol'denburg, remained in the capital and played a key role in the negotiations with the Commissariat of Enlightenment. Initially the Academy felt no less hostile towards the Bolsheviks than did the universities and learned societies, but it had a stronger tradition of political obedience and privileged proximity to autocratic power. After debating the matter, the Academy leadership decided to refrain from issuing an open protest against the Bolshevik coup, and a few months later concluded an agreement with the Soviet government.[2]

The Bolsheviks wanted the Academy to embark on research related to the "rational development of industry and the rational distribution of the country's economic forces" and were less interested in its traditional emphasis on "pure theoretical sciences." Grounds for the compromise were found in the activities of KEPS, the Commission for the Study of Natural Productive Forces. In return for material support and respect for its self-administration, the Academy expressed its willingness to work on "topics originating from the needs of state-building, while acting as the organizational center of the national research effort" (Bastrakova 1973, 127, 141). On 12 April 1918 the Soviet government, Sovnarkom, approved funding for KEPS that allowed the commission to develop activities on a much broader scale. The first two institutes proposed by KEPS—on physico-chemical analysis and on platinum and precious metals—started working by the summer of 1918 and were followed by the institutes on hydrology, ceramics, radium, and soil (Kol'tsov 1999, 80–85).[3] In Vernadsky's absence the work of KEPS was coordinated by his former student, and the commission's secretary, Aleksandr Fersman (1883–1945). The Academy quickly promoted the 35-year-old geologist to full membership, and in May 1920 Fersman departed for the Kola Peninsula at the head of a complex scientific expedition to study the Russian North (Fersman 1968). The expedition found rich deposits of phosphate minerals in the Khibin Mountains along with other ores and resources and marked the beginning of the Soviet scientific and industrial cultivation of the unexplored northeastern territories, which would later become one of the chief economic priorities for the Stalinist regime and its GULAG system of forced labor.[4]

---

[2] See the correspondence between the Academy and Narkompros in February–April 1918 in (Bastrakova, Ostrovitianov et al. 1968, 103–128).

[3] For the documents on the institutes of physico-chemical analysis, platinum, clay, and hydrology, see (Bastrakova, Ostrovitianov et al. 1968, 129–142, 150–165). A general report on KEPS activities throughout the Civil War years was published in (Lindener 1922). KEPS's publishing activity rose despite the catastrophic economic situation from 85 printed pages in 1915, through 732 in 1916, 416 in 1917, to 893 in 1918, 2038 in 1919, and 2264 in 1920 (Blok 1921).

[4] On other expeditions organized by KEPS during the Civil War—the salt expeditions to Solikamsk and Kara-Bogaz, the bauxite expedition, the West-Siberian expedition, and the exploration of the Kursk magnetic anomaly, see (Bastrakova, Ostrovitianov et al. 1968, 177–193; Man'kovsky 1960).

Competition between Soviet commissariats further stimulated the growth of new research institutes. The Commissariat of Enlightenment, Narkompros, managed to monopolize the country's educational system but could not prevent other emerging Soviet bureaucracies—in particular economic (VSNKh), medical (Narkomzdrav), agricultural (Narkomzem), army and navy (Narkomvoenmor), and communications (Narkompochtel')—from developing their own research and development empires. The large majority of the new institutes originated from proposals or activities that went back to World War I or slightly preceded it. What used to be the experimental factory of the Russian Physico-Chemical Society's War Chemical Committee became the Institute of Applied Chemistry under VSNKh. The industrial commissariat also nationalized the private institute Lithogaea into the Institute of Applied Mineralogy and reorganized the research laboratory on automobiles into the Automotive Institute (NAMI). By 1920 VSNKh controlled 16 research institutes, including the Central Aero-Hydrodynamic Institute (TsAGI), which worked on the design of airplanes, and the institutes on metrology, food, fuel, fertilizers, and pure chemicals (Bastrakova 1973, 178–186; Bastrakova, Ostrovitianov et al. 1968, 283–311). The nationalization of banks deprived the Moscow Society for Scientific Institute of its financial assets, but under Narkomzdrav its recently completed physical and biological institutes became, respectively, the Institute of Physics and Biophysics and the Institute of Experimental Biology (Bastrakova 1973, 200). Two other formerly private institutions, Lesgaft Biological Laboratory (renamed the Institute for Natural Sciences) and Bekhterev's Institute for Brain Research, found a new patron in Narkompros (Bastrakova, Ostrovitianov et al. 1968, 251–258). Narkompochtel' undertook the commission to organize research on radio and in 1919 established the Central Radio Laboratory in Nizhnii Novgorod. According to various estimates, 40 to 70 research institutes were created in the country by the end of the Civil War (Bastrakova 1973, 221–231, 162; Bastrakova, Ostrovitianov et al. 1968, 8).

The Bolshevik government was the most important, but by no means the only, political authority that supported scientific research in the Civil War-torn territory of the former Empire. The revolutionary chaos and armed conflict led to the decentralization of the political sphere and created more de facto room for local initiatives by scientists and activist groups. Dozens of local proposals of new universities and institutes of higher learning that had been either turned down or tangled in red tape by the rigid Tsarist bureaucracy received a much easier approval—even if sometimes only on paper and with the primary hope of attaining political legitimacy—from the unstable Civil War administrations and rival regimes. Only ten universities had been established in the country during the preceding 150 years of the monarchy, and no fewer than 24 new ones appeared in just the four years of the revolution and the Civil War. The new schools were founded by a variety of administrations, both Red and White, in the approximate proportion of two to one (Chanbarisov 1988, 39–55). Not all of them survived the turbulent years, but the majority did and changed forever the national scene of higher learning.

Vernadsky's Civil War odyssey is particularly illustrative, since he managed to work under nearly all possible regimes, pursuing with reasonable success the same line of reform in science. At first, he organized KEPS and drafted proposals for a number of research institutes at the Imperial Academy of Sciences in Petrograd; he then continued the same activity under the Provisional Government. After he moved to Ukraine in early 1918, Vernadsky founded a local KEPS in his native Poltava province, and

V. I. Vernadsky with his wife and daughter in 1921. After his Civil War travels and efforts to promote scientific research and institutions under temporary administrations of diverse political orientations, Vernadsky returned to Bolshevik Petrograd in 1921 and resumed his position at the Academy of Sciences. As the creator of KEPS, the Commission for the Study of Natural Productive Forces, he helped reorient the Academy towards a program of major geographical and geological explorations, especially in the underdeveloped northern, eastern, and southern territories. The activities of KEPS contributed to the Soviet, and later Stalinist, industrial modernization and construction projects.
[*Courtesy*: V. I. Vernadsky Museum, Moscow.]

he later played a key role in the organization of the Ukrainian Academy of Sciences (with its own KEPS and research institutes) in Kiev, the capital of the newly proclaimed independent nation. The vulnerable Ukrainian Academy managed to survive all changes of political and military power in Kiev, from the German-backed regime through local nationalists and the Soviet Red Army and White occupations to the Ukrainian Soviet Republic. All except the White Russians, who were the most hostile to Ukrainian independence, granted the Academy some kind of political patronage. During the White occupation in 1919, Vernadsky traveled south to meet with generals and plead on the Academy's behalf. As a result of contracting typhus and changed military fortunes, he got stuck in the White Crimea. There he was elected rector of the recently founded Tauride University and organized—of course—one more KEPS, this time to study the natural productive forces on the Crimean peninsula. Once the Red armies overran the last White stronghold in late 1920 and many academics—including his son George—sailed off from the Crimea towards permanent emigration, Vernadsky returned to Petrograd to work under the Bolsheviks at the Russian Academy. In late 1921 he transformed the former Radium Commission of his original KEPS into a full-scale Radium Institute. Shortly thereafter, Vernadsky delegated the institute directorship to his deputy, radiochemist Vitaly Khlopin, and accepted a visiting professorship in Paris with the official permission of the Soviet government.[5]

In the course of the Civil War, Bolshevik commissars fully accepted as "socialist" the idea of specialized research institutes separated from higher education, which had been developed by Russian academics shortly before the revolution. It corresponded well to their own bureaucratic and political interests and eventually became the dominant institutional form of Soviet science. Most of the newly established research units would be called "institutes," and many older ones—laboratories, bureaus, experimental factories, and stations—were renamed and reorganized into institutes. Bolsheviks especially welcomed research proposals that combined utilitarian goals with modernist, revolutionary overtones and thus were particularly receptive to the novel undertakings that had started during World War I and reflected its needs. The chaotic and utopian mood of the revolutionary period allowed many proposals to win easy approval without long and careful consideration by the authorities. A number of

---

[5] On Vernadsky's travels, see his Civil War diary (Vernadsky 1994); for documents on the first years of the Ukrainian Academy of Sciences, see (Istoriia 1993); for academic life during the Civil War in the Crimea, see (Vernadsky 1921); in the Russian North, see (Zhdanko 1922).

the institutes either remained on paper or did not survive the social difficulties of the Civil War and the financial shortages of the post-war transition to the New Economic Policy. Several dozen, however, proved viable throughout all the turmoil and determined the future directions of the Soviet scientific effort. The period of incessant wars, revolutions, and international isolation lasted overall from 1914 till 1921. Once it was over, Russian scientists could once again exchange information with foreign colleagues, occasionally travel to Europe, and compare their accomplishments with what had been achieved elsewhere. But by that time Soviet Russia had already established the foundation of its distinctive system of research and development. One of the best illustrations and also one of the most important examples of this process is the story of optical glass.

2. From Optical Glass to the Optical Institute

In the 1930s Sergei Vavilov, physicist, historian of science, and coordinator of Soviet optical industry, lamented that after a great start in the late 18th century and the important inventions by Euler, Lomonosov, and Aepinus, research in applied optics and optical instrumentation was neglected in Russia for almost a hundred years (Vavilov [1935] 1956, 147). Serious work had to start again from scratch in the early 20th century. The young military engineer Yakov Perepelkin, a graduate of the Artillery Academy, arrived in 1899 at the Obukhov Steel Works in St. Petersburg with a commission from the Navy Ministry to organize the production of modern optical sights for naval and coastal artillery. By 1905—during the war with Japan—he established at the factory a special optical workshop that manufactured artillery sights of several types from imported optical parts and from locally designed mechanical parts. Perepelkin invited two military scientists as consultants: the applied mathematician and naval engineer Aleksei Krylov and the physicist Aleksandr Gershun. In 1906 Gershun quit his teaching job at the Artillery Officers' Class in Kronshtadt to become the workshop's full-time employee and later director. By 1912 the optical workshop manufactured a wider assortment of military optical devices, including binoculars and range-finders, and employed some 150 workers, though these expansions still did not satisfy the army's needs even during the period of peace (Chenakal 1947, 161–162).

The workshop's staff expanded dramatically during the Great War and reached almost 700 in 1917. Meanwhile, shortly before his death in 1915, Gershun managed to start another, private optical factory under the newly organized Russian Optico-Mechanical Joint-Stock Company. Both facilities were mobilized to work on military needs, together with a few smaller private workshops and two sequestered Riga-based subsidiaries of the German optical firms Zeiss and Hertz. The main wartime bottleneck, however, was the shortage of optical glass. In the pre-war decades, Germany alone had possessed the advanced know-how and had practically monopolized the world production of high-quality optical glass. Both Russia and her allies had depended on the imports of such glass from Germany and had to undertake urgent measures during the war to develop their own production. On behalf of the Chief Artillery Administration, Lieutenant-General A. L. Korol'kov inspected all existing optical facilities and estimated the total amount of optical glass available at the start of the war to be somewhat below one ton. Korol'kov then turned for help to the administration of the Imperial Porcelain and Glass Factory in St. Petersburg.[6]

The situation became critical in the fall of 1915 as optical factories were about to stop production because of the shortage of glass. Attempts to import glass from England and France were not very successful since the allies were experiencing shortages themselves. Meanwhile, the first series of experiments on the production of optical glass at the Porcelain Factory had convinced the military of the necessity to conduct extended research and to "bring into the effort scientific and technical experts."[7] The factory's chief engineer, Nikolai Kachalov (1883–1961), appointed a young chemist, Ilia Grebenshchikov (1887–1953), head of the optical glass workshop and invited as consultants several academic scientists, including the university physicist Dmitry Rozhdestvensky (Chenakal 1947, 163; Grebenshchikov and Rozhdestvensky [1918a] 1993).

The son of a historian and high-level official at the Ministry of Enlightenment, Rozhdestvensky (1876–1940) graduated from St. Petersburg University in 1900 and was subsequently employed there as assistant at the university's Physics Institute. Together with his friends

---

[6] "Perepiska...o zakupkakh, dostavke i proizvodstve opticheskogo stekla i priborov..., 1914–1916," (VIA, 504-10-8, pp. 2–39). On the German optical industry and the wartime problem of the production of the optical glass and military optical equipment in other belligerent countries, see (Feffer 1994; Hagen 1996; MacLeod 1975) and David L. MacAdam, "Early Rochester History of the Optical Society. An Inquiring Sketch," unpublished manuscript, 1964 (Niels Bohr Library, American Institute of Physics-hereafter AIP).

[7] "Perepiska...o zakupkakh, dostavke i proizvodstve opticheskogo stekla i priborov..., 1914–1916," (VIA, 504-10-8, pp. 81–83, 99, 300, 305–306). See alo "Perepiska...o zakazakh opticheskogo stekla inostrannym firmam, 1915–1917" (VIA, 504-10-64).

Paul Ehrenfest and Abram Joffe, Rozhdestvensky founded *kruzhok* (the circle) of young scientists who were trying to establish the new research style in St. Petersburg physics. Ehrenfest (1880–1933) had studied theoretical physics with Ludwig Boltzmann in Vienna, married a Russian student, mathematician Tatiana Afanasieva, and settled with her in Russia in 1907. Joffe (1880–1960) obtained his Ph.D. in experimental physics in Munich with the discoverer of X-rays Wilhelm Röntgen. Their *kruzhok* met independently and in secret from senior professors of physics at St. Petersburg University, who, unlike their colleague Lebedev in Moscow, were oriented towards excellence in teaching rather than advanced research. The young Turks' career advancement was hampered by their challenge to older professors and by a number of bureaucratic problems. They fought for several years an uphill battle against university mathematicians, who controlled one of the formal examinations for the Russian academic degree in physics and kept the requirements so high that passing it was a struggle even for physicists with doctoral degrees from top German universities and mature researchers in their own right. Ehrenfest—an Austrian Jew without any official religious affiliation—could not find a regular position in Russia and eventually left the country in 1912 to succeed Hendrik Antoon Lorentz at the University of Leiden. Rozhdestvensky and Joffe became professors in 1915, after a long wait. Rozhdestvensky received the university chair, while Joffe failed to garner enough votes to be elected to the second chair in physics and left the university for a new professorship at the Petrograd Polytechnical Institute.[8]

At the time of his appointment, Rozhdestvensky was already one of the world's leading experts on spectroscopy and the author of an ingenious method of studying anomalous dispersion, which provided important results for the emerging quantum theory of the atom. He simultaneously became the director of the university's Physics Institute and was also invited to consult about the production of optical glass at the Imperial Porcelain Factory. He readily acquiesced to the idea of working on war-related needs alongside scientific research and started attending a vocational school to learn the professional technician's skill of glass cutting and

---

[8] (Obreimov 1973, 34). On Rozhdestvensky's biography and his studies of anomalous dispersion, see (Frish and Stozharov 1976; Gulo and Osinovsky 1980) and the essay by S. E. Frish in the volume of selected papers by Rozhdestvensky (1951). On Ehrenfest see biographies by Frenkel (1977) and Klein (1970). On Joffe, see books by Josephson (1991) and Sominsky (1964), and a volume of recollections by his colleagues and students (Ioffe 1973).

polishing (Gulo and Osinovsky 1980, 67). Meanwhile, Grebenshchikov traveled to Britain to learn the Allies' know-how. With his and other consultants' guidance, the optical glass workshop delivered its first quantities of glass to instrument factories in May 1916. By the time of the revolution a year later, the workshop had produced approximately three tons of glass with an expanding assortment of types but still unstable and of barely acceptable quality. To solve this problem, Rozhdestvensky set up a special research group at the university and also added a calculating bureau to design optical instruments. In their ambitious plans, Russian scientists envisioned an entire research-cum-production complex with an advanced laboratory, a specialized optical glass factory, and a factory for manufacturing optical devices (Grebenshchikov and Rozhdestvensky [1918b] 1993, 22; Frish and Stozharov 1976, 28; Gulo and Osinovsky 1980, 67–68).

The February revolution and the end of the monarchy initially inspired huge public enthusiasm, but the mood quickly turned sour as the new Provisional Government was unable to control persistent conflict and revolutionary chaos spread rapidly across the country. By the end of 1917, as a result of unrest among workers and the new administration's change of priorities, the production of optical glass at the Porcelain Factory was disorganized, and cadres of qualified workers fled. Rozhdestvensky saved the research part of the enterprise by transferring it in May 1918 to Vernadsky's KEPS, where it was reconstituted as the Department of Optical Technology. In the spring of 1918, KEPS became one of the first academic institutions, if not the very first, to begin receiving subsidies from the new Bolshevik government and submitted proposals for several research institutes. Among other plans discussed within KEPS during the summer of 1918 was an Optical Institute (Gulo and Osinovsky 1980, 69).

Meanwhile, Rozhdestvensky received another offer from Petrograd medical circles. Since 1913 Mikhail Nemenov (1880–1950) and several other activists among medical doctors were trying to organize a Russian Society of Roentgenologists and Radiologists. They finally succeeded in 1916 and in December of that year convened the society's first all-Russian congress in Moscow to discuss the uses of X-rays in war medicine (Troitskaia 1920). According to Nemenov's recollections, in 1913 he submitted the first proposal to turn his X-ray laboratory at the Women's Medical Institute into a radiological institute, but with the start of the war the project was shelved. In March 1918 Nemenov turned with the same idea to the Commissar of Enlightenment in the new Bolshevik government, Anatoly Lunacharsky (1875–1933), and with some astonishment saw his proposal approved within a few days. Acting quickly in the revolutionary manner, Nemenov appropriated or usurped for the new institute several buildings of the former Lyceum, a school for nobles, and medical equipment from the community war clinic, where he served as the chief physician (Nemenov 1920; 1928, 2–6). Because of tensions with other physicians and especially with the administration, the State Roentgenological and Radiological Institute (GRRI) separated completely from the Women's Medical Institute in October 1918, thus also separating teaching and research.

In September Lunacharsky approved the organization of the institute's second division, Physico-Technical, under the directorship of Joffe. The division's task was to study physical phenomena associated with X-rays, to test, repair and produce X-ray tubes and other equipment for Nemenov's Medico-Biological division. After his 1915 appointment as professor at Petrograd Polytechnical Institute, which was an engineering school, Joffe's professional philosophy moved away from purely academic research in physics towards the idea of a special relationship between physics and engineering. Such views resonated strongly with the general wartime changes in academe and with the Bolsheviks' ideology with regard to science and technology. Joffe gladly accepted his new appointment and started building up the Physico-Technical division as the center of research on the physics of X-rays and crystals. To train researchers for his new institution he organized in 1919 a new teaching department—Physico-Mechanical—at the Polytechnical Institute, whose graduates received diplomas with the dual title 'physicist-engineer.' Besides Physico-Technical, two more research divisions were envisioned at the GRRI, according to the institute's statute adopted late in 1918: on optics—with Rozhdestvensky in mind—and on radium (Bastrakova, Ostrovitianov et al. 1968, 232–237; Nemenov 1928, 6–8; Sominsky 1964, 195–198).[9]

Obviously, in the atmosphere of sweeping revolutionary changes, there was more than one way that could lead to the establishment of an optical

---

[9] The optical division within GRRI never materialized, while the Radium division remained largely inactive after the early death of its director, L. S. Kolovrat-Chervinsky, until it was absorbed into Vernadsky's Radium Institute in 1922. In late 1921 GRRI split into two separate institutes, Medico-Biological and Physico-Technical. The latter, directed by Joffe, subsequently became so much more prosperous and famous under the name of Leningrad Physico-Technical Institute that historical narratives often overlook its early status as a division within the more complex, multidisciplinary GRRI. The exceptions are two detailed studies of the Physico-Technical Institute and of the activities of its director A. F. Joffe (Sominsky 1964; Josephson 1991).

institute, but Rozhdestvensky definitely preferred a third opportunity that arose almost simultaneously. With the start of the Civil War in the summer of 1918, the all-but-abandoned military production became important once again to satisfy the needs of the newly created Red Army. On 18 October the Chief Artillery Administration under the new Red leadership requested funds from the Bolshevik government for the restoration of the optical industry.[10] On 15 November Rozhdestvensky received a letter from Nikolai Gorbunov, in which the secretary of the Soviet government and a chief official at the economic commissariat VSNKh inquired about the possibilities of resuming the production of binoculars for the Red Army. Rozhdestvensky, whose political views were socialist, replied without delay with an ambitious proposal for the revival of the entire optical industry through scientific leadership. Exactly one month later, upon receiving encouragement from Bolshevik officials, he invited twenty colleagues and researchers to the founding meeting of the State Optical Institute (Gulo and Osinovsky 1980, 70).

3. From the Optical Institute to an Optical Industry: Modernization, Soviet Style

A meeting convened by Rozhdestvensky on 15 December 1918 adopted and sent to the Commissariat of Enlightenment a proposal of a novel kind of academic institution "in which science and technology would be inseparably linked together." The document argued that it was impossible for an individual university laboratory to cope with the increased complexity and expenditures of modern theoretical and experimental optics. A larger, separate institution was thus needed to conduct planned research in the field in a centralized and coordinated way. At the same time, the new technological tasks of the optical industry and glass production required the increased involvement of science on an everyday basis. The proposed State Optical Institute promised to meet both these demands by fusing advanced scientific research in fundamental optics with the design and industrial production of sophisticated optical technology (Rozhdestvensky [1918] 1993; Frish and Stozharov 1976, 67).

---

[10] "Perepiska s...Petrogradskim opticheskim zavodom o snabzhenii etogo predpriiatiia opticheskim steklom, 1916–1919," (VIA, 504-10-89, pp. 425–426).

Accordingly, the Institute consisted of two departments—science and technology—with tasks defined as:

I. Scientific Tasks:
   1. Research on optics in the visible, infrared, and ultraviolet regions, in particular with regard to the questions of the constitution of the atom ...
   2. The creation and maintenance of central experimental devices for optical research in the Republic, devices that exceed the capacities of individual university laboratories ...

II. Technical Tasks:
   1. Research on optical instruments and devices ... in the Laboratory of Geometrical Optics;
   2. An Optical Workshop for the production of precision devices ...
   3. A Calculating Bureau for the design of large optical systems (also serving the needs of the State Obukhov and Pig Iron Plants), of national importance;
   4. Assisting the development of optical technology in Russia (Polozhenie [1918] 1993, 52).

The emphasis on collectivism, centralism, and large experimental devices that could not be managed by individual university laboratories and institutes justified the dimensions of the proposed enterprise. Designed with the goal of solving a complex technological problem, the institute ignored traditional disciplinary boundaries in science. It was de facto multidisciplinary in character and employed engineers together with scientists of diverse backgrounds. Thus, while the State Roentgenological and Radiological Institute simultaneously pursued research and new X-ray technology in medicine, physics, and geology, the Optical Institute combined physics, chemistry, mathematics, and engineering.

A separate appendix demanded further that the production of optical glass should be taken from the Porcelain Factory and transformed into a separate optical glass factory to be run by a "scientific collegiate body," made up mostly of representatives of the Optical Institute. An additional factory for the manufacturing of optical devices also had to be established and likewise subordinated to scientists. Scientists thus claimed responsibility not only for research and design of technology but also for controlling regular industrial production:

> The scientific element has to play the leading role in the factory, just as it played de facto before [during the World War]. Seriously, the problem of optical glass is more a scientific rather than a technical problem, and the

factory would thus appropriately belong to the Commissariat of Enlightenment. Rather than just consulting, scientific cadres should have a decisive word in all main issues. Currently, consultants cannot get their recommendations fulfilled ... and the entire production halted. It is clear that the factory administration has to be put under the real and powerful control of a scientific *collegium*, which clearly understands the national importance of the production of optical glass and, together with the factory administration, determines technical plans and their fulfillment (Gulo and Osinovsky 1980, 76–77).

On 26 April 1919 Lunacharsky as the Commissar of Enlightenment signed the official decree establishing the State Optical Institute. The laconic text approved the main principles proposed by scientists—the institute with two departments and with the optical glass factory subordinated to it—but did not specify exactly how the factory had to be administered (Gulo and Osinovsky 1980, 81–82). Without waiting for official approval, Rozhdestvensky and his colleagues started the Institute de facto in just a fortnight after the initial founding meeting. The Institute's first list of staff included 35 specialists (Polozhenie [1918] 1993, 51). In January 1919 they hired the first dozen additional "technicians," selected from among the student body of the University.[11] While continuing to take classes at the physico-mathematical department, students received their first practical assignments at the State Optical Institute along with food rations—a matter of life and death in a freezing and starving Petrograd. The institute itself was initially housed in several rooms of the university's Physics Institute until Rozhdestvensky managed to get hold of the adjacent empty building of a nationalized candy factory. With thick walls, deep cellars, and spacious rooms, the building was found perfectly suitable for conducting experiments in optics. In February 1919 the newly organized institute hosted the first Congress of Russian physicists, which established the national learned society, the Russian Association of Physicists (Kononkov and Osinovsky 1968).[12]

---

[11] Several of them later became famous scientists: Vladimir Fock, Aleksandr Terenin, Lev Shubnikov, Sergei Frish, and Evgeny Gross (Gorokhovsky 1968, 12). Like many other scholars, Rozhdestvensky initially hoped to retain his teaching position alongside his new research job at the institute. In 1919 he pushed forward a major reform of the physics curriculum at Petrograd University. By the mid-1920s, however, he had to quit his university professorship because of disagreements with Narkompros over teaching methods, and he then concentrated on research at the Optical Institute (Frish 1992, 129–130). The episode is symptomatic of the growing administrative separation between advanced teaching and advanced research in the Soviet system.
[12] The subsequent activities of the Russian Association of Physicists were studied in (Josephson 1991).

While creating unprecedented opportunities for radical reforms and institutional novelties, the revolution and the civil war also destroyed most of the material conditions for normal work, especially in Petrograd and other big cities overwhelmed by cold, starvation, and criminal anarchy in the streets. Industrial production was disorganized and skilled workers fled to their native villages, with the result that optical glass production was halted in 1919.[13] By year's end Rozhdestvensky could report only a theoretical piece of work, though one with revolutionary aspirations. Typed invitations to his presentation at the Optical Institute in December 1919 assured Petrograd physicists of an incredible luxury: a heated lecture hall. The audience had to arrive on foot, since the city trams were not running, and listened to a speaker who was feeling "very weak, but with a mind exceptionally clear thanks to malnutrition" (Gulo and Osinovsky 1980, 88). In his presentation, Rozhdestvensky challenged the Sommerfeld theory of 1916, the improvement of the 1913 Bohr atomic model and one of the last pieces of scientific news to reach Russia before the total break of communications with colleagues in Europe. He generalized the elliptic orbital model for alkali metals, succeeded in classifying their spectra, and, most important, concluded that atoms have an inner magnetic field which—rather than relativistic effects—caused the doublet separation of spectral lines (Rozhdestvensky 1920). Having no other means of relating the news to foreign colleagues, Petrograd physicists broadcast a radio message addressed to Lorentz and Ehrenfest in Leiden:

> Professor Rozhdestvensky succeeded in proving that spectral series of all elements correspond to Sommerfeld's ellipses. The normal structure of lithium is established. Doublets in series are caused by the magnetic field of the inner shells. Krutkov is working on the anomalous Zeeman effect and on the normal effect in strong fields. Bursian is studying the electric influence of inner shells. We have no literature since the beginning of 1917. The collegium of the Optical Institute requests information on what has been done on related problems outside Russia. Please address the radio message to Rozhdestvensky, University, Petrograd. The help is kindly requested from the Amsterdam Academy in shipping scientific literature. Greetings from the Petrograd physicists (Levshin 1986, 155–156).

Soviet newspapers published inflated reports of a "world-class scientific discovery," the proof of the divisibility of atoms and of the Soviet

---

[13] The Petrograd population fell from 2.42 million in 1917 to 720,000 in 1920, partly from disease, cold, and hunger, but mostly because people fled to the countryside to be closer to supplies of food.

government's generous support of science. A different ideology—the obsession with world domination—showed up in the misrepresentation of the radio message in British media. The newspaper *Nation* commented on 20 November 1920 that if the information about a Russian scientist's mastery of atomic energy is true, the owner of the secret would be positioned to rule the entire planet. Upon receiving a report from Rozhdestvensky, the Commissariat of Enlightenment approved funding for the special Atomic Commission to work on the mathematical theory of the atom, which marked the beginning of Soviet theoretical physics. For almost two years, the Commission provided twenty physicists and mathematicians—including Vladimir Fock (1898–1974) and Aleksandr Friedmann (1888–1925)—with Red Army military rations, which undoubtedly saved several lives. During this time, Friedmann produced the first really great achievement in Soviet physics: two papers on non-stationary solutions in Einstein's general relativity, the cornerstone of all modern theories of the expanding universe (Friedman 1922; Friedmann 1924; Gulo and Osinovsky 1980, 92–96; Tropp et al. 1993).

As the Civil War drew to a close, the Academy of Sciences sent a report to the Soviet government on the critical situation in Russian science and the urgent need to restore its contacts with the West.[14] Several senior scientists—including Nemenov, Joffe, Krylov and Rozhdestvensky—were sent abroad in late 1920 and early 1921 with the primary goal of purchasing scientific literature and instruments. Rozhdestvensky received the astronomical sum of 200,000 hard currency rubles (an estimated $80,000) to equip the Optical Institute. Inflation in Germany allowed him to use this money very effectively. Amply outfitted with the most up-to-date instruments, the institute counted 86 staff members (more than half of them scientists) in 1922 and embarked on intensive research in several directions: spectroscopy and quantum theory, optical glass technology, geometrical optics and the design of optical devices, physical and later also electron optics, photography, photometry, and photochemistry.[15]

The same team that tried to make optical glass during the war—the engineer Kachalov, the chemist Grebenshchikov, and the physicist Rozhdestvensky—now took on the task of organizing its full-scale production. After a serious fight with the administration of the Porcelain

---

[14] "Zapiska rukovodstva Akademii nauk v SNK ob okazanii material'noi i organizatsionnoi pomoshchi nauchnym uchrezhdeniiam," 22 November 1920, published in (Levshin 1986, 174–178).

[15] On Rozhdestvensky's foreign trip and purchases see (Gulo and Osinovsky 1980, 98–104).

Factory and with industry officials who insisted on continuing imports, Rozhdestvensky pushed forward the decision to separate the optical glass workshop as an independent factory under Kachalov's directorship. Appropriate funding came only at the end of 1923, when the factory changed its affiliation from the Commissariat of Enlightenment to the much more powerful economic commissariat, VSNKh. Production resumed in February 1924 with monthly output rising to six tons by the summer. Problems with quality lasted well into 1926, when Optical Institute scientists developed the original know-how that enabled the Soviet Union to stop importing optical glass after 1927 (Grebenshchikov 1933; Rozhdestvensky 1932; Gulo and Osinovsky 1980, 105–130).

The initial success formed the basis for the Optical Institute's subsequent leading role in the creation of the entire Soviet optical industry. The institute's staff grew in the process to 240 in 1931 and 600 in 1936 (Gorokhovsky 1968, 13, 21) while the Soviet Union developed into an independent producer and later exporter of optical technology. The work done at the institute ranged from Fock's fundamental theories of the atom (the Fock–Hartree method) and Aleksandr Terenin's discovery of the superfine structure of atomic spectra to the design of the entire spectrum of optical devices—lamps, cameras, microscopes, telescopes, sights, etc.—for both civilian and military use. The balance between different aspects of research—fundamental and applied, military and civilian, classified and open—was changing with time and often created problems, especially in the early 1930s, when Rozhdestvensky resigned from the directorship over disagreements with industrial officials.[16] The basic principle of combining science with technology, and physics with chemistry and engineering, however, remained in force for the entire Soviet period.

The experience of the optical and other early institutes set the paradigmatic example for the general Soviet style of industrial modernization, which was driven primarily by science rather than by market mechanisms. The mainstream method of creating new technology in the Soviet Union—from 1920s and at least until the end of the Stalin period—consisted in setting up a special research institute in association with

---

[16] The complicated later history of the Optical Institute remains to be written. For the in-house history of the institute and its fifty years of research in fundamental, applied, and technological directions, see (Gorokhovsky 1968). For revealing informal recollections by one of its leading physicists, see (Frish 1992).

Socialist, or Big, Science

an experimental factory or workshop, whereby the institute and its researchers enjoyed a substantial degree of authority over industrial production, rather than the reverse. As the key element of the Soviet modernization effort, such institutes—often nicknamed "the general staff" of the corresponding industry—shared responsibility for many of its successes and failures.

4. The Transition to Big Science

As the Civil War came to an end, the academic community reviewed in the pages of its new mouthpiece, *Science and Its Workers*, what had happened to science in various parts of the country, in the capitals as well as in Siberia, Ukraine, and the Crimea. In the journal's editorial manifesto, KEPS secretary Fersman formulated the two main tasks facing Russian scientists. The first—international—was to restore contacts with foreign colleagues, to resume the exchange of ideas and information, and to "fight with all forces of persuasion against the chauvinism of some countries that resist contacts between former enemies." The second—domestic—was to foster new collective forms of the organization of research and to build up the "state network of scientific research institutions," especially applied ones. In an exalted style that reflected the characteristic revolutionary utopianism, Fersman anticipated radical changes in institutions as well as in the content of science:

> In these years, when the foundations of a new scientific worldview are laid, when old schemes are broken, when the great new achievements that apparently contradict common sense are growing upon the ruins of what was thought to be the unshakable truth; in these years when the bold flight of creativity of the scientist can only be compared to the fantasy of the poet; in these years new forms of scientific work are forged, and scientific thought imperatively seeks new paths, along which it would victoriously proceed towards great new accomplishments (Fersman 1921, 3).

Vernadsky echoed this message in a private letter from Paris to his émigré friend and former political comrade-in-arms Ivan Petrunkevich:

> It seems to me that here [in the West] they do not recognize the huge cultural task that has been accomplished, accomplished in the face of sufferings, humiliations, destruction.... Of course, in a police state ... freedom

is relative and it is necessary to defend it continually. Much new has been created in Moscow and Petersburg, de facto much, although by comparison with the plans of 1915–1917, little. And curiously enough, much has been created in the provinces.[17]

New scientific forms were, indeed, forged out of the innovative World War I proposals modified by revolutionary visions. Although their development was anything but orderly, there were certain important tendencies. Most of the new institutes were organized by professors and some remained affiliated with universities and colleges, yet increasingly they tended to be institutionally separate from higher education. The favorite form—a state-sponsored research institution directed simultaneously towards advanced research and utilitarian service—was closer to Vernadsky's proposal of 1916 than to Timiriazev's of 1911. The popular organizational principles emphasized centralism, planning, collectivism, and a quantum leap towards largesse in all quantitative measures—personnel, instruments, networks—over the previous generation of research laboratories. The magnificent practical, even if not always realistically achievable, goal reflected the typical revolutionary combination of utopianism and utilitarianism. The hope was often founded upon a radically novel theory or technology with revolutionary symbolism, such as radio, aviation, automobile, genetics, X-rays, or radioactivity. Appropriate research and development enterprises usually had to be multidisciplinary in character. Their tasks often included production—manufacturing X-ray tubes, making optical glass, designing aircraft, or improving the sorting of grain—and many of the institutes had experimental factories or production units closely associated with or administratively subordinated to research, rather then reverse.

Loren Graham has described the birth of the characteristically Soviet system of research institutes as "the combination of revolutionary innovation and international borrowing" (Graham 1975). The present study shows that that formula needs to be modified. The new R&D system was, true, a revolutionary innovation, but one that had started already before the revolution itself and did not originate from a particular political or ideological agenda of a revolutionary party. Instead, it came from scholars who responded to the crisis of World War I and to the general revolutionary situation in the country, which contributed to the radicalism of their

---

[17] Vernadsky to I. I. Petrunkevich, 10 March 1923, quoted in (Bailes 1989, 291).

proposals. After the revolution, Bolsheviks accepted, endorsed, and internalized many essential elements of the proposed reform of science not only because they were almost as strongly pro-science as researchers themselves, but also—and no less important—because they, too, operated in the same situation of the war and revolutionary crisis. The values and interests of scientists and of the Soviet government overlapped, particularly, in the idea of a research institute, which helped materialize the proposed reform and gave it its specific spin.

A kind of international borrowing also existed, but a rather ironic one. Exact borrowing between different cultures is hardly possible even under the most favorable conditions. In the story presented above, Russia's international isolation during the war appears to be of much more importance than the desire of its academics to imitate the "West." Russian scientists sensed and shared the general international trends of the period, such as the growing recognition of science as a profession, the tighter links between research and technology, especially military technology, and the resulting increase of governments' interest in science and science policy. They presented their proposals—sincerely as well as rhetorically—as following and catching up with other major scientific nations. The real information they had about the developments in other countries, however, not only was severely limited but was reinterpreted by them in a rather idiosyncratic way. As a result, Russian scientists designed what was, in fact, a novel system of research and development.

During the same period, scientists in other European countries, including Britain, advanced similar proposals for "centrally planned and government-funded" science, public science policy, the opening of new institutions for linking advanced research to industrial and military needs, and an increased political role for scientists (Turner 1980; Wang 1995; Trischler 1990; Hull 1999). In France, Marie Curie lobbied for the organization of a laboratory-factory, supervised by scientists, funded by the government, and responsible for running the national radium industry (Boudia and Roqué 1997, 253; Roqué, forthcoming). Even in the United States, less affected by the war, George Hale tried to reform the National Academy of Sciences, create centralized national laboratories, and connect pure science with military preparations (Tobey 1971, ch. 2). Some of these initiatives failed, while others succeeded partly, or temporarily for the period of war, fading afterwards. In Civil War Russia, assisted by the revolutionary desire for radical change, scientists managed to realize similar proposals in a much more comprehensive, abrupt way, and to preserve the result during the ensuing period of peace under the notion of a

"Soviet (or socialist) model of research."[18] This institutional model started gaining international influence already in the 1920s, when it was borrowed by the nationalist Guomindang regime in the Republic of China (Greene 1997).

Elsewhere, most notably in the United States, comparable changes occurred later, typically as a result of another World War, and survived afterwards in the form of national laboratories and other institutions, under the unofficial name "big science" (the label "socialist" could not be used during the Cold War for obvious reasons). Despite the difference in names, the two phenomena are homologous and apparently are parts of the same social process. They share most of the essential features, such as gigantomania, state support, the cult of science in society, (con)fusion between science and engineering, multidisciplinary research, collective or team work, complex bureaucracy, and militarization. Historians have already referred to Soviet science as the ultimate example of 'big science,' although without explicitly pointing out its historical precedence.[19] The mentality and propaganda of the Cold War period paid more attention to ideological contrasts and oppositions between the rival systems while usually inventing different names for or turning a blind eye towards similarities. Looking back now at the experience of the past century from a less partisan perspective reveals ever more common trends.

---

[18] The episode is not an isolated phenomenon but part of a more general trend, in which the Russian revolution served as a springboard for jumping from backwardness to utopia, pioneering and testing (sometimes prematurely, or ahead of better-prepared nations) many of the essential social and economic innovations of the 20th century (Carr 1947; Hoffmann and Holquist, forthcoming).

[19] See for example (Graham 1992). The basic historical account of big science, however, overlooks the existence of an earlier Soviet version (Galison and Hevly 1992).

Chapter 3

Freedom, Collectivism, and Electrons

The revolution was a powerful experience that transformed the views and perceptions of the world for most of those who lived through it, regardless of their political beliefs. The depth of its impact on the mentality of its contemporaries is perhaps hard to fully appreciate today, in a stable and relatively uneventful world with its correspondingly different, conservative intellectual atmosphere. Then, in the midst of the Civil War, the visions of revolutionary writers, artists, and many ordinary people were as grandiose and exultantly utopian as those of a religious eschatological sect, only for them catastrophic events signified not merely the end of the world, but the simultaneous birth of a completely new one. They saw the entire planet, not just Russia, as undergoing a radical transformation and often metaphorically characterized events from a cosmological perspective, as an astronomical cataclysm and the creation of a new Universe. It is thus befitting that those years also saw the first draft of a new cosmological theory in science—one in which Universe was not stable and static but born and expanding over time, the theory subsequently nicknamed "the Big Bang."

The expanding Universe's first appearance was actually not grandiose at all, but very modest, consisting of two short papers published by the Petrograd mathematician Aleksandr Friedmann in the German journal *Zeitschrift für Physik* (Friedman 1922; Friedmann 1924). Friedmann found that the main equation of Einstein's new general theory of relativity allowed a non-stationary mathematical solution for the dimensions of the entire Universe. At the time, this solution was seen primarily as a rather imaginary scenario, a mere mathematical possibility. Yet the basic preference for a stable or non-stable Universe was, of course, not just mathematical, but bordered on metaphysical, as reflected in the fact that Einstein's initial reaction to Friedmann's first paper was a certain disbelief and suggestion that there must be a mathematical mistake in it (Frenkel 2002). Although Einstein conceded the correctness of non-stationary solution a year later, he also proposed to modify the basic equation of general relativity in order to restore the cosmological stability. Perhaps nobody could foresee at the time that forty years later Friedmann's solution would become the basis of mainstream cosmological theory.

To what extent abstract formulae coming out of a mathematician's mind can be influenced by events in the outside world is, of course, hard to investigate. Whether there could be a real, not merely symbolic, connection between the world around Friedmann in the spasms of violent transformations and his mathematical vision of a dynamic, expanding Universe remains a challenge for some future study. This chapter presents a case in which a similar question can be answered with sufficient confidence. The case is that of some revolutionary ideas applied to the description of the behavior of atoms and electrons. The origins of the collectivist approach in the quantum theory of solid state were also rather humble in 1920s and '30s, but forty years later they, too, developed into the mainstream foundation of the modern quantum physics of condensed matter and many-body systems.

1. Freedom as Problem

What kind of freedom do scientists have in mind when they say that an electron or another particle is "free"? The most common model of a system of free particles is an ideal gas, in which atoms are rare and move unfettered, interacting only when they directly collide. Yet some physicists, such as Yakov Frenkel, whose specialty was the quantum theory of matter and whose political views were socialist, were sensitive to the fact that in this presumably ideally free system atoms are free but electrons are not. They are, on the contrary, enslaved by free atoms. If, however, atoms are packed together closely into a solid body, they lose most of their freedom and become largely confined to specific loci of a crystal, while in this same process electrons gain in freedom as they become liberated from individual atoms.

Liberated into exactly what state? The answer is not simple. In a metal, for example, electrons transport electric current and therefore are apparently free enough to move through the solid body. On the other hand, they are still subject to very strong forces from the atoms of the crystal lattice as well as from other electrons. This complex situation led to differences in opinions and approaches developed by physicists. Some physicists based their calculations on a simplified model of electrons as, in the first approximation, belonging to their proper atoms and only occasionally switching their allegiances. Others envisioned electrons as almost free, an ideal gas of their own, and downplayed the possible impact of interactions, treating them as minor disturbances. Frenkel and some similar-minded colleagues felt that both alternatives simplifying assumptions—free and bound electrons—were inadequate as representations of the more complex actual state of the electrons' freedom in dense bodies. He also had a special word in his vocabulary to refer to this state, calling it 'collectivist.' The idea came from leftist political language and social theory.

Disagreements over the large issue of freedom played a particularly important role during the early formative stages of the quantum physics of the solid, liquid and plasma states of matter—or condensed matter in current usage—from the 1920s through the '50s. At stake were not only the language proper but also the mathematical models and conceptual foundations of an emerging scientific discipline. A variety of specific approaches and theories that existed and competed during that period rested on their authors' conflicting intuitions regarding the freedom of particles. In their attempts to conceptualize these intuitions in physical and mathematical terms, physicists often used social metaphors, implicitly as well as explicitly, consciously as well as unconsciously. These metaphors reflected their varying interpretations of the general concept of freedom, their political philosophies, and also—no less important—personal and often incompatible existential experiences of social life in different countries and regimes.

The present chapter describes the origins of one important line in this debate: the collectivist approach in the early history of condensed matter physics. This approach was initially proposed by socialist-minded physicists, who were trying to design mathematical methods for the description of the collective behavior of particles and relied in their search on collectivist metaphors and on the collectivist understanding of freedom.

Their efforts led to the invention of a new fundamental concept in contemporary physics, or, if one prefers, to the discovery of new fundamental objects in the world: a family of physical models that are currently called 'quasiparticles' or 'collective excitations' (Kojevnikov 1999).[1]

'Collectivism,' like 'freedom,' was used and abused so much in political propaganda that it seems necessary to start with some clarifications of its meanings. As if to remind us that words sometimes have tricky trajectories, 'collectivism' as a political term originated with opponents of Marxism. It referred to the theory that the means of production should be owned neither by private individuals nor by the state but by free associations of laborers. The conflict between Mikhail Bakunin and other proponents of this view on the one hand and Karl Marx and his followers on the other led to the split of the First International in 1872. The new political movement that formed thereafter accepted "collectivism" as its program and "anarchism" as its name (Nettlau 1996, ch. 8). The anarchists were the Marxists' main rivals within the international workers' movement, having strongholds in France, Switzerland, Spain, and Italy. They continued to be a powerful force within the European left until their last important organizations were exterminated in the 1930s, especially during the Spanish Civil War.

Much earlier, however, by at least 1900, anarchists had lost their monopoly over the term 'collectivism.' Although they and some others continued to use it in defense of workers' freedom against both private and state property, the word was also appropriated by virtually all socialist factions. Its meaning changed, too, as it began to be used ever more often as a vague synonym for anti-capitalist values, while its critique of étatisme weakened. Socialists accepted collectivism as an alternative to the liberal individualist concept of freedom, noting that the latter often went hand in hand with exploitation and slavery.[2] Collectivism, for them, was the true strategy of liberation for the oppressed and their only way to succeed in the struggle for freedom.

The next major change of meaning occurred owing to collectivization, a violent reform of Soviet agriculture around 1930 that deprived peasants of their private plots, uniting them into big collective farms, or *kolkhozy*. Legally, the Soviet *kolkhoz* was a cooperative of peasants, not a state enterprise—a large concession from the point of view of hard-core

---

[1] For more information on the history of other main approaches in the physics of condensed matter, see (Hoddeson *et al.* 1992).
[2] On this and other antinomies of freedom see for example (Bauman 1988).

Marxists. The *kolkhoz* thus had some formal resemblance to the original anarchist program, although it in fact was anything but a free association. The results of collectivization were notoriously disastrous. Among other, more important things, they also damaged the reputation of collectivism, showing that what had been conceived as a liberation concept could also become a method of enslavement. In this as in a number of other important historical cases, the pursuit of a new type of freedom in the political realm turned into the emergence of a new type of dictatorship. A similar pursuit in the realm of physical models produced a different outcome, leading to the discovery of new kinds of natural objects.

New objects started to appear in solid state physics around 1930, when all the meanings of collectivism discussed above were still in wide circulation. Their current name, quasiparticles, is of relatively recent origin, having been first introduced after World War II. In their early decades, these new objects were most commonly referred to as 'collectivized' particles or 'collective excitations.' A basic idea indicative of their name may be illustrated by two simple examples. Consider a string of connected atoms, one of which is in an excited state of higher energy. Since it interacts with neighboring atoms, it can give this energy to one of them; the excitation can thus move from one neighbor to another though the atoms themselves do not move. The excitation's movement along the string can be described mathematically in a fashion very similar to the movement of an ordinary particle, and it received the particle-style name 'exciton.' Or consider a crystal lattice in which one place is unoccupied. If a neighboring atom receives some extra energy, it can jump to the vacancy, leaving behind an empty space into which another atom can move, and so on. The vacancy will thus be traveling through the lattice in a manner similar to the movement of a particle. In this case, actual atoms move, too, but rather than analyzing the behavior of thousands of atoms, it is much more convenient to describe the process of their collective movement by means of just one, albeit fictional, particle, the 'hole.'

Given quasiparticles' fundamental importance as the basic language in many fields of modern science, surprisingly little has been published about the history of the concept. A short version of the story was told by two physicists, Charles Walker and Glen Slack (1970), driven mainly by curiosity to find out "who named the -ON's" (both quasiparticles and ordinary particles). They did not restrict the inquiry to names but tried to determine for each particle who was the author of the concept and who invented the name for it. Their findings for quasiparticles are presented in the table. Not all their attributions are beyond doubt, nor is their

EARLY QUASIPARTICLES	CONCEPT	NAME IN PRINT
Phonon	Tamm, 1930	Frenkel, 1932
Magnon	Bloch, 1930	Pomeranchuk, 1941 Referring to Landau
Exciton	Frenkel, 1931	Frenkel, 1936
Polaron	Landau, 1933	Pekar, 1946
Roton	Landau, 1941	Landau, 1941 Referring to Tamm
Plasmon	Bohm and Pines, 1951	Pines, 1956
Polariton	Fano, 1956	Hopfield, 1958

Early quasiparticles and the physicists who invented and named them. Until around 1950, the approach was developed mainly by Soviet authors.
[*Source*: (Walker and Slack 1970). A similar list can also be found in (Kaganov and Frenkel 1981, 28–35, 51–54).]

list of quasiparticles complete, but it gives a useful first approximation to important places and names. With the exception of Felix Bloch, all of the pre-1950 names on their list belong to Soviet physicists. Indeed, the method of collective excitations was developed primarily within the Soviet Union until the late 1950s, when it lost its national specificity and gained worldwide acceptance.

This chapter describes the earliest stage in the history of quasiparticles, introducing the two pioneers of the concept: Yakov Frenkel, the initiator of the entire collectivist approach and the author of the model of 'collectivized particles,' including the hole and the exciton, and Igor Tamm, whose phonon later became the paradigm for a somewhat different model, 'collective excitations.'

2. The Origin of the Collectivist Metaphor

Yakov Il'ich Frenkel (1894–1952) grew up in the family of a Jewish revolutionary in the last years of the Russian Empire. His father, Ilia, had been a member of "The People's Will," an underground organization that prepared terrorist attacks against leaders of the monarchical regime. After a six-year exile in Siberia, Ilia Frenkel withdrew from active politics and became a small merchant. He continued to support illegal revolutionaries (Frenkel 1996, 1–4), however, and sympathized primarily with the Socialist Revolutionaries, or SR, a radical non-Marxist political party oriented towards the peasants rather than the nascent industrial proletariat. SR envisioned Russia's way to a future socialist society as being through the *mir*, the traditional communal organization of the Russian village, in which peasant families owned the village land collectively and periodically redistributed plots among themselves. The first national elections after the March 1917 revolution and the fall of the monarchy gave the majority of votes to socialist parties in general and to SR in particular, yet power in the industrial capitals fell to the Marxist Bolshevik party in the November uprising. SR's leftist faction initially formed a coalition with Bolsheviks in the first Soviet government but revolted later in 1918. Many other SR members joined forces with opponents of the Bolshevik regime during the unfolding Civil War.

Frenkel the son had just graduated from Petrograd University at the time of the revolution and on the day of the Bolshevik coup in the capital was busy there taking his major exam in physics, a formal prerequisite for obtaining the first teaching position at a university. The collapse of the *ancien* regime allowed local initiative groups to establish dozens of new universities all over Russia. Frenkel, who had already published papers in leading physical journals, was offered a junior position at the first of these post-revolutionary schools, Tauride University in the Crimea, and moved there in early 1918. He shared his father's political sympathies and was critical of the Bolshevik government, but allegiance to the cause of the revolution was more important to him than distinctions among socialists. Although regretting that the revolution came to be headed by Marxists, Frenkel concluded in a letter to his father that it was "late now to struggle with the Bolsheviks; we have to help them diminish the negative results of their policy and enhance the positive ones. On the other hand, ... I am too far from active politics ... and am not at all inclined to exchange my science for it" (Frenkel 1996, 20–21).

In the Crimea, Frenkel combined university teaching with membership in the governing board of the Commissariat of Enlightenment of the local Crimean Soviet.[3] Political power in the south alternated many times in the course of the Civil War. As the Whites made the Crimea their stronghold in the summer of 1919, Frenkel was jailed for having worked in the Red administration, and only the fact that he was an academic, a profession

---

[3] His post, according to one account, was deputy commissar of the enlightenment of the Crimean republic (Frenkel 1970, 8).

somewhat respected by both sides in the cruel conflict, saved his life. Not yet knowing that he would be so lucky, he tried to comfort his mother, philosophizing about sitting in jail:

> "I am not pining in the least; rather I am occupied with reading Drude and Grave.... If one doesn't give oneself up to thoughts about what could be ... then it's just like living in clover, like being in a sanitorium. The whole difference is that in a sanitorium there are usually rooms that lock from the inside, and in prison, the cells lock from the outside" (Frenkel 1997, 204).

The books Frenkel mentions here are a textbook on higher algebra by Kiev mathematician Dmitry Grave and the treatise on optics by the German physicist Paul Drude (1900).

Drude's treatise of 1900 contained his famous electron theory of metals, which rested on the physical assumption that electric current consisted of the movement inside the metal of a gas of free electrons (then recently discovered entities). Drude's model gave a satisfactory quantitative explanation of the ratio between the electrical and thermal conductivity and, in an improved form given by H. A. Lorentz a few years later, was the best available pre-quantum theory of metals (Hoddeson *et al.* 1992, 27–31). Reading Drude's book while sitting behind bars perhaps helped Frenkel to realize that electrons inside lattice cells can hardly be free. (Incidentally, the Russian word for "bars" of a jail and for "lattice" of a crystal is the same, *reshetka*.) He verified the idea with a calculation based on the virial theorem, demonstrating that electrons are contained within the body of the crystal by bonds stronger than those inside an individual atom. Whether Frenkel developed his model in the Crimea or a couple of years later is hard to document, since he was able to publish it only in 1924, after the Civil War had ended, Russian academic life and publishing had resumed, and he had returned to Petrograd to work with Joffe at the newly founded Physico-Technical Institute (Frenkel 1924a; 1924b).

Frenkel proposed to replace the classical Drude–Lorentz model with a theory based on quantum ideas. Quantum mechanics remained to be formulated; he relied in this first attempt on Bohr's atomic theory, according to which electrons circled around the atomic nucleus on elliptical orbits.[4]

---

[4] Bohr's theory of 1913 managed to arrive in Russia despite the interruption of scientific contacts and literature caused by World War I and the revolutions. Frenkel reviewed it for the journal of the Russian Physico-Chemical Society and used it in his first original paper in 1917. Some scientific journals from Germany arrived in the Crimea during the short period in 1918 when the peninsula was occupied by the German army (Frenkel 1970, 10). Supplies of foreign scientific literature to Soviet Russia resumed in 1921 and by 1924, when Frenkel was writing his paper, he must have had adequate access to recent journals.

In metals, Frenkel calculated, atoms are forced so close to each other that their outermost orbits overlap. Before completing the full ellipse, an electron would come close enough to a neighboring atom and jump over onto an elliptical orbit around another nucleus. In Frenkel's model, the electric current in a metallic body was represented by electrons gliding from one atom to another in a chain, passing from one owner to another like land plots in the Russian village commune. These electrons were no longer owned by individual atoms, as in the gaseous state of matter, but neither did they become absolutely free, as in the ideal gas of the Drude–Lorentz model. Frenkel called their more complex state of freedom 'collectivist' and summarized the essence of his model in metaphorical terms borrowed from the language of the revolutionary era:

> In this way, valence electrons become "free" electrons, contributing to the electrical conductivity of metals. It must be noted that they are not free in the real sense of the word. On the contrary, they are bound to the body of the metal stronger than within isolated atoms. But they have become emancipated from the domination of particular atoms; they no longer belong to individual atoms but to the entire collective formed by these atoms. The quantum character of their motion can only be described, strictly speaking, as 'collectivist' (Frenkel 1958, 2: 57–58).[5]

The use of metaphors, including far-reaching anthropomorphic ones, characterized Frenkel's style of scientific creativity. In a student paper on the photoelectric effect in 1913, Frenkel had described electrons as "emigrating" from the surface of the metal at the same time as he himself—unsure whether the quota for Jewish students would allow him to study at a Russian university—was considering emigration to America (Frenkel 1996, 10, 15). In a letter of 1924 he referred to "inanimate objects, such as molecules, atoms, and electrons," as "microscopic inhabitants of the animate universe" and praised physics as being "not so much exact science as ... a drama or comedy of the life of atoms and electrons" (Frenkel 1996, 66). With these words, he expressed his admiration for Paul Ehrenfest, but

---

[5] The above is my translation from the Russian version of the paper. In its German version, the same passage reads as: "In solcher Weise werden die Valenzelektronen "freie" Elektronen, welche für die elektrische Leitfähigkeit der Metalle verantwortlich sind. Man muß beachten, daß sie nicht 'frei' in eigentlichem Sinne werden. Vielmehr sind sie noch stärker mit dem ganzen Metallkörper verbunden als in isolierten Atomen. Aber sie werden von der Alleinherschaft bestimmter Atome emanzipiert; sie gehören nicht mehr den individuellen Atomen an, sondern dem von diesen gebildeten Kollektiv. Was nun den Quantencharakter ihrer Bewegung anbetrifft, so kann dieser, streng genommen, nur 'kollektivistisch' bestimmt werden" (Frenkel 1924a, 218–219).

they actually tell more about Frenkel's own preferred style of doing physics. Throughout his career, he developed analogies between phenomena from very distant areas of physics, and not only physics, as long as this helped to make sense of things. Colleagues, especially Lev Landau, were often critical of Frenkel's imaginative visualization, occasional sloppiness in calculations, and promiscuous creation of models, preferring more dry, precise and consistently solid work. Some of the critics, however, later acknowledged that, although a number of his metaphors did not survive, others proved to have important value for physics, and praised Frenkel as "a generator of new ideas," even though he often left to others the critical work of checking, perfecting, and justifying them.[6]

In 1958, the editors of a posthumous edition of Frenkel's selected works acknowledged that his collectivist metaphor of 1924 had initiated a "profound development in the modern quantum theory of the solid state." They even observed further that the "collectivization of valence electrons takes place in all crystals," not only in metals (Frenkel 1958, 2: 58, 72). By that time, collectivist terms in physics had lost most of their direct political connotations. They had acquired a life of their own in professional parlance, textbooks, reviews, and technical papers. The initial metaphor had undergone, in Sabine Maasen's terms, not only *transfer* from political to scientific discourse, but also *transformation*: it had been furnished in its new setting with language, formalism, and meanings specific to the discipline of physics (Maasen 1995). This transformation, however, had not been easy or straightforward. Physical models and mathematical apparatus capable of describing the complex collectivist state of freedom were not available at the start, and it took quite some time and effort to develop them.

3. Liberated Holes

In a 1926 paper in which he was trying to develop a classification of various kinds of thermal motions in solid and liquid bodies, Frenkel reached from his model of collectivist electrons to the concept of the hole. Even in crystals, atomic ions are not bound absolutely to their positions in the lattice, as is apparent from the phenomenon of solid diffusion.

---

[6] See a description of Frenkel as a "romantic" and a characterization of his controversial style in (Tamm 1962, 174); and an expression of belated regret for youthful underestimation of Frenkel by A. B. Migdal, one of Landau's students, in (Frenkel 1986, 216–218).

George von Hevesy had proposed a mechanism for diffusion in solids as two (or occasionally more) ions simultaneously exchanging their positions in the lattice (*Platzwechsel*). In 1923 Joffe put forward a different hypothesis, according to which ions in a crystal enjoyed the greater freedom of occasionally leaving their proper places and wandering within the interatomic space, thus adding an ionic contribution to the total electric current. Frenkel now developed the idea of his Leningrad colleague and institute director one step further, by noting that the "dissociation" of an ion from its proper place would also liberate an "empty space" ("*ein leerer Platz*"), a vacancy in the lattice that would behave like a particle (Frenkel 1926).

Frenkel's liberated empty spaces of 1926 traveled through the lattice in the same way as his collectivist electrons of 1924, by jumping from one atomic position to a neighboring one, and thus enjoyed a similar degree of freedom. An elementary act in this process can be described either as an ion jumping to a nearby vacant place or as a vacancy moving in the opposite direction. Frenkel thus characterized an empty space as a "negative atom," an "ion of the opposite sign." It can travel through the lattice until it meets a liberated wandering ion, with which it can recombine into an ion properly fixed in the lattice. In later Russian-language publications Frenkel referred to vacancies in the lattice as "*dyrki*" (holes), which by that time was also the Russian term for the electron holes, or anti-electrons, of Dirac's famous hole theory of 1929 (Dirac 1930). The two concepts, one in solid state physics and the other in quantum electrodynamics, have much in common, indeed, and it is also very likely that Dirac had known of Frenkel's lattice holes at least since their conversations on board a steamer in 1928, during the week of the Volga congress of Russian physicists.[7]

Having played a heuristic role in the initial development of the idea of holes, the analogy with collectivist movement became less necessary once the concept was in place. Frenkel could operate both with and without the collectivist metaphor when he discussed holes in his multiple papers and books. Dirac was probably not aware of it at all; at least he did not use collectivist terminology in his 1930 papers on the hole theory, which made the concept widely known, if not widely accepted, among physicists. Dirac's theory had many opponents who, in particular, did not like its

---

[7] On Dirac's hole theory see (Moyer 1981; Kragh 1990, 87–105). On his visit to the Soviet Union in 1928 (Dirac and Tamm 1993–1996).

Ya. I. Frenkel and P. A. M. Dirac on board a steamship, Volga, 1928. The seven national congresses organized by the Russian Association of Physicists between 1919 and 1930 reflected the bourgeoning strength of the Soviet physics community. Perhaps the most remarkable and successful of these was the 6th congress in August 1928, which started in Moscow and continued on board a ship steaming down the Volga river. The meeting was attended by a large group of prominent foreign physicists, including Dirac, whose recently discovered relativistic equation for the electron was the major topic of discussions at the congress, and Alfred Landé (also on the photo).
[*Credit*:AIP, Emilio Segrè Visual Archives.]

'metaphysical' postulate of an unobservable vacuum filled with an infinite number of negative-energy electrons, in which only the vacancies, the holes, were supposed to be observed. Eventually, the hole model lost its popularity in high energy physics to the mathematically equivalent representation of anti-electrons, or positrons, moving as free particles in empty space. But it became universally recognized in condensed matter physics, where the existence of a medium was obvious.

Although Werner Heisenberg (1901–1976) was among the critics of Dirac's approaches to quantum electrodynamics, he appreciated the idea of the electron hole and in 1931 returned the concept to the physics of the solid state. Heisenberg considered holes ("*Löcher*") as vacancies in either electronic shells of an atom or nearly filled electronic bands of a crystal, vacancies that behave like positively charged particles (Heisenberg 1931). Since Dirac and Heisenberg, the term "hole" has become most commonly, though not exclusively, associated with electronic vacancies. Frenkel's original ionic vacancies, too, have become common knowledge in physics, but under a different name. In the 1930s they were incorporated by Friedrich Wilhelm Jost, Carl Wagner, and Walter Schottky into the general theory of defects in solids, in which the pair made of a wandering ion and a vacancy in the lattice is called a "Frenkel defect" (Brown *et al.* 1995, 3: 1528–1529; Hoddeson *et al.* 1992, 255–264).

4. The Collectivized Electron and the Bloch Electron

Frenkel's model of electrical conductivity of 1924 offered a solution to the main difficulty of the Drude–Lorentz theory, that of specific heats. Had electrons in metals been free, their thermal motion would have added a significant contribution to the value of specific heat of metals, which was not observed in experiments (Hoddeson and Baym 1980). Frenkel's electrons passed from one atom to another with velocities much higher than thermal ones and therefore did not have to contribute to specific heats. Despite this promising feature, the proposal found little support among physicists. Even Ehrenfest, who was very sympathetic to the young author, thought that his paper contained "many clever ideas and a big confusion."[8] Although Arnold Sommerfeld (1868–1951) once referred to Frenkel's theory as "well known" (Sommerfeld 1927a, 828), Frenkel himself found exactly the opposite situation when he tried to communicate further developments of his ideas to the cream of the international physics community at the Volta memorial conference held in in Como, Italy, in 1927 (the same congress at which Bohr presented his first version of the complementarity interpretation of quantum theory). Frenkel started boldly: "In the classical theory of electrical conductivity of metals, the so called "free" electrons are regarded as particles of gas that move

---

[8] Ehrenfest to A. F. Joffe, 24 November 1924, in (Ehrenfest and Joffe 1990, 176).

with constant velocities between collisions with positive ions.... A few years ago I showed that this conception is totally wrong." After presenting his arguments once again, he concluded:

> The only type of freedom electrons could obtain under these conditions is, so to speak, the freedom to change their master, or the atom to which they belong. While in the gaseous state each electron belongs to its proper atom, in the liquid or solid state it becomes 'the slave of the collective' formed by all atoms, and enjoys a rather relative freedom of constant transition 'from hand to hand,' i.e., from one atom to another.[9]

At the same meeting, Sommerfeld announced a different modification of the Drude–Lorentz theory. He avoided the difficulty of the specific heats by applying to electrons the new quantum statistics of Pauli and Fermi (Sommerfeld 1927b).[10] With only this modification, the ideal gas model of electrons was taken over from the classical theory. Sommerfeld wrote of these electrons as free, even though he admitted that quantum statistics imposed some additional restrictions on them. By prohibiting more than one electron from occupying the same phase cell, it created a relative deficiency of available phase space, making electrons "a nation without space" ("ein Volk ohne Raum") in Sommerfeld's metaphor borrowed from the contemporary ideology of German nationalism (Sommerfeld 1928, 375). In the following year, Sommerfeld's students and assistants developed and applied his theory further to other problems of the solid state. Some of them, however, were troubled by the difficulty about freedom. As Hans Bethe recalled:

> Sommerfeld... recognized that Drude's main difficulty, namely the large specific heat of the free electrons, would be eliminated by applying Fermi statistics... However, it was very unsatisfactory that Sommerfeld had to assume completely free electrons. How could the electron be considered as free in the presence of the obviously very strong variations of the potential energy inside the metal? I thought that this was a very major objection against the Sommerfeld theory; Sommerfeld himself seemed to

---

[9] "Le seul genre de liberté, que les électrons puissent acquérir sous ces conditions—c'est pour ainsi dire la liberté de changer leur maître, c'est à dire l'atome, auquel ils peuvent être censés d'appartenir. Tandis que dans l'état gazeux chaque électron appartient à un atome déterminé, dans l'état solide ou liquide il devient l'esclave du collectif, formé par tous les atomes, juissant de la liberté très relative d'ailleurs de passer incéssamment "de main en main" c'est à dire d'un atome à l'autre" (Frenkel 1927, 66, Russian translation in Frenkel 1958–1959, 2: 71–72).

[10] For Frenkel's 1928 review of the conference and the comparison of his and Sommerfeld's theories see (Frenkel 1970, 253–256).

Freedom, Collectivism, and Electrons

Ya. I. Frenkel talking to H. A. Lorentz in Como, Italy, September 1927, probably trying to explain the difference between his idea of collectivized electrons and Lorentz' model of free electron gas inside metals. The International Conference in Physics, organized by the Mussolini government to commemorate the centenary of Alessandro Volta's death, gathered almost all living luminaries of European physics. The conference's main event was the first public presentation by Niels Bohr of his complementarity interpretation of quantum mechanics.
[*Credit*: V. Ya. Frenkel. *Courtesy*: AIP, Emilio Segrè Visual Archives.]

be less concerned. The problem was solved by Felix Bloch in his famous Ph.D. thesis (Bethe 1980).

Frenkel and Sommerfeld had initiated the application of quantum ideas to metals, but it was Bloch (1905–1983), in 1928 a student of Heisenberg's in Leipzig, who found how to do this consistently on the basis of the fundamental Schrödinger equation. Bloch's landmark accomplishment explained why electrons can move through the lattice with apparent ease despite the strong internal forces acting on them. The key turned out to be the lattice's periodicity. In the version of his theory of 1927, Frenkel estimated the mean free path of electrons by considering the scattering of corresponding de Broglie waves on an atom. Reasoning once again by analogy (this time with the scattering of light waves), he concluded that in a perfectly periodic crystal electrons would propagate without scattering. Resistance to the current arose only from impurities, defects, and the thermal movement of the lattice ions (Frenkel 1927; Bloch 1929, 555–556).

Bloch proved this thesis in 1928 by rigorous calculation: "I had gotten the essential idea that a periodic arrangement is not really an obstacle for waves, but it's only the thermal vibrations. Heisenberg was very pleased. I told it to him only in the one-dimensional case, in a very primitive way. And he said, "That is explained now. Now I understand."[11]

Sommerfeld's school at first perceived Bloch's approach as "a quantum theoretical analog of Frenkel's theory."[12] This view had some basis, since Bloch's electrons were neither a free ideal gas as in Sommerfeld's model, nor bound to particular atoms as in Heisenberg's 1928 theory of ferromagnetism. Bloch understood the problem of the complexity of the electron's state of freedom and distanced himself from both these contrasting alternatives. He referred to his approach as "intermediate," since it took into account at least one kind of strong force inside the metal: the periodic potential. However, Bloch had no special notion in his vocabulary to characterize this state of intermediate freedom, and he did not use Frenkel's collectivist terminology and model, perhaps because of their specific political undertones. The intermediate state was also hard to describe mathematically, and in order to perform calculations Bloch had to resort to rough approximations. He considered two limiting cases (*Grenzfälle*): near freedom, with electron as almost free and the potential treated as a small perturbation, and tight bonding to a single atom, whence the influence of other atoms was treated as small disturbance. Although aware that neither of these approximations correspond to more complex reality, he thought they could be relied upon insofar as they both provided qualitatively similar results (Bloch 1929; 1931, 882).[13]

Pursuing this approach further, in 1929 Bloch and Rudolf Peierls (1907–1995), another student of Heisenberg's, described the strange and counterintuitive (*unanschauliche*) properties of electrons that were neither bound nor free: their momentum **k** was not conserved but could change by a quantum (*Umklapp* processes), and their kinetic energy E deviated from the usual quadratic function of the momentum (which Peierls used to explain the anomalous Hall effect in metals). Not only were electrons able, in a periodic potential, to jump from one atom to another despite

---

[11] Felix Bloch, oral history interview conducted by Thomas Kuhn, 14 May 1964, p. 21. Archive for the History of Quantum Physics (hereafter AHQP).
[12] William Houston reporting to his mentor about Bloch's yet unpublished theory (Houston to Sommerfeld, 30 May 1928, AHQP).
[13] As Bloch admitted later, he "had never understood how anything like free motion could be even approximately true [for electrons in a metal]" (Bloch 1980, 25).

their energies being lower than the potential barrier separating the atoms, but also, as Peierls found further, there were certain regions of higher electron energy above the barrier level within which the electrons, contrary to classical intuition, were prohibited to move (forbidden bands) (Peierls 1929a; 1929b; 1930). Peierls thus formulated his own notion of quantum freedom in a review of the state of the theory in 1932: "The difference between 'free' and 'bound' electrons, which is important in the classical theory and for which it is decisive whether or not the electron's energy suffices to overcome the potential barrier between atoms, is largely erased in quantum mechanics" (Peierls 1932a, 265).[14] In his own review, Frenkel welcomed Bloch's and Peierls' new achievements from the point of view of his understanding of freedom. To him, they had proved mathematically "that all valence electrons are free with regard to individual atoms ... and at the very same time bound with regard to the collective formed by all these atoms" (Frenkel 1932a, 182).

Meanwhile, in 1929 the word "collectivization" became familiar to everybody in the Soviet Union as the main political slogan of the year of the "Great Break" and of the collectivization campaign in agriculture. The on-going tragedy of villages was not perceived in full by those who lived in the cities. While teaching quantum theory to students, Frenkel presented its unusual concepts in collectivist terms, which were familiar to and popular among his young audience. Students sensed (in their own understanding of collectivism) the political connotations of the concept. As one of them, Oskar Todes, recalled later, "Frenkel's formulation of the 'collectivization' of free electrons in metals emerged from the events of contemporary political life. It enabled us to remember for our entire lives the main physical ideas about the behavior of these electrons" (Frenkel 1986, 94). Todes himself, together with two enthusiastic classmates, organized a student *kolkhoz* for collective work on the curriculum, yet the authorities criticized their initiative as a "collective of kulaks," since the three were regarded as the best students in the class.

After several years of use, the notion of "collectivized electrons" was no longer just a figurative metaphor for Frenkel, but the appropriate description of the complex reality within metals and the basis of his understanding of the physics of solids. This conviction led him in the 1930s into disagreement with further developments in the field, the so-called band theory of solids, and caused his alienation from the mainstream of the

---

[14] See also a review of the state of the theory at that juncture by Bloch (1933).

solid state community. It also motivated his search for alternative approaches. The one that Frenkel eventually found, the exciton, had profited from a proposal of the elastic quantum, or phonon, suggested in 1930 by his Moscow colleague Igor Tamm.

5. The Phonon and Quantum Individuality

The son of a railroad engineer, Igor Evgenievich Tamm (1895–1971) was accustomed to moving. He was born in Vladivostok in the Far East of the Empire and grew up in Elizavetgrad, a provincial town in the Ukraine. Like many Gymnasium students of the pre-revolutionary decade, Tamm read forbidden political literature voraciously and developed a serious interest in politics. He became involved in a Marxist study circle, participated in workers' demonstrations and illegal meetings, which made his parents worry. Trying to distract the boy from politics, they convinced him to go abroad, to a university in quiet Edinburgh rather than politically active London.[15]

Having spent an uneventful year in Edinburgh, Tamm was getting bored when the outbreak of World War I prevented him from continuing his studies abroad. He transferred to Moscow University, where his academic interests shifted from mathematics and chemistry to physics despite the poor quality of physics teaching in Moscow after the Kasso affair of 1911. At twenty, Tamm hoped and feared that he might live the life of a revolutionary, regarding a possible career in science as philistinism. In 1915 he joined the Social-Democratic Workers Party as a member of its Menshevik faction. During the subsequent years of wars and revolutions Tamm alternated between Elizavetgrad and Moscow, Kiev and Odessa, between pursuing his academic studies and participating in the turbulent political life under alternating political regimes. After the collapse of the monarchy in the capitals in the spring of 1917, Tamm turned political agitator and left Moscow for Elizavetgrad to assist in the revolution there. In April he was elected a member of the city Soviet and in June he represented his home town at the First All-Russian Congress of Soviets in Petrograd. A leftist among Mensheviks, he shared the Bolsheviks' uncompromising opposition to the imperialist war and voted

---

[15] Tamm's diary specifically mentions his reading of a brochure on "Collectivism" by Jules Guesde, a French socialist and one of the founders of the socialist party. See (Vernsky 1995, 99–100).

with them on this crucial issue against the new offensive on the front (Vernsky 1995, 103–105).

Unlike the easy overnight coup in Petrograd, the Soviet takeover of power in Moscow in November 1917 took a week of heavy fighting. Although Tamm, caught in the crossfire, despised the Bolsheviks' "fanaticism," he was not too far from them politically and initially entertained some hopes for a cooperation between leftist Mensheviks and the new regime. With the start of the Civil War in the summer of 1918, however, the Bolsheviks declared all other political parties illegal. By that time Tamm was already concentrating more on his academic studies. He did not fight in the war, although his sympathies were definitely on the Red side, and managed to graduate in the fall (Tamm 1995, 143–145; Feinberg 1995). He then spent 1919/1920 as an assistant at Tauride University in the Crimea, where his lifelong friendship with Frenkel began. Like Frenkel, Tamm experienced arrest and the menace of execution at least once, in the summer of 1920, when he tried to pass secretly across the front between the Whites and the Reds, from the Crimea to his fiancée in Elizavetgrad. Caught by the Reds with no documents, he was ordered shot as a White spy and escaped death only by proving his identity as a mathematician. To check this, the Red commander, who also happened to be a former mathematics student, demanded that Tamm derive the Taylor series.[16]

In 1922 Tamm returned to Moscow and soon began teaching as a lecturer at Moscow State University. Some of his best friends and former colleagues-in-arms joined the Bolshevik party, but Tamm remained formally unaffiliated. His disagreements with communists were philosophical rather than political: instead of believing in dialectical materialism, Tamm was influenced by Mach and the positivists. He was the first Moscow theoretician to use new quantum mechanics after it appeared in 1925. Ehrenfest noticed him and invited him to Leiden in early 1928, where Tamm met Paul Dirac (1902–1984), the British genius and one of the leaders of the young quantum generation. Although reputedly unsociable, Dirac developed a close friendship with Tamm, Frenkel, and a few other Soviet physicists. This friendship and Dirac's leftist sympathies made him interested in the Soviet experiment. Almost every year between 1928 and 1936, altogether seven times, Dirac visited the Soviet Union. For

---

[16] The following detail that could not have been invented by a scientist gives verisimilitude to the story: Tamm confessed that, having struggled all night, he failed to reproduce the full derivation, whereupon the Red leader also admitted that he had forgotten most of his college math and postponed the execution (Vernsky 1995, 105–108).

Tamm, he became the main authority in physics, alongside Leonid Mandelstam (1879–1944), Tamm's mentor and revered colleague at Moscow State University (Dirac 1990; Mandel'shtam 1979).

Both Mandelstam's and Dirac's influences are evident in the paper of 1930 in which Tamm introduced a new hypothetical particle, later named the phonon. Tamm was following up a discovery by Mandelstam and Grigory Landsberg, who in early 1928 had observed a new effect while studying the scattering of light in quartz. They found two distinctive additional frequencies in the scattered light, one higher and one lower than the frequency of the incoming light, which they called "molecular" or "combinational" scattering and interpreted as a combination of the original electromagnetic oscillation with the elastic oscillations of the solid body.[17] Tamm complemented Mandelstam and Landsberg's semiclassical explanation with a strict theory based on Dirac's form of quantum mechanics and electrodynamics.

Since the Moscow team understood that the new effect could not be caused by scattering by independent atoms, Tamm considered the solid body "as a whole," as a system bound together by strong interactions between the atoms. Elastic vibrations of these atoms had already been treated in the earlier quantum theories of Einstein, Debye, Born and von Karman, in which quantization of the energy of vibrations helped to explain the specific heats of solids. Dirac developed more advanced methods of wave quantization in his quantum theory of radiation (1927). Tamm used these methods in 1930 to describe electromagnetic radiation in his theory and also extended the same methods of quantum electrodynamics to the treatment of the solid state, for the quantization of elastic waves in the crystal. For the physical interpretation of quantized electromagnetic waves, Dirac relied on the corpuscular model of light quanta, or photons. In addition to them, Tamm introduced analogous "elastic quanta," particles corresponding to quantized elastic waves, with their own momentum, direction of propagation, and energy. He thus interpreted the change of light frequency during scattering in the solid as the process of absorption or emission of one particle by another, of an elastic quantum by a quantum of light (Tamm 1930, 345).

Unlike Frenkel, Tamm avoided using metaphors in his published papers. The analogy with the photon alone was sufficient for him to

---

[17] At about the same time, the effect, currently known as Raman-effect, was also observed in liquids by C.V. Raman and R.S. Krishnan in Calcutta. On the history, see (Fabelinskii 1978; Brown et al. 1995, 2: 996–997; Singh and Riess 2001; Singh 2002; Venkataraman 1995).

Freedom, Collectivism, and Electrons

P. A. M. Dirac and I. E. Tamm in Leiden (first and third from left), spring 1928. Tamm was in Leiden during the first half of 1928 on the invitation of Paul Ehrenfest, while Dirac came to visit on his European tour. Their meeting marked the beginning of a scientific collaboration and life-long friendship between the two theorists. Also in the photo are Soviet physicists O. M. Trapeznikova and I. V. Obreimov, who were working in Leiden's low-temperature laboratory, for more than a decade the only place in the world where one could work with liquid helium. Later that year Obreimov was summoned back to the Soviet Union and appointed organizing director of the new Physico-Technical Institute in Kharkov, Ukraine. The institute, UFTI, opened in 1930 with the first low temperature laboratory in the Soviet Union, which in 1933 also started experimenting with liquid helium.
[*Credit*: AIP, Emilio Segrè Visual Archives.]

justify the idea of the elastic quantum in print. Yet the proposal had an additional value-laden meaning related to Tamm's attitude towards Bose–Einstein statistics. In a 1926 review article, Tamm had discussed physical interpretations of the new quantum statistics, according to which particles appeared indistinguishable. In his view, this lack of individuality left two possible choices: either to admit that individuality is fundamentally lost on the microscopic quantum level or to save individuality by postulating some kind of a physical cause, as Einstein had hinted, in an unknown interaction between particles. Tamm expressed his hope that indistinguishability would be found to be only a formally valid, phenomenological description of what at a deeper level of reality would be an ensemble of distinguishable but interacting particles. Having to choose between free particles without individuality and particles with individuality but without absolute freedom, he preferred the latter (Tamm 1926;

see also Tamm 1928). The issue of maintaining individuality in a collective with strong interaction had been one of the main principles of Mensheviks as political movement. The split of the Russian Social-Democratic Workers' Party in 1903 into Menshevik and Bolshevik factions, which would have so many tragic consequences, occurred along exactly this line. The Bolsheviks insisted on the clause in the party statutes that established strict discipline and subordination of individual members to their party organizations, while the Mensheviks defended a more liberal clause with more room for individual rights and choices.

By 1930 developments in quantum theory definitely favored the conception of photons as free but fundamentally indistinguishable particles. Yet Tamm's proposal of elastic quanta offered the hope of saving individuality. Mathematically, the new quanta were exactly like photons, an ideal Bose gas of particles, whose number was not conserved, but behind the appearance of free, indistinguishable elastic quanta was the reality of strongly connected and collectively oscillating atoms with individuality but without much freedom. Frenkel apparently shared Tamm's values and immediately welcomed the new particles in a paper of 1931, calling them "sound quanta" and "heat quanta" (Frenkel 1931b, 1289). In his textbook on wave mechanics (Frenkel 1932b), he suggested the name "phonon," which became the standard term. Frenkel, too, interpreted phonons phenomenologically. They were for him fictive particles, whose very usefulness cast doubt on the physical reality of their close prototype, photons.[18] In his philosophy of physics, Frenkel was inclined towards positivism even more strongly than Tamm, and he dared to criticize dialectical materialism openly at a conference in 1931, when it was not politically safe to do so. This public act would not be forgotten by Soviet philosophers and made Frenkel a frequent target of their criticism, which grew especially militant in the late 1940s. Among other charges, attackers cited Frenkel's use of phonons as "fictive particles" instead of "real" sound waves as the proof of his positivistic heresy.[19]

---

[18] "It is not in the least intended to convey hereby the impression that such phonons have a real existence. On the contrary, the possibility of their introduction rather serves to discredit the belief in the real existence of photons" (Frenkel 1932b, 267–268). The name 'phonon' simultaneously appeared in (Frenkel 1932c).

[19] Positivism, and Mach in particular, was popular not only among non-Marxist Russian socialists (Shreider 1923) but also among Mensheviks and some less orthodox Bolsheviks, which was why Lenin attacked this philosophy in his *Materialism and Empiriocriticism* (1909). On accusations against Frenkel, in particular by the Marxist philosopher Mikhail Omel'ianovsky, see (Sonin 1994, 139). On Frenkel's critique of dialectical materialism, see (Frenkel 1996, 225–227).

But phonons, like their close analogs photons, could be interpreted from many different perspectives, which no doubt helped them gain wide and quick acceptance. If for Tamm and Frenkel they were a phenomenological description of the collective of strongly interacting atoms, for Peierls, who used them in a paper of 1932 under the name *Schallquanta*, they were merely synonyms for quantized sound waves and, as such, useful but not conceptual novelties. These nuances of interpretation did not make a serious difference for the mathematical formalism because the formulas were all equivalent and could be easily and fully translated into one another. They did make a huge difference, however, for the generalization of the concept of the phonon into a broader model of "collective excitations."

Tamm did not publish further on phonons. Having done important work on the solid state in the early '30ies, he returned to quantum electrodynamics, nuclear physics, and particle physics, the fields that interested him the most. He participated in the further development of the concept of quasiparticles mostly indirectly, as a discussant at conferences and seminars, as the mentor of younger students, and through his theoretical colloquium at the Physical Institute of the Academy of Sciences. But other physicists who shared his interpretation of phonons transformed it into a much more general hypothesis: that not only for oscillations of atoms in solid body, but for any kind of collective system with strongly interacting particles, there should exist a phenomenological description in the form of an ideal gas of fictive particles, 'collective' or 'elementary' excitations. Led by this assumption, physicists tried to describe complex properties of dense bodies by postulating new kinds of particles that resembled phonons but possessed novel and strange properties and did not necessarily have well-defined classical analogs. As a general strategy, this way of dealing with condensed matter would be advocated and pursued with tremendous success by another Moscow group, Lev Landau and his students starting in the late 1930s.[20] In a more rudimentary form, a similar generalization of Tamm's elastic quantum helped Frenkel in 1931 to introduce a new hypothetical particle, the 'excitation quantum,' or exciton.

6. Shared Excitation

Frenkel spent the academic year 1930/31 as a visiting professor at the University of Minnesota and anxiously followed political developments at

---

[20] See Chapter 10.

home. Newspapers and letters delivered mixed news: the disappearance of food from the stores and the introduction of rationing, propagandistic accounts of class war and the collectivization campaign in the countryside, and promises of a bright future. In his letters home Frenkel mentioned the contradictions of the Soviet life—its idealism and cruelty, great successes of industrialization and agriculture, and the lack of most basic goods—while approving wholeheartedly the Soviet system in general and, with somewhat more reservation, the collectivization of agriculture (Frenkel 1966, 251).

In his major paper written during that year, Frenkel considered another kind of excitation of a solid body: not the displacement of an atom from its equilibrium position which leads to elastic oscillations, but the absorption of a light quantum by an atom which puts the atom (actually, one of its electrons) into an excited state. In the former case, oscillation of one of the atoms does not remain localized but spreads across the whole solid in the form of sound waves (classical physics) or phonons (quantum physics), whereby all atoms participate in the collective movement. The same thing, according to Frenkel, must also occur with the excitation of an atom by a light quantum. In a gas, such an excitation remains the property of an excited atom until it either reemits a light quantum or collides with another atom and transforms the energy of excitation into the kinetic energy, or heat. But in the solid body, owing to interactions between the atoms, the excitation would not remain localized on one atom but should be shared by the entire collective. In order to describe this sharing mathematically, Frenkel constructed a wave function for an excited crystal as a superposition of wave functions corresponding to excitations localized at different atoms and found the stationary states of such a system in the form of "excitation waves." The latter were similar in form to Tamm's quantized elastic waves, but did not come about as a result of the quantization of a classical wave. Although their classical analog could be found post hoc, it did not provide so direct a basis for reasoning as sound waves did for phonons (Frenkel 1931a).

In the follow-up paper, Frenkel introduced a new particle corresponding to excitation waves which he called the "excitation quantum," making explicit the analogy with Tamm's "elastic quantum" and describing its important role as the intermediary in the process of the absorption of light by solids and its further transformation into heat. The direct process would have required a simultaneous transformation of a light quantum into hundreds of elastic quanta (because the energy of a photon is typically a hundred times larger than the energy of a phonon), and, according to quantum mechanics, would have had a very low probability of occurrence. The process can take place much more easily through an intermediate excited state, "visualized from the corpuscular point of view as the transformation of the incident light quantum into an excitation quantum having the same energy and momentum," which would then live for some time in the solid body and gradually dissipate its energy into the energy of elastic quanta, or heat. If realized in nature, Frenkel's excitation quantum or exciton would imply the existence of narrow lines in the absorption spectra of solids (Frenkel 1931b, 1284–1285).

Experimental physicists had not yet seen these narrow bands. Although Frenkel could point to some experimental data that indirectly supported his proposal, the excitation quantum remained a largely hypothetical entity. Initially, other quantum theorists were mostly critical of it. While in Minneapolis, Frenkel learned from a letter from his wife, Sarra (and she apparently from Peierls, Pauli's assistant, who had come to Leningrad in spring 1931 to marry his fiancée), that Wolfgang Pauli (1900–1958) had rejected the excitation quantum in his usual "gentle" way. In response Frenkel wrote her:

> You mention that my long paper in *Phys.Rev.* was severely criticized in Zurich and they decreed it to be wrong. I believe that this opinion is undoubtedly incorrect—not only on the basis of my own discussion with American theoreticians, but also according to my personal conviction. The fact that Pauli considers my work *Falsch* proves only, in my opinion, that it is not *Trivial*.[21]

Pauli suggested that Peierls meet with Frenkel in Europe and discuss their disagreements before publishing the critique (Pauli 1985, 89), but it is not exactly clear who was more effective in persuading whom in the end. In Peierls' alternative version of the theory of light absorption in solids, he accepted parts of Frenkel's theory, including phonons and the intermediate excited state of the crystal. He objected, however, to the idea that the excitation would be distributed evenly over the whole crystal, arguing instead that it should be localized in a small area. Mathematically, Peierls managed to treat only the case of such an excitation bound to a particular atom (Peierls 1932b, 916, 942; 1932c). Until the early 1950s, when exciton spectra were observed, Frenkel's proposal found very few supporters. The main problem that hindered its acceptance, however, was not the criticism by Peierls but its perceived incompatibility and rivalry with the emerging

---

[21] Frenkel to his wife, 1 April 1931, quoted in (Zakharchenia and Frenkel 1994, 470).

mainstream approach in solid state theory, the band theory that was developed in Britain and the United States during the 1930s and assumed that electrons in solid bodies are almost free (Hoddeson *et al.* 1992, ch. 3).

Frenkel did not accept the band theory and continued to criticize its basic model, arguing that electrons are in a more complex, collectivized state of freedom, though at times his collectivist alternative looked as a marginal, almost one-man crusade. It took gradual efforts of the next generation of Soviet physicists, in particular Lev Landau and his school, to develop additional, more powerful methods, which will eventually bring collectivist approaches to triumph in the 1950s. For these later developments, which will be described in Chapter 10, to occur, a new discipline of theoretical physics had to establish itself institutionally in the Soviet Union, producing a massive following and training younger students.

Chapter 4

Lev Landau's *Wanderjahre,* or Theoretical Physics in the Context of Cultural Revolution

As an academic discipline, theoretical physics barely existed in the Russian Empire, but it rose quite spectacularly during the post-revolutionary decades and eventually became, alongside mathematics, the most advanced field in all of Soviet science. Perhaps nothing illustrates the change in its status better than the following tale of two emigrations. As a young student in Moscow in 1910, Paul Epstein (1883–1966) wanted to study the new theories of relativity and quanta. His senior colleagues advised him to study abroad, in Germany, where those theories had originated. Had he wanted to learn experimental physics, they said, he could have studied with Piotr Lebedev, who had established a school for training advanced researchers at Moscow University. No comparable opportunity to learn theoretical physics, however, existed in the country.

There was, to be sure, a stellar theorist in St. Petersburg, the Austrian Paul Ehrenfest, but he did not have a regular academic position and was not likely to find one in Russia. In 1912 Ehrenfest finally obtained a professorship as successor to the famous H. A. Lorentz at the University of Leiden and moved to the Netherlands (Frenkel 1977; Klein 1970). As for the young Epstein, he followed his colleagues' good advice and went to study with Arnold Sommerfeld in Munich. As a Russian subject, he was interned in Germany during the First World War and accomplished some very important work on the development of the Bohr–Sommerfeld atomic model. After the war he was offered a professorship in theoretical physics at the newly established California Institute of Technology and moved to the United States, a country that at the time also lacked a tradition of its own in theoretical physics and was eagerly importing scholars from Europe.[1]

A different kind of emigration of theoretical physicists from Russia to America occurred around 1990. This time senior researchers came by the dozens, if not hundreds, along with more numerous junior scholars and postdocs. At the century's end, American theorists no longer felt inferior to the Europeans and were not concerned with catching up, but the reputation of Soviet theoretical physics stood so high that its representatives landed prestigious positions at Princeton and MIT, Argonne and Livermore, Stanford and Minnesota. Back home, their advanced expertise was no longer needed, or so at least sounded the message from political authorities in the new post-communist Russia, who were busy devaluing science along with everything else that their predecessors, the communist authorities, had seemed to value.

Unlike many other products of the communist regime, however, Soviet science proved a highly competitive international commodity. Professionals in the field had become so used to the decades of Soviet excellence in theoretical physics that they often regarded it as a kind of natural Russian tradition, like chess. As with many similar claims about national "traditions" elsewhere, such a perception lacked historical perspective. Theoretical physics is generally quite young, having first established itself in Germany around 1900 (Jungnickel and McCormmach 1986). In Russia, it rose to prominence during the 1920s and early 1930s, supported by a rather unusual constellation of circumstances. In its early years, Soviet theoretical physics profited from being an 'adopted child' of two unlikely partners, international Rockefeller philanthropy and Bolshevik cultural revolution. This chapter describes how that cooperation became possible by following the early career of Lev Landau and his struggles to establish the new scientific discipline in revolutionary Soviet society.

1. Education of a Soviet Scientist

Lev Davidovich Landau (1908–1968) belonged to the first truly Soviet generation of scientists, those educated immediately after the revolution. He was born in Baku, the Transcaucasian oil capital, to the middle-class Jewish family of a petroleum engineer. As a child, Landau was a prodigy

---
[1] Epstein's autobiography, P. S. Epstein papers, California Institute of Technology.

and an *enfant terrible*, and he retained some characteristics of immaturity well into adult life. By the age of thirteen he had learned calculus, contemplated suicide, and was about to be expelled from the classical Gymnasium because of his stubborn and rebellious character (Bessarab 1978, 11–12). It would be hard to imagine him doing well in a regular, disciplined system of education, but, luckily for him, the system with those properties was dismantled after communist power reached the Caucasus in 1920.

The new revolutionary schools were more chaotic and subject to all kinds of radical reorganization and pedagogical experiments, motivated by the desire to give to members of the lower classes full access to education. Reverse class privileges in education,—'affirmative action,' as it were— now favored workers and peasants, especially those who were politically active.[2] To stimulate their advance to the highest educational levels, the new school system relaxed the formal barriers between stages of education. In particular, it refused to grant or recognize diplomas and academic titles, which were all abolished as 'relics of the medieval past.' One could thus, in principle, enrol in a university without graduating from a high school, start a graduate program without a formal university diploma, and be hired as a professor without a Ph.D.

For someone like Landau the relaxed system proved advantageous, allowing him to skip some formalities. He never wrote a Ph.D. thesis or its equivalent, for example. In 1920, as his classes at the Gymnasium were interrupted by the revolutionary chaos, he studied at home for a year, after which he spent another year at a vocational school in the same class with his elder sister. In 1922, at the age of fourteen and apparently without formal graduation, he enrolled at the newly organized Baku University. The new college, itself a product of hasty revolutionary transformations and initiatives, started as a university in name but not yet in substance. Two years later, again following his sister's example, Landau transferred to a school with a long and glorious scholarly tradition, Leningrad University (Bessarab 1978, 22). There he took regular classes and found himself in the company of young theoretical physicists.

St. Petersburg–Petrograd–Leningrad University (the name of the city changed twice during the turbulent decade of 1914–1924) was the alma mater of the majority of the first generation Soviet theoretical physicists.

---

[2] On the applicability of the present-day term "affirmative action" to Soviet policies see (Martin 2001). The first educational reforms of the Bolshevik government are described in (Smirnova 1979; 1984).

Ehrenfest, before leaving the country in 1912, had introduced to the field two young students, Yury Krutkov (1890–1952) and Viktor Bursian. Yakov Frenkel and Vladimir Fock completed their university studies during the subsequent years of war and revolution. In the post-revolutionary early 1920s, they were joined by a group of younger students of theoretical physics, including George Gamow (1904–1968), Dmitry Ivanenko (1904–1994), and the even younger Landau. The latter three, a.k.a. "Jonny," "Dimus," and "Dau," formed the core of the so-called "physical jazz-band," a gang of classmates famous (or infamous) for making fun of others, especially senior colleagues (Gorelik and Frenkel 1994, ch. 2).

In this circle of new friends, Landau overcame his earlier loneliness and timidity, although—being several years junior to the rest—he did not attract the attention of the female part of the company. The absence of any real experience did not preclude him from demonstrative theorizing about the opposite gender and the science of courting. Their company, the "jazz-band," was not merely a student gathering whose members excelled at reciting poetry, making practical jokes, blatantly disrespecting good manners, and posting satirical "Physikalische Dummheiten" ("Physical Nonsense") on university walls. The group was also in some meaningful ways a byproduct at the radical student culture of the 1920s with its trappings of a student rebellion, the main thrust of which was transposed from the explicitly political into scientific spheres (Hall 1999, ch. 6). While some of Landau's politically engaged university classmates staged street demonstrations in support of Leon Trotsky and the leftist Bolshevik opposition to the Politburo, he and his friends remained formally unaffiliated in politics, but took special pleasures in challenging authorities in their own discipline of physics, senior professors whom they dismissively called "*zubry*" (bisons).

Landau did not lack interest in politics as such. On the contrary, he entertained strong opinions, usually to the left of the official Soviet line, and he may well have also sympathized with Trotsky.[3] The adolescent Landau idealized militant revolutionaries and dreamed of becoming one (Bessarab 1978, 14–15), but he found his personal revolutionary initiation in the new and radical theory of quantum mechanics. News of this latest breakthrough in physics arrived from Germany when he and his friends were university students. To Landau, the new theory that overturned the foundations of physics and made obsolete many of the basic beliefs and

---

[3] In his later hatred of Stalin, Landau shared at least some of the Trotzkyites' left-wing criticism of Stalinism. See also (Pavlenko *et al.* 1998, 215–216).

skills of his professors was in many ways as revolutionary as the political and social upheaval that was going on in his country. In his eyes, the two revolutions—the social and the scientific—complemented and reinforced each other.

This affinity, also perceived by some other young Soviet physicists, contributed to their immediate and enthusiastic embrace of the new quantum mechanics. Student members of the "jazz-band" learned the new theory as it developed, through every fresh issue of the German journal *Zeitschrift für Physik* arriving in Leningrad libraries. In the meantime, the somewhat older Frenkel and Fock were already preparing courses on the new physics. In the anti-hierarchical, anti-traditional atmosphere of the 1920s, the teenaged Landau and his friends did not need special permission or encouragement from their professors to start pursuing research in the new field entirely on their own and writing papers for the *Zeitschrift für Physik* even before they finished taking regular classes at the university.

Soviet science had extremely tight connections with Germany during the 1920s, with German journals opening their pages to the rising tide of research papers coming from the USSR. At its peak in the mid-1920s, up to 16 percent of the articles published in the *Zeitschrift für Physik* were by Soviet authors (Josephson 1988a). The first significant Soviet contribution to quantum mechanics came from Fock in the summer of 1926, shortly after the publication of the first papers on wave mechanics by Erwin Schrödinger. Independently and almost simultaneously with Oskar Klein, Fock developed and studied the relativistic generalization of the Schrödinger wave equation (Fock 1926). On the heels of Fock's paper came two short notes by Gamow and Ivanenko (1926) and Ivanenko and Landau (1926) that dealt with the derivation of the "Klein–Fock equation." Later that year Landau completed a more mature paper, in which he used the methods of the Heisenberg–Born–Jordan matrix mechanics to calculate the rotator and the spectra of the two-atom molecule (Landau 1926).

The pattern of the initial reception of quantum mechanics in the Soviet Union was similar to that in the United States. In both countries the theoretical physics communities were young, emergent, and particularly receptive to new trends coming from abroad. They responded to quantum mechanics immediately and vigorously, and what the initial papers by young students may have lacked in quality and professionalism they made up in sheer numbers and the enthusiasm of converts. By comparison, established physics communities in Great Britain and France reacted to the new German theory with more caution, producing papers by only a few authors but of generally much higher quality, such as those by Louis de Broglie and P. A. M. Dirac (Kozhevnikov and Novik 1989, 131–132). For Landau and his friends, quantum mechanics became an integral part of their youth culture and offered an additional reason to speak dismissively of senior colleagues who were not as quick in mastering the newest theories. One of their favorite targets was the senior experimentalist Abram Joffe.

In the mid-1920s, Joffe emerged as the most powerful among directors of major physics institutes, having outpaced others in gaining support from Soviet commissariats, raising funds for research, and creating new positions. Underlying his success was the establishment in 1924 of close ties between his Physico-Technical Institute in Leningrad (LFTI), supervised by the Commissariat of Enlightenment, and the much more powerful economic commissariat VSNKh, a linkage that offered the possibility of drawing funds from both ministries (Obreimov 1973, 44). In a few years, the research staff of his institute grew from several dozen to several hundred, while Joffe himself effectively became the chief spokesman of the physics community in the eyes of Soviet officials and the public (Sominsky 1964).

This expansion created positions for Landau and other "jazz-band" theorists. In 1927 Landau completed his coursework at Leningrad University (diplomas were still not required) and started working as *aspirant* at Joffe's LFTI. '*Aspirant*' is usually translated as 'doctoral student,' but since theses and titles had been abolished in 1918, the position in practice amounted to that of a junior research worker, an apprentice learning the skills of the academic trade without the requirement of writing a thesis. The outbreak of the so-called cultural revolution in 1928 brought with it a new and intensified wave of radical, haphazard reforms in the system of higher education and greatly increased the demand on academic institutions to open up even more positions of this sort and educate huge numbers of engineers and scientists in the shortest time possible (Fitzpatrick 1979). The new cultural policies created additional opportunities and disturbances in the regular progression of academic careers, so that it was not uncommon for a university student to work simultaneously as *aspirant* at a research institute or for an *aspirant* to teach simultaneously as a professor at some other institution of higher learning (Livanova, 1978, 16; Gorelik and Frenkel 1994).

The *aspirant* Landau joined the expanded theory group and seminar at LFTI, which were headed by Frenkel. As is reflected in the photo, the seminar's main activities and identity centered on quantum mechanics

Theoretical Physics in the Context of Cultural Revolution

Frenkel's theoretical seminar at the Leningrad Physico–Technical Institute, early 1929. The group then included a good portion of the nascent Soviet theoretical physics community. Inscriptions on the blackboard (Anno Quanti XXIX)) identify the date as the 29th year of the quantum era and the year of the Dirac equation. The slogan above the blackboard points to the year of the Cultural Revolution and the Stalinist Great Break in the Soviet Union ("For solving the tasks of the reconstructive period, for keeping the pace of the country's industrialization, for the socialist reorganization of agriculture, forward!") From left to right: L. E. Gurevich, L. D. Landau, L. V. Rozenkevich, A. N. Arsenieva, Ya. I. Frenkel, G. A. Gamow, M. V. Machinsky, D. D. Ivanenko, and G. A. Mandel. Gamow had just returned from Europe international celebrity; Landau would soon depart for his own foreign trip.
[*Credit*: V. Ya. Frenkel. *Courtesy*: AIP, Emilio Segrè Visual Archives.]

and its applications. One of the papers Landau published during this period—introducing the density matrix (Landau 1927)—would be valued by him even late in life. He also wrote a paper with Ivanenko on the relativistic generalization of the wave equation with spin, but before they were able to bring their project to a completion Dirac brilliantly resolved the problem with his famous Dirac equation for the electron (Iwanenko and Landau 1928; Dirac 1928). The next level of professional maturity, together with international experience, came to Landau with the help of a Rockefeller postdoctoral fellowship.

2. Rockefeller Philanthropists and Bolshevik Russia

In the judgment of the former president of the Rockefeller Foundation, Raymond Fosdick, Russia was "[t]he one large country with which the Foundation never established a satisfactory relationship" (Fosdick 1952, 286). Indeed, contacts between the icon of world capitalism and the icon of world revolution were limited both in scope and in time, being confined to several years around 1930. But Fosdick had additional reasons to downplay the foundation's one-time modicum of cooperation with the Bolshevik regime, as his recollections appeared at a time when American anticommunist and espionage paranoia was at its highest. Even the official directory of Rockefeller grant recipients published in those years omitted the majority of the former fellows from the Soviet Union, listing mainly those few who either came from the Baltic republics or decided to stay in the West (Rockefeller 1953). Among the conspicuously absent names were several famous physicists who were—or were believed to be—playing important roles in the Soviet atomic bomb project.

These very fellows on the other side of the Iron Curtain were equally uninterested at the time in drawing attention to their earlier connection with the Rockefeller name and money, owing to an analogous upsurge of Cold War spy mania in the Soviet Union. Apart from a few muted hints in the biographical literature, the connection remained largely forgotten until the end of the Cold War, when several historical studies called attention to it (Rosenbaum 1989; Frenkel and Josephson 1990; Kozhevnikov 1993). In overall financial terms, the Rockefeller Foundation's contribution to Soviet science was miniscule in comparison to their philanthropic investments elsewhere or to the level of the Soviet government support of science. Yet it managed to leave an important mark, in particular on the development of Soviet theoretical physics.

The cooperation was not easy on either side. In theory, the Bolsheviks rejected private philanthropy on the general ideological grounds that it tried to substitute cosmetic improvements of the status quo for real (e.g. revolutionary) ways of solving social problems. However, the post-Civil-War social and humanitarian situation was so catastrophic that the Bolsheviks had to leave high principle aside and accept emergency aid from Herbert Hoover's American Relief Administration. Rockefeller philanthropists, for their part, had no love lost for the communist regime and no desire to help its survival. They also generally preferred to abstain from emergency relief work in favor of more stable and long-term constructive building of social infrastructure and were thus reluctant to get involved in Russia, at least until the most acute crisis was over and the general political situation was somewhat stabilized.[4] In mid-1923, having received several applications for aid, including one from the Bolshevik Commissariat of Public Health, the Foundation revisited its policy towards Russia:

> [P]resent conditions in Russia ... included the following salient features: 1. Famine conditions will continue over wide areas at least until the next harvest, about July 1. From 5,000,000 to 10,000,000 people are likely to be near starvation during this time. 2. The present government seems likely to remain in power since there is nothing to take its place. Progress may come through infiltration of new personnel and adoption of new ideas by the existing government rather than through another political revolution ... 3. Health conditions throughout the country are indescribably bad due to war, disorganization, revolutions, and counter revolutions, isolation from other countries, poverty, collapse of exchange, dearth of medical men brought about in part by heavy mortality in fighting plagues and epidemics. 4. The present condition of the universities and other institutions is pitiable. Laboratories and hospitals are lacking in every form of scientific equipment and in most elemental supplies. 5. While famine conditions still continue, many observers believe that much constructive help can be given in the near future by co-operating in rehabilitation of medical education and in counsel and demonstrations in health programs.[5]

No formal action was taken and the consensus remained that it was still too early to send anything except some emergency help through the American Relief Administration. Ultimately, the Rockefeller Foundation's involvement with Soviet science came about not so much through a conscious policy decision but as part of a much broader project supporting European science that was initiated in 1924 by Wickliffe Rose. Rose, who had worked at the International Health Board, was promoted in 1922 to head another branch of the Rockefeller philanthropy, the General Education Board (GEB). He suggested extending GEB operations outside U.S. borders, which led to the creation of an additional philanthropic institution, the International Education Board (IEB), also under Rose's directorship (Gray 1941). In 1924 Rose toured major academic centers in Europe, choosing locations for institutional grants and designing a system

---

[4] George E. Vincent to John D. Rockefeller Jr., 9 August 1921 (Rockefeller Archive Center—hereafter RAC—RF-1-1-785-2-15).
[5] Minutes of the Rockefeller Foundations re: Policy with respect to Russia. (RAC, RF-1-1-785-2-15).

of international postdoctoral fellowships to allow promising students to visit foreign institutions for advanced training. At each place he visited he consulted with senior scientists, one of whom was Paul Ehrenfest.

From his base at the University of Leiden, Ehrenfest did everything possible to restore the contacts and academic exchanges with Russian science that had been broken during the years of war and revolution. Once correspondence resumed in 1920, he exchanged letters with his close friend and former colleague Joffe in Petrograd, suggesting, in particular, that Joffe send a young theoretical fellow to Niels Bohr in Copenhagen to learn at first hand the newest developments in quantum theory (Ehrenfest and Joffe 1990, 140). Ehrenfest was one of the first scientists from the West to visit post-revolutionary Russia, traveling there in the summer of 1924 to participate in the congress of the Russian Association of Physicists. Impressed by the spirited quest for knowledge amid economic catastrophe, he contacted Robert Millikan and Paul Epstein at Caltech, urging them to find

Paul Ehrenfest's visit to Leningrad, 1924. Ehrenfest, who had lived in Russia before, was one of the first scientists from Western Europe to visit the Soviet Union and he did much to restore academic contacts interrupted by seven years of wars and revolutions. Here he is shown with leaders of the State Physico–Technical Institute/Laboratory in Leningrad. From left to right: I. V. Obreimov, N. N. Semenov, P. Ehrenfest, A. F. Joffe, A. A. Chernyshev. [*Credit*: V. Ya. Frenkel. *Courtesy*: AIP, Emilio Segrè Visual Archives.]

ways to fund a visit to the United States by Joffe and the young theoretical physicist Krutkov. In his reply to a subsequent inquiry from Millikan to the Rockefeller Foundation, Rose expressed reluctance to support a short visit by a senior scholar but was much more open to the idea of a longer fellowship for a junior researcher, which resonated with his general plan of supporting international exchanges in European science.[6] It probably did not help Joffe's case that foreigners often confused him with the prominent Bolshevik of the same last name. As an American consulate official asserted, "Professor Joffé is a militant communist, of the left branch, or the party of Zinovieff, and, as such, would of course be liable to exclusion from the United States under the Act of June 5, 1920."[7]

In accordance with their general preferences, the Rockefeller Foundation declined most requests and proposals to support scientific institutions and scholars in the Soviet Union. The 1927 visit by Alan Gregg to Moscow also did not result in any institutional grant or investment in scientific infrastructure, except for some book donations (Solomon 2003.) But Rockefeller philanthropists granted fellowships to individual Soviet scientists who traveled abroad. With the notable exception of the famous physiologist Ivan Pavlov (Windholz and Kuppers 1988), most of the recipients were younger scholars who received postdoctoral fellowships for up to a year's stay at a foreign, usually European, center of advanced research. In total, IEB and the Rockefeller Foundation authorized about fifty such fellowships to scientists from the Soviet Union between 1925 and 1932, which were distributed almost evenly among four disciplines: mathematics, physics, biology, and medicine (Kozhevnikov 1993).

Applications did not come directly from postdoctoral students because both the IEB and the Rockefeller Foundation relied instead on nominations by a few respected academics in Western Europe and America with whom they had established personal contact. For most of the Soviet physicists and for some mathematicians, such nominations came from or were arranged by Ehrenfest. He initially proposed two theoretical physicists from Leningrad, Krutkov and Frenkel, who received IEB fellowships in 1925 and spent the year mostly in Göttingen with Max Born, just in time to witness the discovery of quantum mechanics and to help communicate the news back to Russia.[8] In 1926 Ehrenfest proposed Fock and Tamm for IEB

---

[6] Ehrenfest to Millikan, 17 December 1924; Millikan to the IEB, 7 January 1925; Rose to Millikan, 16 January 1925 (RAC, IEB-1-3-52–818, 836).
[7] Donal F. Bigelow, American Consul in Paris, to Augustus Trowbridge, IEB, 12 November 1925 (RAC, IEB-1-3-52–818).
[8] RAC, IEB-1-3-53–836; RAC, IEB-1-3-49–733.

fellowships, but the board only approved a fellowship for Fock for a one-year stay in Göttingen. The reason for Tamm's rejection apparently was that Max Born, who seconded both nominations, was asked to rank the two candidates and placed Fock somewhat higher on the basis of his publication record.[9] Ehrenfest was eager to correct the bureaucratic 'mistake' and saw to it that Tamm received a Dutch fellowship from the Lorentz Fund that enabled him to travel to the Netherlands and Germany in 1928.[10]

At a December 1926 meeting with the Rockefeller representative in Europe, Augustus Trowbridge, Ehrenfest and Joffe learned of a change in procedures requiring fellowship candidates to be interviewed by a Foundation officer.[11] This requirement created a serious complication for Soviet candidates, who had to arrange to travel to Europe (usually Paris) prior to the approval of their fellowships. Joffe used his influence within Soviet commissariats to get funding for such foreign trips for his nominees, experimental physicists Kirill Sinel'nikov and Dmitry Skobeltsyn. Both received IEB fellowships in 1928 for work on nuclear physics in Cambridge and on cosmic rays in Paris, respectively.[12] The same year Gamow went to Göttingen with funding from Narkompros and there wrote an extremely influential paper that pioneered the application of quantum mechanics to the atomic nucleus and radioactive decay (Gamow 1928). An impressed Niels Bohr arranged for Gamow's stay in Copenhagen with the help of a Danish Rask–Ørsted fellowship and also introduced him to a visiting IEB officer. Gamow divided his eventual IEB fellowship of 1929/30 between Cambridge and Copenhagen.[13]

Starting in 1928 most of the international fellowships were handed out by the Rockefeller Foundation instead of the IEB. Priorities also shifted towards the biological and medical sciences, but physicists and mathematicians, including Landau, still received some fellowships. According to some archival reports, the Foundation reached an informal agreement with the Soviet medical and educational commissariats Narkomzdrav and Narkompros, which were together expected to nominate up to ten candidates annually.[14] The actual numbers never reached this level, but the granting of fellowships continued until the end of 1932. After that date, the collaboration stopped, apparently without any formal decision,

---

[9] RAC, IEB-1-3-48-725.
[10] On Tamm's travels and scientific experiences in Europe see (Dirac and Tamm 1993).
[11] RAC, IEB-1-2-40-566.
[12] RAC, IEB-1-3-59-979, 980.
[13] RAC, IEB-1-3-49-741.
[14] Robert A. Lambert to Stephen P. Duggan, 20 January 1934 (RAC, RF-1-1-785A-2-16).

but mainly because after the Nazi victory in Germany Soviet authorities became increasingly reluctant to allow Soviet citizens to travel abroad (Kozhevnikov 1993). Though less abruptly, international cooperation and scientific exchanges generally declined in Europe throughout the 1930s owing to the rising international tensions, and the Rockefeller program of international fellowships gradually faded away by 1938 (Gray 1941).

Overall, the period when Soviet scientists were allowed to travel abroad proved rather short, and the actual number of fellowships given to them was relatively small, but timing, choices, and serendipity were on the side of the young theoretical physicists. The window had opened just as the new discipline was forming and allowed its future leaders to acquire some crucial international experience and contact with colleagues in other countries. For the small community of Soviet theoretical physics, four such leaders would play crucial roles during the subsequent two decades—Frenkel, Fock, Landau, and Tamm, three of whom were Rockefeller fellows, the fourth, Tamm, being a Lorentz fellow. It would be left to them to guide the field through the twenty years of Stalin's dictatorship, when Soviet science worked in virtual international isolation, with practically no foreign travel, visits, personal communications, conferences, or correspondence, and when most contacts with the rest of world science would be reduced to exchanges of printed works.

3. A Quantum Rebel

Landau departed for his first foreign trip in October 1929. In the first half year, supporting himself with a stipend provided by Narkompros, he visited Berlin, Leipzig, and Zurich, working with Heisenberg and Pauli. In April 1930 he arrived in Copenhagen for a small conference of quantum theorists at Niels Bohr's Institute of Theoretical Physics. Around that time Rockefeller officers, acting upon Frenkel's recommendation, granted Landau a nine-month fellowship, which he subsequently divided among Cambridge, Copenhagen, and Zurich.[15]

He frequently complained of having come a little too late to fully partake in the quantum revolution (Livanova 1978, 21). By 1927, when he finished university, the two-year extraordinary rush for quantum mechanics was essentially over, and physicists started working more on

---

[15] Lev Landau fellowship card (RAC. Rockefeller Foundation. 10. Fellowship Cards USSR. DME+MS+NS).

applications of the new theory to solid state and the atomic nucleus and on generalizing it into the relativistic domain, or quantum electrodynamics. Yet, with his inclination to overthrow foundations and with criticism as his best-developed faculty, Landau was most strongly motivated when he saw a possibility of exposing a limitation of existing approaches or somebody's mistake or oversight. His best-known accomplishment of the time was the proof that electrons in a magnetic field demonstrate not only spin paramagnetism—orientation of spins along the field lines, explained by Pauli—but also a smaller but opposite effect, diamagnetism, resulting from the quantization of finite circular orbits of electrons in the field (Landau 1930).

Another paper of the same year can also be viewed as a criticism of Pauli, though in this case encouraged by Pauli himself. Together with Pauli's assistant Rudolf Peierls, Landau designed a corpuscular analog of the Heisenberg–Pauli quantum electrodynamics of 1929 (Heisenberg and Pauli 1929–1930; Landau and Peierls 1930). Their calculations were based on the model of photons as quantum particles rather than on the quantized wave field. Pauli was curious to see whether such a corpuscular approach would prove any better at solving some of the grave difficulties that he and Heisenberg had encountered in their version of the theory, yet he probably expected the answer would be negative. Indeed, Landau and Peierls' version proved equivalent and equally plagued by problems, which at the time seemed terminal for quantum electrodynamics. Pauli grew so pessimistic that he was ready to declare that the path towards relativistic quantum theory had been ill-chosen and that entirely new radical ideas were needed. He was certainly not alone among the small community of first quantum theorists in thinking this way, and the resulting disturbance became known as the 'new crisis of the quantum theory.'

The first crisis, which occurred in 1923–1925, had mostly had to do with the difficulties of the so-called old quantum theory of Bohr and Sommerfeld and was eventually resolved by the new path found in quantum mechanics (Jammer 1966). This time, the situation appeared to be repeating itself, only now the methods of quantum mechanics itself were found wanting in the domain of relativistic phenomena. In both episodes, the feeling of crisis in science coincided with periods of acute worsening in the general economic and social situation of post-World War I Europe, symbolized by the 1923 hyperinflation and the 1929 stock collapse. The feeling of a general crisis influenced the way physicists reacted (and overreacted) to difficulties in their own special field and the readiness with which they were prepared to abandon existing theories and postulates.

The difficulties in quantum electrodynamics around 1930 included the 'negative energy solutions' of the Dirac equation, a number of calculations yielding infinite results, and problems associated with electrons as part of atomic nuclei (as was commonly assumed then). Writing in a Soviet popular science journal, a newer member of the Leningrad "jazz-band," Matvei Bronstein (1906–1938), a.k.a. "Abbot," described the mood in the following words:

> After four triumphant years, the clouds over quantum theory are stormy once again. In spring 1930, a council at Niels Bohr's Copenhagen institute brought together some of the main experts of quantum theory, including Pauli and Heisenberg. The meeting parodied a ceremonious procedure: Pauli was holding a horn and blowing it every time he sensed the speaker was getting into a mental quagmire or a new unresolved difficulty. Unfortunately, he had to blow his horn way too often. The situation was rendered hopeless, and in a joking way, those present adopted a resolution to never again deal with quantum theory.[16]

Bronstein was not present at the Copenhagen meeting, and must have relied in his somewhat overdramatized description on a personal communication from Gamow or Landau, or both. Landau took the situation in earnest as the coming of another revolution, which he more than welcomed. Some of the young theorist's radical proposals went a little too far even for condescending authorities such as Bohr. With the enthusiasm of converts, Landau and Peierls applied Bohr's complementarity argument to the difficulties in quantum electrodynamics. In their second paper, finished by the time of the next Copenhagen conference in the spring of 1931, they argued that uncertainty limitations on measurement were much stronger in the relativistic domain than in non-relativistic quantum mechanics (Landau and Peierls 1931). Bohr was also very pessimistic about quantum electrodynamics, but in his view Landau and Peierls' call for a new revolution in fundamental physical concepts chose the wrong target—the notion of the electromagnetic field—as the main source of existing problems and the main candidate for sacrifice. In addition, Landau was probably unaware that Copenhagen, unlike the revolutionary Soviet Union, still respected academic hierarchy and that high debate over philosophical fundamentals was the privilege of a few recognized masters

---

[16] (Bronstein 1931, 38–9). The editors of the Soviet journal were pleased to hear scientists talking about a crisis in physics, which, they insisted, had to be interpreted more generally and politically as the 'crisis of bourgeois science' and a sign of the general revolutionary situation in the West.

Theoretical conference at Bohr's Institute of Theoretical Physics, Copenhagen, spring 1930. Note the horn and toy cannon that were used as part of the performance inaugurating the outbreak of a new crisis in quantum theory. First row, left to right: O. Klein, N. Bohr, W. Heisenberg, W. Pauli, G. A. Gamow, L. D. Landau, H. A. Kramers.
[*Courtesy*: Niels Bohr Archive, Copenhagen].

there, not apprentices in the field. It took Bohr and Leon Rosenfeld three years to write a response that validated the notion of the electromagnetic field in relativistic quantum theory and persuaded most physicists (Bohr and Rosenfeld 1933). Landau remained unconvinced, but he kept his objections to himself.

He returned to Leningrad in March 1931, twenty-three years of age, a mature scholar, not yet a mature person, and feeling that his task was to establish the discipline of 'theoretics,' as he dubbed his profession, in the Soviet Union. Back home, the cultural revolution he had seen starting as he was preparing to depart for foreign travel, was still in full swing.

4. Theoretics and the Cultural Revolution

The cultural revolution was part of the more general social revolution that occurred in the Soviet Union between 1928 and 1932 and also included the collectivization of agriculture and a major drive towards industrialization. It trumpeted the end of compromise with "bourgeois specialists"—engineers and scientists who had collaborated with the Bolshevik regime while distancing themselves from its values—and the demand for new "Red" specialists who would sincerely support Communist policies. Radical leftist experiments in education accompanied a crash attempt to train a new technical elite in huge numbers (Fitzpatrick 1978; 1979). The new science policy demanded close linking of education and science with the campaign of rapid industrialization, thus forcefully rejecting the notion of "pure science." The same policy also dramatically increased the importance attached to science and engineering and brought about a great expansion in financial support, the number of positions, and research institutions. Altogether, it was a short period of chaotic institutional changes, radical reforms, and social mobility.

The established leaders of the scientific community found themselves in an insecure position. They were vulnerable to attacks from many sides (from the security police and the political authorities from above, and from militant junior members of their institutes from below) and could be accused of various misdeeds ranging from mistakes in management to political crimes (the favorite political charge was "wrecking," or sabotage). Many senior scientists and engineers failed to survive these attacks and had to give way to a new generation (Bailes 1978, part 2). Among the variety of conflicts the cultural revolution let loose, it specifically provoked an acute generational conflict in academe in which the political authorities tended to side with younger radicals, especially those who had received their education during the previous decade under Soviet rule. The conflicts in which Landau became involved upon his return to Leningrad were, if not the most typical of the cultural revolution, still unmistakably products of the time.

Unlike most senior scientists, Joffe could feel sufficiently secure throughout the turmoil. Though a 'bourgeois specialist' in looks and manner, he managed to surf the revolutionary wave and turn the events to his, his institute's, and the entire discipline's advantage. Joffe was regarded as clearly pro-Soviet among his generation of scholars and continued to enjoy support and protection from Soviet commissariats. His reputation and influence were particularly strong at the economic commissariat, VSNKh, which with the start of the campaigns of industrialization and cultural revolution took over from the Narkompros the role of the main authority in science policy (Fitzpatrick 1985; 1992, ch. 5; Josephson 1988b; Strekopytov 1990). In the late 1920s, as a member of VSNKh's ruling collegium, Joffe advised the government about opening new physico-technical institutes in other parts of the country—Kharkov and Dnepropetrovsk in the Ukraine, Sverdlovsk in the Urals, and Tomsk in Siberia. If not necessarily creating as such the major expansion of physics into new places and institutions, which was likely to have happened anyway, he certainly greatly contributed to it and shaped its direction, eventually acquiring a reputation as the 'dean' or 'father' of Soviet physics.[17]

Still, as an institutional authority during iconoclastic times, Joffe had to deal with challenges presented by rebellious academic youth and with the generational conflict in his own institute. He also proved susceptible to the spirit of revolutionary times in his desire to have his institute contribute something substantial to the Soviet industrialization drive. His pet idea at the time, a high-profile project to develop thin-film insulators, was making newspaper headlines and attracting considerable investment from the Soviet government and foreign firms. Upon his return to the Soviet Union, Landau attacked his institute director's high-stake project as misconceived and ridiculed Joffe for not having sufficiently mastered theoretical physics. His behavior was rude, but his judgment—as far as physics was concerned—proved correct. Eventually, Joffe would have to concede that his proposal was based in part on wishful thinking—not unlike many other high-flying ideas of those extravagant years—and in part on trusting results of preliminary measurements that turned out to be inaccurate (Hall 1999, ch. 8).

Landau's own high-stakes project of the year was equally misconceived and utopian, though in the social rather than the physical domain. He and Bronstein made splashes trying to orchestrate the election of the 27-year-old Gamow to the Soviet Academy of Sciences and to transform the Academy's Physico-Mathematical Institute (which had only a nominal physics department, most of whose dozen members also held positions elsewhere) into a separate Institute of Theoretical Physics, under Gamow's leadership. Gamow was the first member of the "physical jazz-band" to gain international recognition for his 1928 theory of nuclear decay. He and his friends wanted to capitalize on that fame and on the authority conferred on younger academics by the cultural revolution to demand concessions from their seniors. Their main revolutionary demand was the recognition of theoretical physics as a separate discipline, and in their generational arrogance they did not include the slightly

---

[17] On Joffe and his Institute see (Sominsky 1964; Josephson 1991). For a number of documents pertaining to Joffe's role in establishing new institutes for physics research see (Ioffe 1980).

older Frenkel, who was the first theoretical physicist elected to the Soviet Academy as correspondent member in 1929 (Gorelik and Savina 1993).

In more ordered times, Landau and his friends would not have felt entitled to make such claims, nor would the Academy have paid any attention to a suggestion pushed by a few youngsters. It was a sign of how strongly the cultural revolution had shaken the traditional academic hierarchy that in 1931 the idea was seriously discussed at all. Still, Joffe, as the Academy's senior physicist, managed to block the proposal. Joffe was an old-fashionedly polite, mild-mannered man who apparently thought he had suffered enough from the impudent "jazz band" gang. He also subscribed to a more traditional view that the role of theorists in physics institutes was to advise experimental workers, rather than build separate institutes. He agreed with other senior physicists at the Academy, experimentalists Rozhdestvensky and Mitkevich, that the Academy should establish a full-size physics institute, but one headed by an experimentalist. Gamow and Fock, two theoreticians, were elected corresponding members in 1932, but the idea of an independent institute of theoretical physics was abandoned by the Academy until almost forty years later.[18]

Another scandal of 1931 involving Landau concerned Marxist philosophy. Landau called himself a Marxist, but his views were far from conformist. He regarded the historical and social theory of Marxism—"historical materialism" in Soviet terminology—as an ultimate example of scientific truth. At the same time he dismissed the Marxist philosophy of 'dialectical materialism' (and any other philosophy of nature for that matter) as complete nonsense. One of the practical jokes at which the jazz-band so excelled misfired when they sent a mocking telegram to one of the country's leading Marxist philosophers of science, Boris Hessen (1883–1938). Joffe was particularly upset by this prank, for he regarded Hessen as one of the physicists' most important allies among philosophers. Hessen firmly supported Einstein's relativity and the new physics in general, both as a philosopher, arguing that modern advances in physical theory corresponded well to Marxism, and as an administrator, in his position as director of Moscow State University's Research Institute of Physics. For Landau and Bronstein, the episode led to a public reprimand at a meeting of the Leningrad Physico-Technical Institute for "antisocial" hooligan behavior and to suspension of their part-time teaching contracts (Gorelik and Frenkel 1994, ch. 3.8).

---

[18] After Landau's death, his students would organize in the Academy the L. D. Landau Institute of Theoretical Physics (Khalatnikov and Mineev 1996).

After this string of incidents, Landau's relations with his institute's director were inevitably strained, and in August 1932 he moved to Ukrainian Physico-Technical Institute (UFTI) in Kharkov, then the capital of the Ukrainian Soviet Republic.

5. The Kharkov School

UFTI became a natural rallying place for those younger Leningrad physicists who preferred to realize their professional ambitions independently of Joffe's patronage and authority. Itself a product of the cultural revolution, UFTI was one of the first and arguably the best of the new generation of physical institutes that Joffe helped organize around 1930. The economic commissariat VSNKh, soon to be reorganized as the Commissariat of Heavy Industry (NKTP), approved the decision to build a physico-technical institute in Ukraine in 1928, and the new building officially opened in 1930 (Ioffe 1980, 110–113). Its research staff combined local Kharkov physicists and a group of young physicists from Leningrad, mostly recent *aspiranty* from Joffe's institute. The possibility of leaving for Kharkov actually helped diminish some of the generational conflicts at LFTI. As recently educated scientists—some also politically active—they could expect cultural revolution momentum and authorities to support their initiatives at the new place.

Several of them, still in their twenties, soon began to play leading roles in the new institute as heads of its main laboratories, and there they realized projects that for one reason or another they had felt they could not pursue at LFTI. While the Leningrad institute continued to specialize primarily in X-rays and solid state research, UFTI leaders chose the newer and hotter fields of low-temperature and nuclear physics. Kirill Sinel'nikov, Aleksandr Leipunsky, and Anton Val'ter started building the country's first artificial accelerator of atomic particles, the electrostatic Van de Graaf machine. Lev Shubnikov (1901–1937, co-discoverer of the Schubnikow–de Haas effect) organized a low-temperature laboratory, the first in the Soviet Union to have a helium liquefier and work with liquid helium temperatures (Val'ter 1933; Shubnikov 1990; Pavlenko, Ranyuk, and Khramov 1998).

In the spirit of the time, the chief rhetoric justifying the establishment of a new institute in the Ukraine was that of service to local industry, hence the leading role of the Commissariat of Heavy Industry as the main sponsor and supervisor of the physico-technical institutes (and also those in

Theoretical Physics in the Context of Cultural Revolution

Scientist tuning the apparatus used for the splitting of lithium nuclei, Ukrainian Physico–Technical Institute, Kharkov, 1932. Revolutionary imagery associated with the idea of smashing atomic nuclei made the project a public relations coup for the young institute. The success of the experiment was reported in an open letter from physicists to the Soviet political leadership, published by *Pravda* and other Soviet newspaper.
[*Courtesy*: Russian State Archive of Cinema and Photographic Documents, Krasnogorsk, hereafter RGAKFD.]

chemistry and geology) during the cultural revolution and industrialization drives.[19] The institutes benefited greatly from NKTP's financial and material resources but at the same time came under increased pressure to produce industrially applicable knowledge and had to persuade the

---

[19] One can compare industrialization-related plans for the development of physico-technical institutes in Leningrad and Kharkov in (GFTL 1929) and "Stenogramma zasedaniia prezidiuma VSHKh USSR, 21 marta 1929 g.," (P. L. Kapitza Archive, IFP. Moscow).

officials of the direct industrial potential of some of their more fundamental research projects. In the case of UFTI, the Commissariat proved willing to support, alongside industrially useful research, some unprofitable fields such as nuclear physics, which attracted much attention and advanced the international prestige of Soviet science.

UFTI grew so rapidly that in a few years its size and budget began to rival those of the parent institute in Leningrad (NKTP 1934; 1935). Among the country's physics institutes, it also developed the most extensive international contacts. Many of its leading researchers spent considerable periods studying and working in some of the most advanced European laboratories. The institute lured foreign specialists with generous contracts and hosted a string of visitors from abroad, including such luminaries as Dirac and Bohr. In 1932 UFTI started publishing a major physics journal in foreign languages (with articles in German, English, and French), the *Physikalische Zeitschrift der Sowjetunion*. It quickly surpassed the German *Zeitschrift für Physik* as the main channel through which Soviet physicists made the results of their research known internationally.

UFTI's first director, Ivan Obreimov (1894–1981), also prided himself on having understood better than Joffe the rising importance of theoretical physics and was willing to support the new discipline more seriously (Obreimov 1973, 45, 53). Already in 1929, before the institute opened officially, UFTI hosted the first All-Union conference on theoretical physics. The institute's first theoreticians, Ivanenko and Lev Rozenkevich, came from Leningrad. Obreimov tried to persuade Ehrenfest to move to Kharkov, but his hopes for a senior theorist did not materialize. Ehrenfest visited UFTI a couple of times and tried to convince Joffe and other experimentalists that Landau, in spite of his youth, terrible character, and dogmatic and inconsiderate opinions, had the right potential (Ehrenfest and Joffe 1990, 282–291). Indeed, soon after his move to Kharkov Landau established himself as the lead theorist there.

For Landau as for others, escape to Kharkov offered an opportunity to realize the cultural project that he, as a young radical rebelling against his seniors, had pushed for back in Leningrad. UFTI directors Obreimov and Leipunsky shared his visions for the discipline and supported his ambitions in building up, if not a separate institute, but a strong theoretical group as one of the institute's main divisions. What later became known as the Landau school started in Kharkov, as Landau began attracting a following of talented students, including Aleksandr Kompaneets and Evgeny Lifshitz from Kharkov, Isaak Pomeranchuk from Leningrad, and Aleksandr Akhiezer from Kiev. Himself now in a position of some authority, Landau did not encourage students to become rebels as he had been, but to acquire technical skills and calculate concrete effects. In pedagogy, he also started requiring more order and discipline. Prior to accepting a new student, Landau subjected him to a special set of exams, later codified and glorified as the 'Landau minimum.' Eventually, these exams were supplied with a series of textbooks, the famous *Course of Theoretical Physics*, written by senior students (Lifshitz, primarily) under Landau's direction (Kaganov 1998; Khalatnikov 1989; Hall, forthcoming).

The times were also changing, and after 1932 the sheer enthusiasm and radicalism of the cultural revolution gave way to a renewed emphasis on solid training and disciplined expertise, the marks of the emerging ordered society of high Stalinism. The restoration of a more conservative, traditionalist order in higher education took several years, until 1936 or so. Beginning in 1934, Soviet *aspiranty* had to write theses and earn degrees equivalent to the Ph.D. (*kandidat*). A higher degree (*doktor*, somewhat similar to the German *Habilitation*) was instituted for senior scholars. Scientists who had accomplished their careers and attained professorial level during the interim sixteen years when these degrees were abolished were granted academic titles *post factum*, on the basis of their cumulative research accomplishments. Together with a number of other senior colleagues, Landau received his title of *Doktor of Science* in 1935.

He also became more reserved and solid in his papers, although hardly more polite and considerate in behavior. The string of papers in which he was critical of the foundations of quantum theory ended with a 1932 work on the theory of stars, which mentioned Bohr's dangerous idea of energy non-conservation (Landau 1932). Landau shared Bohr's hope that deviations from strict energy balance might help explain the unresolved problems in relativistic quantum theory. The suggestion seriously upset not only the majority of physicists but also Soviet philosophers, who regarded the principle of energy conservation as a sacred cornerstone of philosophical materialism. They would occasionally criticize Landau for publishing such heresy, but he did not insist on the idea for long, because some of the difficulties it was supposed to resolve were soon dealt with by the introduction of new elementary particles, the positron, discovered in 1932, and the neutrino, first proposed by Pauli in 1930 and developed into the elaborated theory of weak interactions by Fermi in 1933. Other difficulties in quantum electrodynamics remained, in particular the ever more frequent infinities in various calculations, but the general feeling of a major crisis in fundamental theory became much more subdued. After 1932 and for the next thirty years or so Landau abandoned foundations and directed his efforts mainly towards applications of quantum theory.

"Think less about foundations," he instructed his students accordingly (Hall forthcoming). They and Landau calculated various effects in quantum electrodynamics—on stopping, pair creation, and the scattering of light on light—but his main attention turned increasingly towards the solid state and condensed matter. From 1934 on UFTI's Lev Shubnikov developed a thriving experimental research program on superconductivity and other low-temperature effects in metals (Shubnikov 1990; Kriogenika 1978; Yavelov 1985). Landau's close collaboration with the experimental group at UFTI helped him to predict the phenomenon of antiferromagnetism (Landau 1933c). His first attempt to attack the riddle of superconductivity with the hypothesis of spontaneous currents was not successful (Landau 1933b), but in 1937 he developed a theory of the intermediate state as consisting of superconducting and normal layers (Landau 1937a) and much later, with Vitaly Ginzburg, he authored the correct macroscopic theory of superconductivity (Ginzburg and Landau 1950). The crowning achievement of the Kharkov period was the thermodynamic theory of the second-order phase transitions, in which the state of the system changes continuously but its degree of order, or symmetry, switches abruptly (Landau 1937b).

In the socio-political domain, the adaptation of Landau and other former radicals to Soviet society's phase transition from the chaotic revolutionary to the more ordered Stalinist state did not proceed so smoothly. As state policies many aspects of the cultural revolution ended by 1932, but habits acquired earlier did not die instantaneously. In posture, pronouncements, and relationship to at least some authorities Landau continued acting as a rebel. As late as 1936 he and Leipunsky were still publicly attacking Joffe at a special meeting of the Academy of Sciences that reviewed LFTI's performance and significantly undermined Joffe's reputation and power (Leipunsky 1990, 115–122; Vizgin 1990–1991). As Stalinist purges drew nearer, such behavior would entail ever more risks for both the attackers and the attacked.

As the international situation worsened and expectations of war grew stronger, Soviet science and the entire society became increasingly disciplined and militarized. In 1935 security guards began to regulate entrance to UFTI and some of its buildings, formerly of unrestricted access. Landau with some of his institute's friends protested and mocked the new security regulations as well as the redirection of some laboratories towards classified projects. They demanded from the commissariat

Theoretical Physics in the Context of Cultural Revolution 97

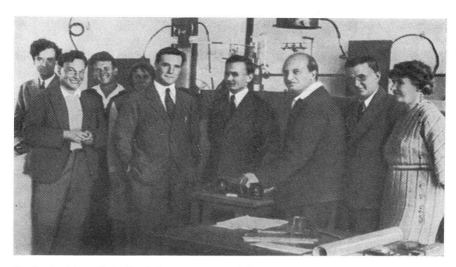

Kapitza's visit to Shubnikov's low temperature laboratory in Kharkov, summer 1934. Still a happy time, before things took a turn for the worse. Kapitza had not yet learned that he would have to remain in the Soviet Union instead of returning to Cambridge from this visit. Nor had administrative conflicts and purges yet hit the flourishing Ukrainian Physico–Technical Institute. From left to right: L. D. Landau, A. I. Leipunsky (UFTI's director), Yu. N. Riabinin, O. N. Trapeznikova, P. L. Kapitza, (?), L. V. Shubnikov, B. N. Finkel'shtein, (?). Three years later, Shubnikov would be executed as an "enemy of the people"; Landau and Leipunsky would be arrested, but later released.
[Source: (Leipunsky 1990), 192 ff.]

the removal of UFTI's new director, Semion Davidovich, who had instituted those measures. Davidovich was replaced with Leipunsky, but during the struggle one of Landau's students, Moisei Korets, who had written an article in the institute's newspaper against classified research, was arrested and tried. He was later released, but the legacy of the institutional conflict and continuing tensions turned into a mortally dangerous mix and contributed to tragic arrests at UFTI in 1937, when the scale of repression escalated and the cruelest waves of the Stalinist political purges started roaming across the country (Pavlenko, Ranyuk, and Khramov 1998, 170–198).

The story of Landau's escape from Kharkov illustrates how even a minor incident could become life-threatening in the context of the Great Purges. In December 1936 Landau had a personal quarrel with the rector of Kharkov University, where he was teaching part-time. Returning from the meeting, he announced to his friends that he was about to be fired. In order to put some pressure on the rector, seven of his colleagues and students submitted resignations from their part-time teaching jobs.[20] A month later, in January 1937, a wave of arrests of "Trotskyites" reached the university, and a number of officials, including the rector, disappeared forever. In the poisonous atmosphere created by those arrests, which saw much soul-searching, finger-pointing, and paranoid vigilance against "enemies from within," speakers at UFTI meetings denounced the act of collective resignation as an "anti-Soviet strike." Whether Landau was simply terrified or grasped intuitively that a quick change of location could increase one's chances of survival, he also suddenly disappeared. Several weeks later his friends in Kharkov received a message from him that he had decided to take a job in Moscow, in the new Institute of Physical Problems directed by Piotr Kapitza.

A great many projects and trends that had strongly profited from the cultural revolution suffered a particularly bad luck during the purges of the late 1930s. Whether this had more to do with the controversial status of the projects themselves or with the iconoclastic behavioral patterns of their protagonists, one way or another theoretical physics, after making great institutional advances in the early 1930s, fell on hard times by the decade's end, not only in Kharkov but also at Leningrad University and in Tamm's group in Moscow.[21] As a discipline, it would retreat somewhat and have to regroup before making further progress several years later. Landau abandoned his thriving theory group in Kharkov and went to Kapitza's institute, where he would be the sole theoretician for two years until his student Lifshitz was also hired. Eventually, the Landau school would reconstitute itself both in Kharkov and in Moscow by students who either stayed or followed their teacher to a new place. For a few years, however, the chances for its survival, and for Landau's personal survival, appeared dismal. New Stalinist times and political realities required changes in behavior and a new style of relationship between scientists and the authorities. Landau was not adapting well to the transition, but his new institute boss, Piotr Kapitza, managed most remarkably to find his own solution to the situation and establish a special connection to Stalin's government. This would also help him save Landau's life.

---

[20] See recollections by A. K. Kikoin and A. I. Akhiezer in (Khalatnikov 1989); L. V. Shubnikov's personal dossier, Archive of Kharkov State University, personnel files for 1933–1941, fold. 347.
[21] On the temporary dissolution of Tamm's theory group at Moscow's FIAN see (Hall, 1999, 184–190). A center of physics that suffered almost as severely as UFTI during the purges was Leningrad University. Among theoretical physicists in Leningrad Bronstein, Bursian, Fock, and Krutkov were arrested (Kosarev 1993, 143–145).

Chapter 5

Scientist under Stalin's Patronage: The Case of Piotr Kapitza

Piotr Leonidovich Kapitza (1894–1984) gained recognition not only for his scientific achievements, which won him the Nobel Prize in 1978, but also for his public activities. These activities, however, have received very different interpretations. While some American journals called Kapitza a "dissident," the official Soviet press portrayed him as a true Soviet scientist, honored by the government because his work had served the socialist state. Post-Soviet authors usually praise him as a non-conformist who resisted Stalinism, battled the ruthless chief of security police, Lavrenty Beria, and was persecuted for his defiance (Kedrov 1984; Spruch 1979; Rubinin 1994). All these portraits have some factual basis. Kapitza was an elite and very influential academician—he won the Stalin Prize twice, the Order of Lenin five times, and the title Hero of Socialist Labor twice. During the Stalinist purges of late 1930s, he bravely defended some persecuted scientists and saved several lives, and in 1946 he was dismissed from all his official positions and disappeared from public view for several years. Rumors in the West held him responsible for the construction of the Soviet atomic bomb, but later reports claimed, on the contrary, that he was disgraced after his refusal to take part in the bomb project. Sovietology, a field that specialized in presenting rumors and prejudices as results of scholarly research, had both these mythical versions printed:

> After 1937 [Kapitza became] the leading member—de facto physical director—of the Supreme Atomic Energy Panel of the Soviet Union. In

the beginning of 1946 Kapitza was nominated Soviet Lieutenant-General but later was removed from office by Stalin, because of his attempts to exchange information on the results of atomic research with his foreign colleagues. He worked now only within the framework of pure scientific research, not being allowed into the Super-Complex. In 1949, after insistent advice from professor Hertz and professor Joffe, he was again appointed as atomic chief with all honors. In 1953 he finished this cycle of his research in the atomic field with the completion of Soviet hydrogen bomb (Biew 1954, 286);

[In 1946, Kapitza] was arrested and sent to the Gulag for refusing to work for the military who wanted him to carry out research into the use of nuclear energy. After Stalin's death he was released immediately and brought back to Moscow. (*A Biographical Dictionary of the Soviet Union 1917–1988*. London 1989.)[1]

Although some archives related to military and atomic weapons research remain classified, a wealth of documentary information about Kapitza has become accessible since the start of Gorbachev's perestroika in the mid 1980s, thanks, in particular, to publications prepared by P. E. Rubinin (Kapitza 1988–1998; 1989a; 1990; 2003). These sources help dispel much of the folklore surrounding Kapitza and make it possible to piece together the seemingly incompatible parts of the puzzle into a complex, but coherent life trajectory of a scientist who lived through the challenges and contradictions of Stalinist society. Although many features of Kapitza's career and the choices he had to make were unique, they nevertheless help reveal some general characteristics of Soviet science, its role in industrial modernization, and very specific patterns of patronage granted to science by the top political leaders.

1. Expatriate

Piotr Kapitza, the son of a military engineer and general of the Russian army, was born on 9 July 1894 in Kronstadt, an island navy fortress off the coast of St. Petersburg. Expelled from a classical *Gymnasium* for slow

---

[1] Another completely fictive "eye-witness account" of Kapitza's secret activities can be found in the collection of oral history interviews with emigrants from the Soviet Union (Hoover Institute Archive, B. I. Nicolaevsky collection, box 412. folder 9). In contrast, (Shoenberg 1985) provides an example of how, by refusing to deal in speculation and rumor, one could write an extremely reliable biography even before archival documents became accessible.

Scientist under Stalin's Patronage: The Case of Piotr Kapitza

Joffe's student seminar at the Petrograd Polytechnical Institute—the nucleus of the future post-revolutionary Leningrad Physico–Technical Institute—in 1916. Joffe is standing on the right, while sitting on the far left are Kapitza, Frenkel, and Semenov.
[*Credit*: AIP, Emilio Segrè Visual Archives; Frenkel collection.]

progress, Piotr completed his secondary education in a *Realschule* in 1912. At the time, a Gymnasium-level examination in Latin was still required for enrolment in Russian university, but one could matriculate without Latin at engineering schools. Kapitza began studying at the St. Petersburg Polytechnical Institute with a view to becoming an engineer, as he probably would have, had he not met the professor of physics Abram Joffe. In 1916 Joffe established a physics seminar for advanced students at the Institute, and Kapitza, who had just resumed his studies after two years of military service at the front, became one of the seminar's first members along with several other future luminaries of Soviet science: Nikolai Semenov, Yakov Dorfman, and Yakov Frenkel (Kokin 1981, 10–25). That same year the student Kapitza published his first two short papers in the journal of the Russian Physico-Chemical Society.

Kapitza and other student members of Joffe's seminar subsequently became the basic research staff of the Physico-Technical Division, which Joffe organized at the State Roentgenological and Radiological Institute in 1918 and which in 1921 became a separate Physico-Technical Institute (LFTI). Throughout the Civil War years and the economic disaster that followed, Kapitza and other researchers struggled with a lack of even the most basic supplies and, together with the rest of the Petrograd population, suffered from hunger, cold, and epidemics. In the course of just a few months, influenza and other illnesses took the lives of Kapitza's father, his two-year-old son, and his young wife with a newborn baby. In order to help Kapitza out of a deep depression, Joffe took him as assistant on a trip to Europe, where he went in the spring of 1921 with the goal of purchasing equipment and scientific literature for the new institute (Kapitza 1990, 106–107).

The mission was also an early effort in science diplomacy. The Soviet regime was not yet recognized as a legitimate government by most European powers and was viewed internationally with much hostility and suspicion. After a month-long waiting period in Estonia, Kapitza received a British visa and went to England in May 1921. Joffe, who traveled by way of Berlin, arrived a little later. While visiting Cambridge in early July, Kapitza managed to win Ernest Rutherford's consent to study and work at the Cavendish Laboratory through the winter, a decision also approved by the Soviet diplomatic representative in London (Kapitza 1989b, 2–5). Although initially only a short stay was envisioned for Kapitza, it turned into a full-scale academic career in Cambridge. The enculturation was anything but easy, as Kapitza described in a letter to his mother:

> I shall try to give you a general idea of my situation. Imagine a young man who arrives at the world-famous laboratory in the most aristocratic and conservative English university... who is not known to anybody, who speaks bad English and has a Soviet passport... Why was he accepted? I still have no idea. I once asked Rutherford. He burst out laughing and said: "I surprised myself by agreeing to accept you, but in any case I am very glad I did." The first thing this young man encounters there is the following declaration from Rutherford: "If instead of research you are going to engage in Communist propaganda, I am not going to tolerate it." Everybody is keeping away from this young man, everybody is afraid of compromising their own reputation by consorting with him. I realize that my best strategy is to take up this challenge. I start a most difficult project with hardly any faith in its successful outcome... The truth is I often worked until I nearly fainted. I did break through, however. And this is certainly a piece of good fortune. It cost me dearly, though (Kapitza 1989b, 47).

After this initial success, Rutherford suggested that Kapitza extend his stay at the Cavendish. With time, Kapitza began to view the English-style schooling he was receiving at Cambridge as superior to the German style in which Joffe had been trained, but more than anything else he valued that "the opportunities for work here are such as I never dreamed of in Peter[sburg] even in peacetime." "To go back to Petrograd—to struggle without electricity and gas, with the lack of water and equipment—is impossible. I have only now come into my own. I find success exhilarating and my work inspiring. This is all I have left after the death of my family... Judging by what Abram Fedorovich [Joffe] told me about the situation at the... institute, I concluded that their thermometers barely rise above the freezing point, although Abram Fedorovich tends to embellish everything. His personal view is that I should work here for another year" (Kapitza 1989b, 51, 88). Kapitza thus decided not to return to Soviet Russia until later, as "a mature man capable of doing true science, not an ersatz margarine" and able to dictating his own conditions, though that decision was causing some tensions with institute friends back home (Kapitza 1986, 204, 206). He actually remained in Cambridge for thirteen years.

At Rutherford's suggestion, Kapitza started a study of the passage of alpha particles through matter and proposed to measure the energies of alpha particles by placing the Wilson chamber in a magnetic field that curved their tracks. This required generating magnetic fields far stronger than those that could be sustained by magnetic coils. Kapitza circumvented

P. L. Kapitza in Cambridge, 1920s. Kapitza's generator of strong magnetic fields and other machinery pushed the Cavendish Laboratory to the limits and initiated its transition from tabletop experiments to large-scale research.
[*Credit*: AIP, Emilio Segrè Visual Archives.]

that difficulty by designing a special generator of magnetic pulses that lasted only a thousandth of a second, not long enough to destroy the apparatus but sufficient for taking photographs of particle tracks. To help him build this generator, Rutherford agreed to bring to the Cavendish a mechanic from Estonia, E. Ya. Laurmann, who had worked with Kapitza in Petrograd. After three and a half months of intensive work, Kapitza proudly wrote to his mother that he had obtained the first three curved tracks of alpha particles (Kapitza 1989b, 60; Kapitza 1923). Having at his disposal the equipment for producing magnetic field of unprecedented strength, Kapitza gradually shifted his interest away from Rutherford's favorite nuclear radiations towards studying magnetic properties of matter in extremely high fields.

Rutherford strongly supported his new pupil. In 1923 Kapitza was officially admitted to the Cavendish as a research student for the Ph.D. degree. He received the degree later that year and became a fellow of Trinity College. In 1925 he was named Assistant Director of Magnetic Research at the Cavendish, and he was elected fellow of the Royal Society in 1929, at a time when, according to David Shoenberg, the Society was relatively closed to foreigners (Shoenberg 1985, 340). Although Kapitza never learned to speak English properly, he brought more life into the Cavendish by his eccentric manners and by introducing into the reserved British academic culture a livelier mode of discussion, leading a Russian-style seminar, *kruzhok*, for younger physicists, which became known as the Kapitza Club. Rutherford got his famous nickname 'Crocodile' from Kapitza, who approached his boss with a mixture of fear, admiration, and informality.

According to Rutherford, Kapitza, "if not a genius, had the brain of a physicist and the ability of a mechanician, a combination so rarely wedded in one brain, that it made him something of a phenomenon" (Joravsky 1985, 33). Indeed, Kapitza's main scientific accomplishments typically involved the construction and operation of novel and powerful experimental machines capable of producing extreme conditions: very strong fields, very low or very high temperatures. In 1929 he reported that in strong magnetic fields the electric resistance of metals changes from a quadratic to a linear function of the field, which became known as the Kapitza law of magnetoresistance (Kapitza 1929). Since many properties of substances become much more distinctive at lower temperatures, he started planning experiments with liquid hydrogen and helium. At the time, only a handful of laboratories in the world—Leiden, Berlin, and Ottawa—had the capacity to liquefy helium. Kapitza began building a new type of a liquefier based on the adiabatic expansion of helium in a cylinder with a piston. The principle of adiabatic expansion was of course well known, but the freezing of lubricants did not allow its use at extremely low temperatures. Kapitza decided to get rid of the lubricant altogether. He allowed helium to escape freely through the gap between the cylinder and the piston but made the expansion so rapid that the loss of gas was small. Helium itself performed the function of the lubricant (Kapitza 1934).

With Kapitza's new devices, the style of research at the Cavendish was shifting from tabletop experiments to working with big machines and towards contacts with industry and with the Department of Scientific and Industrial Research. In 1930 Rutherford persuaded the Royal Society to provide funds from the Ludwig Mond bequest for building the Royal Society Mond Laboratory in the courtyard of the Cavendish. The Mond Laboratory opened in 1933 with Kapitza as its director. In April 1934 he obtained his first liquid helium, though it was not the first liquid helium in Britain. The Oxford team beat him slightly by using more conventional techniques and apparatus for liquefaction (Shoenberg 1985).

As his British career progressed, Kapitza was also developing contacts with colleagues and officials back home. After spending a vacation in the Soviet Union in the summer of 1926, he paid visits there almost every year. Soviet officials who directed the science department at the economic commissariat VSNKh were offering him prestigious positions and conditions at home.[2] He agreed in principle to return to a permanent position in some unspecified future, meanwhile accepting an appointment in 1929 as the official consultant to the newly organized Ukrainian Physico-Technical Institute (UFTI) in Kharkov, where Ivan Obreimov and Lev Shubnikov were establishing the first Soviet low-temperature laboratory. Kapitza also offered his assistance in instructing younger Soviet physicists in England. Taking advantage of the relative ease of foreign travel between 1929 and 1932 and with some help from Kapitza, several students visited Cambridge and worked there. Among them were Yuly Khariton, later the scientific director of the main Soviet nuclear weapons laboratory, Aleksandr Leipunsky, and Kirill Sinel'nikov, two subsequent directors of UFTI and prominent nuclear scientists, and Lev Landau,

---

[2] Curiously enough, politicians who held this post in sequence had once been powerful communist leaders: Leon Trotsky, Lev Kamenev, and Nikolai Bukharin. Having lost their struggle for power to Stalin, they were on downhill career paths within the Soviet bureaucratic apparatus. Kapitza was in personal contact with the latter two.

theoretical genius. In 1929 Kapitza, together with Frenkel and Semenov, two of his colleagues and friends since the days of Joffe's seminar at the Polytechnical Institute, was elected corresponding member of the USSR Academy of Sciences.

Having obtained liquid helium in 1934, Kapitza was planning an ambitious program of experiments in the newly built Mond Laboratory. But he was never to perform those experiments in Cambridge. In August 1934 he went on one of his usual summer trips to the Soviet Union, visiting his mother in Leningrad and Shubnikov's low-temperature laboratory in Kharkov. As he was preparing to return to England at the end of September, he learned from Soviet authorities that he would not be allowed to leave the country.

2. Prisoner

While Kapitza remained in Leningrad, his wife Anna, whom he had married in 1927, returned to Cambridge to their two young children and stayed there another year. After all the efforts that she, Rutherford, and other British colleagues undertook to negotiate Kapitza's release failed, she rejoined her husband in Moscow. During their separation they corresponded frequently; Kapitza alone wrote more than a hundred letters recording at length his initial experiences in the Soviet Union.[3]

At first, he felt surrounded by hostility. State officials talked to him with suspicion, and police agents openly followed him (Kapitza to Molotov, 7 May 1935, in Kapitza 1989a, 39–43). Since the Nazis had taken power in Germany, Soviet society was bracing itself for another major war; fear of foreigners and spies permeated the nation, and great purges of the alleged fifth column of "spies" and "enemies of the people" loomed on the horizon. Kapitza discovered that he did not even have the sympathy of his closest colleagues, including Joffe and Semenov. They were probably afraid to befriend someone under surveillance, but even in private conversations, according to Kapitza, his colleagues tended to approve of his detention (Kapitza to Anna Kapitza, 5 April 1935, in Badash 1985, 67). Kapitza found more understanding in several older scientists, Ivan

---

[3] During his first year in the Soviet Union, Kapitza was not allowed to write letters abroad except to his wife in Cambridge. The translations done for Rutherford were the basis of (Badash 1985). The originals are in the Archive of the Kapitza Institute of Physical Problems, Moscow (hereafter IFP).

Pavlov, Aleksei Krylov (his father-in-law), and Aleksandr Bakh. Although of different political orientations, they had more empathy for his situation and eased his loneliness. His main complaints focused on his inability to continue research: "I can't read papers related to my research because I might become half-mad. ... I could once understand how someone could go mad, but I never thought I would ever be brought to such a condition when deprived of my scientific work" (Kapitza to Anna Kapitza, 21 May 1935, in Kapitza 1989a, 44).

Although Soviet officials justified Kapitza's detention on grounds of the Soviet industry's need for his expert advice and professional research, he had no laboratory or equipment and received no requests for consultation. He even thought of leaving physics for physiology so he could collaborate with Pavlov, but the authorities strongly objected. Finally, after several months of uncertainty and negotiations, the Politburo of the Communist Party adopted a resolution about Kapitza on 21 December 1934. Stalin and other leaders decided to set up a new Institute of Physical Problems (IFP) within the Soviet Academy of Sciences, appoint Kapitza its director, and give him an apartment in Moscow, a personal Buick, and a dacha in the Crimea for his family (Akademiia 2000, 165). The resumption of experimental research still had to be postponed while Kapitza supervised the design and construction of the institute's building and waited for the equipment to arrive. His forced exile from science lasted altogether two years, until the fall of 1936.

Cambridge physiologist E. D. Adrian visited Moscow in August 1935 for the International Physiological Congress and brought back a report for Rutherford about Kapitza's situation:

> Kapitza considers that there were three reasons for his detention: (a) unfounded reports from England that he was doing war work: these reports must have come from Cambridge and from a well informed source. (b) Gamow: when Gamow was out of Russia he wrote to Molotov asking for the same standing as Kapitza used to have, and he made this a condition of his return to Russia. (c) the reason that his abilities would be valuable during the war (Kapitza 1990, 267).[4]

As far as we know, Kapitza did no military research at Cambridge, though he did consult with industries and had grants from the

---

[4] A similar memorandum also exists in P. A. M. Dirac's handwriting (IFP). Dirac was in Moscow in August 1935, visiting with Igor Tamm and Kapitza (Dirac and Tamm 1996). Before returning to Cambridge in September, he memorized the text because he was apparently afraid that authorities might confiscate the document on the border.

Department of Scientific and Industrial Research (Kapitza 1969). His abilities, indeed, proved useful to the Soviet government later during the war. It is not certain whether Gamow had explicitly referred to Kapitza in his negotiations with Soviet authorities, but they could easily see the connection. Gamow decided to remain in the West after taking part in the 1933 Solvay Congress in Brussels (Gamov 1970, 118–133). Kapitza advised him to request a one-year official extension of his foreign trip from the Soviet Academy and hope that the authorities might eventually get used to his prolonged stay abroad and let it continue as a "chronic illness."[5] Gamow was granted such an extension until September 1934, but as the international political climate continued to darken in the 1930s, scientists such as Kapitza and Gamow had fewer and fewer ways of negotiating their exterritorial status. Gamow eventually settled in the United States and did further important work in physics, in particular on Big Bang cosmology. His Soviet citizenship and his status as the Soviet Academy of Sciences' corresponding member were officially revoked in 1938.[6] Gamow had warned Kapitza against returning to the USSR in the summer of 1934 and wrote to Bohr after he learned of Kapitza's detention:

> You may have heard also that Kapitza is captured [in the] USSR just as a proton by the carbon-nucleus. Dirac got recently a letter from Cambridge saying, that [the] Soviet government will not let him go under any conditions, and I have seen in [a] Moscow newspaper that he is appointed as director of the new In[stitute] of Physical Research of the Academy of Sciences. I hope he will not feel too bad, he was playing himself this dangerous game and just missed it.[7]

Although Gamow's case could on its own have prompted Soviet leaders to reconsider the cases of other Soviet scientists who worked abroad, general policy changes in the Soviet Union pointed in the same direction. Obtaining permission to travel abroad, though never a trivial matter, became increasingly complicated from 1933 on, and by 1938 such trips had stopped completely, as had visits to the Soviet Union by foreign scientists. Soviet scientists stopped attending conferences abroad, and no more international conferences were held in the country. In September 1936, the permanent secretary of the Academy of Sciences, Nikolai Gorbunov, wrote to

---

[5] Kapitza to Niels Bohr, 15 November 1933 (Niels Bohr Archive, Copenhagen, hereafter NBA).
[6] Minutes of the meeting of the General Assembly of the USSR Academy of Sciences, 29 April 1938 (Archive of the Russian Academy of Sciences, Moscow, hereafter ARAN).
[7] Gamow to Bohr, 20 January 1935; see also Gamow's letters to Bohr of 24 November 1933, 13 June 1934, and 1 June 1935 (AHQP).

another Soviet scientist, chemist Vladimir Ipatieff, who had been working abroad since 1930, in terms that made the situation clear:

> Dear Vladimir Nikolaevich! It has been six years since you have been absent from the USSR and have not been participating in the enormous tasks for the construction of socialism. You are a Soviet citizen, a prominent scientist, and a full member of the Academy of Sciences. Our country needs you. Therefore, on behalf of the Presidium of the Academy of Sciences, I ask for your clear and sincere answer to the following question: do you think that you have an obligation to work totally for the sake of your country, the Soviet Union, for its strength and prosperity, and if you do, are you ready to draw from this the necessary practical steps? This is a legitimate question, since your voluntary separation from our country has been so protracted. If you answer the above question positively, you must return to the USSR as soon as possible to continue your scientific work. The Academy of Sciences will take all necessary steps to create favorable conditions for your scientific work and for your living conditions. In the event that you decide otherwise, the Academy of Sciences and the whole nation will have to draw the proper conclusions about your attitude toward the Soviet Union. We wait for your prompt response, and we hope that you will return soon.[8]

Even if Kapitza had remained in Britain in 1934, he would have had to choose a couple of years later whether he would return completely to the Soviet Union or remain in exile forever. Intermediate positions would not have been possible.

3. Missionary

Trying to obtain his release, Kapitza argued that he could not continue his research without the advanced laboratory instruments left in Cambridge. In December 1934 the Politburo decided to inquire about the possibility of purchasing the equipment of the Mond Laboratory, and the Soviet government allocated free currency for this purpose. At first, the negotiations with Cambridge University were difficult, perhaps because Kapitza was not permitted to take part in them and Rutherford still hoped for his release (Badash 1985, 20–36). Dirac and two Cambridge physiologists, A. V. Hill and E. D. Adrian, who visited Moscow in August 1935 in connection with the International Physiological Congress, made a last attempt

---

[8] ARAN, 459-3-10.

to plead with Soviet authorities on Kapitza's behalf. But they returned to England with only a long memorandum in which Kapitza told Rutherford what he would need to continue his work in Moscow. In October, Kapitza was allowed to write directly to Rutherford, and a new Soviet official whom Rutherford trusted took over the negotiations. In the fall of 1935 the Senate of Cambridge University agreed to sell copies of all the principal equipment of the Mond Laboratory. In October 1936, Kapitza could report to Bohr that the institute was practically finished and equipped and that he expected the resumption of experimental work in few weeks. He added some general considerations about Soviet science and his strategy:

> In general the position of science and research people is somewhat peculiar here. It reminds me of a child with a pet animal which is tormented and tortured by him with the best intentions. But indeed the child grows up and learns how to look properly after his pets, and make of them useful domestic animals. I hope it will not be long to happen here.
>
> I am very critical here, and make my criticism quite openly, and I think this is the only right way of acting. I even find now that the responsible comrades listen and on [a] number of occasions are quite willing to discuss and change things. Much less sympathy I find [among] my own colleagues, scientists who are mostly interested in the comforts of their personal work and hate to put questions on a broad base. In spite of all this I have a strong conviction that after a number of mistakes and blunders, science will progress here; the general line on which the social life of the country is organized is much more superior and more correct than that of any of the countries of the old capitalist world. And the leaders are people most sincerely devoted to their work and personal, selfish motives exist [at a] minimum, which is inevitable and keeps people human.
>
> It is for the scientists themselves to take [the] opportunity of these circumstances and to find their own proper and useful place for the work in this new system. If this has not yet happened it is mostly due to the attitude of Russian scientists, as I just said, who cannot grasp the opportunity of the future and only grumble [about] small things. Indeed at the moment the conditions of work here are not nearly as good as in Cambridge, but they are rapidly improving. I am trying my best to help people here organize science, and it is my conviction that the injustice done to me must not blind me... During great historical events there are always victims, such is life, and the worst in my case is over. I feel the responsibility of my position, especially having the experience, which I gained in Cambridge. Besides just resuming my work here, I think I must try to organize my Institute in such a way as to show people here all the

healthy and powerful methods of the work in the Cavendish. I will try to follow Rutherford's methods as far as I am capable.[9]

Kapitza thus remained reasonably optimistic, expressing his apparently sincere belief in socialism and its values of rational, planned organization of economic life, social justice, and internationalism. Though he did not overlook discrepancies between proclaimed principles and social reality, he was inclined to downplay them as temporary, in contrast to what he took as the more essential characteristics of the new society:

> What is done by a telephone call in England, requires hundreds of papers here. You are trusted in nothing. ... They only trust paper—that is why paper is so scarce! Bureaucracy is strangling everyone. ... [But] to destroy this bureaucracy will not be an [easy] task ... As far as I can see, this is more a question of education than of organization, and to educate takes years. [Yet] even in spite of my cursing, I do believe that the country will come out of all these difficulties victorious. I believe it will prove that the socialist economy is not only the most rational one, but will create a state answering to the world's spiritual and ethical demands. But, for me as a scientist, it is difficult to find a place during the birth pangs, and as I wrote in my last letter, the time is not yet ripe and that is the tragedy of my position. The only way out is to be like a hot-house plant under the special care of the government. But is this right? ... [L]ots of things are not clear to me. But life will show.[10]

Later life actually demonstrated that bureaucracy persisted in Soviet society while social enthusiasm waned, but Kapitza's personal strategy remained, as stated above, to be a "hot-house plant under the special care of the government" while holding critical opinions. His criticism was not simply a manifestation of independent character but also a conscious position taken as a social duty. In one of his first letters to his wife written from the Soviet Union, Kapitza mentioned that Pavlov once told him: "Piotr Leonidovich, look—I am the only person here who says what he thinks. I will die soon, and you must take my place. It is so important for our country, which I love much more now that she is in a difficult situation." Kapitza observed, "I will not be afraid to say what I think, but I do

---

[9] Kapitza to Bohr, 20 October 1936 (NBA).
[10] Kapitza to Anna Kapitza, 23 February 1935 (Kapitza 1989a, 35–37; Kapitza 1990, 225–227). A British visitor to his institute described Kapitza's attempts to reduce the typical Soviet bureaucracy at his institute (Shoenberg 1942).

not have the same opportunities as he [Pavlov]. He is a recognized leader of a scientific school, and I am here alone, without support or respect."[11]

Public political criticism became all but unthinkable by the mid-1930s as the Stalinist regime solidified. Kapitza managed, nevertheless, to find some ways of expressing dissent by resourcefully using existing cultural traditions of the Soviet society. In general, Soviet citizens were encouraged to write critical letters to public officials or to the newspapers. Such letters could be worded quite strongly against a particular social problem or drawback, or against an individual official, including some highly placed ones, without criticizing the Soviet system as a whole or political leaders in their role as a symbolic representation of that system. Sometimes newspapers printed such grassroots letters, and sometimes such publications could effectively trigger a public campaign or initiate a change in policy. The Soviet interpretation of democracy included a tradition that a lay person could address a letter or written complaint directly to the Central Committee of the Communist party or to Stalin himself. It was, however, less common for an "above average" person of some rank to bypass the usual chain of command and appeal directly to the highest political authorities. Kapitza acted as if he was not aware of any such restriction and conveyed his criticisms to top political leaders. His personal correspondence includes about forty-five letters to Stalin, dozens each to Politburo members Molotov and Malenkov, and several dozen to other political leaders.

In his first letters to politicians Kapitza devoted his energies to defending his dignity and independence. He insisted that he ought to be treated with respect, that his letters ought to be answered, that he ought to be free from ungrounded accusations and suspicions, and that he ought to be seen promptly when he made an appointment. His letters were purposely far from servile and formal. In one letter to Molotov he wrote:

> You better accept me as I am: a bit impudent, a lover of freedom, independent in my scientific work, unable to wag my tail, even if I had one, but certainly committed to the Union and to the work for socialist construction, to which you are also committed. I am sincerely willing...to help establish science in our country. But you better abandon forever the notion of training me like a dog: "being a good child you will receive this and such, and if you behave badly, you stinker, we will not allow you to go to the theater and we will charge you such a fee for a parcel that you will whine." I tell you once and forever, that ever since my school desk

---

[11] Kapitza to Anna Kapitza, 4 December 1934 (Kapitza 1989a, 30–31; Kapitza 1990, 213–214). On Pavlov and his relations with the Bolsheviks see (Babkin 1949; Todes 1995; 2002; Tolz 1997).

> I could not behave like a "good boy." For instance, you told me, when we spoke: "We have lots of Kapitzas." But this again resembles the training of an animal. I know that you think you should take me down a peg. You will see later that I am not overly supercilious. I have a lot of other bad qualities, but not this... If, instead of all this training, you would have tried to involve me in our country's life, which in fact is much more remarkable than you even think, then we could have been friends a long time ago.[12]

Yet Kapitza would adopt the same metaphor of training once he had a chance to have the roles reversed, as he complained to the deputy-chairman of the Soviet government Valery Mezhlauk about an agency that supplied instruments and materials for his institute: "I am sorry for bothering you once more with the problem of 'Tekhnoimport,' but I feel that if it isn't punished and beaten for every stupid and untidy deed, we will never teach it how to work well. You know that when a dog is being tamed, the most important thing is not to relent in the initial stages of training.... 'Tekhnoimport' behaves like a bad, untrained dog. After the first reprimand you gave it, it improved for a while, but now it has reverted to its old ways" (Kapitza to Mezhlauk, 19 December 1936 in Kapitza 1989a, 112–114).

He did not stop short of occasionally offending some of his higher addressees and patrons. His probably most impudent letter to Mezhlauk stated:

> Tell me in general how do you picture the mechanism by which builders in the Soviet Union can be made to observe their schedules?... So I ask myself the question of why you, the Government, can't do anything and I have two possible answers: (1) Essentially, you don't consider my work as a scientist sufficiently important and necessary for the country to provide the necessary tempo of construction to enable me to start work as quickly as possible. But in that case why did you detain me?... (2) The second possibility is no better—this is that you, the Government, are unable to make Borisenko & Co Ltd obey you. But then, what sort of a Government are you, that you can't get a little two-story house built on time and put its ten rooms in order after assembly? In that case you are simply wets!... Well, that's the picture of what's happening. Imagine you saw a violin at your neighbour's and you were able to take it from him. And what do you do with it instead of playing on it? For two years you use it to hammer nails into a stone wall.... And as for the violin you have

---

[12] Kapitza to Molotov, 5 July 1935 (IFP).

taken by force you can't even play 'Chizhik' [a simple children's song] on it (Kapitza to Mezhlauk, 26 April 1936 in Kapitza 1990, 327–328).

It runs counter to the existing stereotypes that high members of Stalin's government would tolerate such affronts from a subordinate in the late 1930s, but in the case of Kapitza they did. Besides an earlier precedent set by Pavlov and the Soviet habit of letter-writing to top leaders, there were a couple of other cultural traditions that helped Kapitza play the role he had chosen. Through his family roots in the Russian military nobility, he must have been familiar with a pattern of seeking the highest patronage through a mixture of subordination and familiarity in behavior. He apparently used similar tactics in Britain to win Rutherford's patronage and trust. Another strategy was available to him because of his long stay abroad. Back in the Soviet Union, he was often seen as a half-foreigner and treated with suspicion, but in some cases this very status could be advantageous in that it allowed him some leeway to behave unconventionally, like someone unaccustomed to Soviet ways. He apparently exploited this possibility quite consciously. Carefully calculated dissent became Kapitza's ingenious method of improving his own connections and standing with the highest politicians. Through this, he made himself known as a person and taken seriously, established a certain degree of sincerity in the dialogue, and generally succeeded in building a reasonably close patron-client relationship instead of the usual one of bureaucratic subordination. It would not be too long until the advantages of such a relationship started paying off.

4. Client

Never too modest, Kapitza felt responsible for improvement of the general conditions for science in the Soviet Union, while claiming that his peers, the official leaders of the Academy, either did not care enough or were not brave enough to raise important issues with powerful political figures. He commented dismissively on the Academy of Sciences and its ruling body, the Presidium, in his words an "obsolete wagon" that had not changed in the course of the revolutionary decades and whose privileged members acted like priests presiding over a sacred ritual in the name of science. His own official rank as a corresponding member of the Academy matched neither his ambitions nor his evaluation of his own accomplishments. In a letter to Rutherford, Kapitza maliciously portrayed members of the Presidium, from the permanently sleepy 90-year old president, geologist

Aleksandr Karpinsky, to his colleague physicists Joffe, Vavilov, and Aleksandr Frumkin. "Probably ... the only person on the Council who has some personality," according to Kapitza, was the communist official Nikolai Gorbunov, who had been recently appointed the Academy's permanent secretary, "[I]n any case, when you talk to him, he does express views and opinions, what others scarcely dare."[13] In his own administrative activities, Kapitza often sidestepped the Presidium and made appeals over its head directly to political officials.

Kapitza first succeeded in establishing close contacts with middle-level politicians such as Mezhlauk and Karl Bauman, head of the Science Department at the Party Central Committee. These officials were able to help him solve some relatively minor problems with the construction of his institute, equipment, and management, even reserving for him tickets to a popular theater performance. Kapitza characterized this method of solving problems of everyday Soviet life with an English proverb: "To use a steam-hammer to crack nuts." Really important issues required connections in higher circles. Kapitza tried to arrange an appointment with the head of the Soviet government, Molotov, and after a number of failed attempts found an occasion to write him a personal letter. In April 1935 the news of Kapitza's forced detainment leaked to British newspapers (Rutherford tried to keep the information from the public as long as he hoped Kaptiza's release was negotiable). Mezhlauk asked Kapitza to announce through the Soviet press that he preferred to work in his own country. Kapitza wrote to Molotov flatly refusing to do what was requested of him: "I cannot say what I don't think. Not only do I not feel better here than in Cambridge, but on the contrary I feel very bad, and the best I can do is to stay silent.... Without my equipment, without my books, without my scientific colleagues and with my work rudely interrupted at a very interesting juncture, I feel at present sad, unhappy, shattered and worthless."[14]

After Kapitza learned that the Senate of Cambridge University had approved the Soviet government's purchase of the Mond Laboratory equipment, he decided the occasion was right for writing to Stalin personally. On 1 December 1935 he communicated the news to the Soviet leader and elaborated on his future research plans. Like many of his other writings to politicians, this was a long letter, almost a minor treatise. Kapitza often started with a particular problem of his own or his institute

---

[13] Kapitza to Rutherford, 26 February–2 March 1936 (Kapitza 1990, 279–286, on 281).
[14] Kapitza to Molotov, 7 May 1935 (Kapitza 1990, 319–323, on 321–322).

and proceeded by linking it to a broader discussion of science and science policy in the Soviet Union. In this way, he tackled the problems of the relationship between basic and applied research, of the proper ways of managing resources, of secrecy in science, and of respect towards scientists and their role in the society. He rarely received written replies and did not necessarily expect them, but he did not fail to notice that his letters sometimes produced actual effects, reflected in the politicians' behavior and actions. Kapiza viewed his writing activity as missionary, as he initially felt himself a half-foreigner in a strange yet familiar land, or pedagogical, as he later grew accustomed to and identified himself with Soviet society.

In word and deed, Kaptiza presented himself in those letters as a pragmatist devoted to the development of Soviet science and industry and the construction of socialism. His proposals often sounded risky, but he did not overstep certain limits, for example always restricting his suggestions to issues of direct relevance to science and scientists, while carefully avoiding more general political questions. The practical results he achieved were also limited, but at least he succeeded in creating exceptionally good conditions for the work of his institute and the related community of researchers. Kapitza's institute was organized differently from other academic institutions. It was significantly smaller—closer to British than to typical Soviet standards—and with a lower proportion of clerical staff. As director, Kapitza enjoyed more leeway in hiring personnel and allocating funds than other directors of the Academy's institutes. His connections with politicians helped him in solving the problems of building and equipping his laboratories. He sometimes even ordered some basic laboratory supplies from British firms, arguing they were more efficient and flexible in filling small-scale orders.[15]

The arrival of major equipment from Cambridge in the fall of 1936 allowed Kapitza to resume his experimental research after a two-year interruption. In February 1937 he obtained the first liquid helium in Moscow, and before the year's end he was able to report probably his greatest scientific discovery—the superfluidity of liquid helium. It was known that liquid helium existed in two phases: helium I, effectively a boiling liquid (between 2.19 and 4.2 K), and helium II (below 2.19 K). In 1936 Willem Keesom in Leiden observed an extraordinary high thermal conductivity in helium II and announced this as a new kind of "superconductivity," in

---

[15] See Kapitza's description of his new institute in (Kapitza 1937). For a comparison between the management of Kapitza's institutes in England and in the Soviet Union, the Mond Laboratory and the Institute of Physical Problems, see (Shoenberg 1942).

addition to the well-known electrical superconductivity of metals under low temperatures. Kapitza suspected that the phenomenon was not a true conduction of heat but convection, an extraordinary high transport of helium particles due to low viscosity. He decided to check the data on the viscosity of helium with greater accuracy and, in order to avoid turbulence, let helium flow through a microscopic gap between two parallel glass plates. His suspicions were overconfirmed, as he recorded no measurable viscosity whatsoever. The possible upper limit on viscosity due to the experimental margin was at least 1500 times lower than the previously accepted value. Kapitza announced his discovery as "superfluidity" of liquid helium in a short note in *Nature* in December 1937. The result was confirmed in the same issue of the journal, in a paper by J. F. Allen and A. D. Misener, who were independently pursuing similar measurements in Kapitza's old laboratory in Cambridge (Kapitza 1938, Allen and Misener, 1938).

A flurry of investigations of the newly discovered phenomenon ensued, during which Kapitza determined that the thermal conductivity in liquid helium was at least 20 times higher than the value reported by Keesom. Even convection could not provide a mechanism for such high conductivity, because helium atoms would have had to travel with impossibly high velocities. In 1940 Kapitza designed an elegant experiment to demonstrate that heat was transported not by irregular convectional movements of atoms but by a directed flow of liquid. A beam of light heated a bulb containing helium inside a liquid helium bath. A directed flow of warmer helium escaped from the bulb through a small capillary and, when falling upon a small vane, pressed it into rotation. Surprisingly, no inward flow could be detected. Cooler helium somehow had to get inside the bulb to replace the outgoing warmer liquid, but its movement did not produce any noticeable mechanical effects. Liquid helium behaved as if it consisted of two fractions with rather different properties. Kapitza could only speculate that helium went inside along a very thin layer near the walls of the capillary. He clearly needed a theoretician's help to understand the paradoxes of low-temperature liquid.

His in-house theoretician Lev Landau, was, however, in a helpless situation himself and needed to be saved from the cruel machine of Stalinist purges. Landau was hired by Kapitza in March 1937 after he ran for his life from Kharkov. Six months later, the chaotic machine of purges picked UFTI in Kharkov for its deadly carnage. Several top scientists, including Shubnikov and Rozenkevich, were arrested, forced to confess the crimes of "espionage" and "sabotage," and executed after a short trial

(Pavlenko, Ranyuk, and Khramov 1998). Landau figured in some of the extorted Kharkov confessions as co-conspirator, but his being in another city delayed his arrest for another half a year. In Moscow, Landau came under surveillance and was arrested on 28 April 1938 together with two friends and a piece of incriminating evidence, a handwritten leaflet for the upcoming May Day. The leaflet was composed in the name of an imaginary "Moscow committee of the anti-fascist workers' party" and called on

P. L. Kapitza as a recently elected member of the Academy of Sciences and director of the Institute of Physical Problems in Moscow, 1940. The stressful time of Stalinist purges was also, for Kapitza, the time of his groundbreaking research on liquid helium, in particular the discovery of superfluidity. Through his letters and memoranda, Kapitza managed to establish a relationship with Stalin and other top politicians, which brought him many rewards and privileges, but also increased risks.
[*Courtesy*: Database of the Institute for Problems of Information Transmission, Moscow, hereafter IPPI.]

comrades to save socialism by resisting the criminal Stalinist clique (Landau 1991; Gorelik 1995a; 1995b). Although some of Landau's students remain unconvinced even now of its authenticity, the leaflet is probably genuine and Landau appears to have dictated or at least approved its text. Still, he was somewhat lucky even in this grave misfortune thanks to the fact that he was not captured until several months after the Kharkov arrests. After Beria replaced Ezhov as the head of the security police in late 1937, the purges began to subside, with investigations and trials becoming less expeditious. This gave Kapitza some time to attempt a rescue.

Kapitza's connections with high politicians proved very effective in the time of the Great Purges of 1936–1938, when he attempted to help several arrested scientists. He could not question the basic presumption of the purges—that traitors were plentiful and spies and enemies of the people existed in the midst of Soviet society—but he was able to plead for a few individuals and ask for their cases to be investigated with care. Kapitza had to be very selective in choosing when to intervene lest his appeals be devalued. His first attempt to interfere with the working of the political purging machine failed. In July 1936 he wrote Molotov expressing doubt about the wisdom of the media campaign of harassing mathematician Nikolai Luzin for his allegedly non-Soviet views and unpatriotic behavior, but his letter came back with a rather rude remark in Molotov's handwriting, "Not needed; return to citizen Kapitza."[16] Fortunately, this did not discourage Kapitza, and his next attempt bore fruit. In February 1937 he wrote two letters, one to Mezhlauk and the other to Stalin, to plead for the arrested Vladimir Fock from Leningrad University. Within few days, the theoretical physicist was released from jail.[17]

In Landau's case, Kaptiza acted immediately with a personal letter to Stalin but received no response. A year later, in April 1939, he wrote to Molotov claiming that he needed the theoretician's help to understand recent discoveries in helium. This time his letter worked. Kapitza was allowed to bail the convicted prisoner out of jail in return for a written promise to prevent Landau from committing further "counterrevolutionary" acts.[18] Kapitza succeeded in a couple more cases of persecuted

---

[16] Kapitza to Molotov, 6 July 1936 (Kapitza 1990, 331–333). Soviet officials typically applied the address form "citizen" instead of the usual "comrade" when talking to politically alien individuals. On the Luzin affair see (Luzin 1999).
[17] Kapitza to Mezhlauk, 12 February 1937, Kapitza to Stalin, 12 February 1937 (Kapitza 1990, 337–339).
[18] Kapitza to Stalin, 28 April 1938; Kapitza to Molotov 6 April 1939; Kapitza to Beria, 26 April 1939 (Kapitza 1989a, 174–179; Landau 1991, 151–154).

scientists—in 1941 he helped Obreimov, the first director of the Kharkov Physico-Technical Institute, gain release from a labor camp—and failed in a few other cases. In all these pleadings, he restricted himself to pragmatic arguments without questioning the justness of judiciary and political decisions. He called politicians' attention to the importance of the work performed by the arrested scientist and asked that the case be considered carefully so as to avoid a costly error. The rewards, at least, were the lives of two of the nation's best theoretical physicists. And in Landau's case, an additional reward was Landau's 1941 theory of the superfluidity in liquid helium, which explained many of the strange phenomena discovered by Kapitza.[19]

5. Minister

When the Soviet government decided to keep Kapitza in the country, he was told that his services were needed by the developing Soviet industries. He was willing to offer consultation and in 1935—while waiting for his research equipment to arrive from Cambridge—visited several industrial plants in Moscow:

> Yesterday morning I went to the industrial plant which deals with low temperature... they are starting some research there. Expected to stay there until 2 p.m., but stayed until 3 p.m. I am very fond of our youth and enthusiasm. This plant (VAT) is one of a new generation of plants constructed after the revolution, and there is a great difference in spirit and atmosphere... This visit to VAT warmed me up. Moreover, the people there are working on topics which are close to my field.[20]

Several engineers from the plant visited Kapitza the following month to discuss the possibility of increasing the efficiency of oxygen production. Their questions interested Kapitza, and he soon designed a new device for that purpose.[21]

Oxygen is industrially produced by extraction from liquid air, with the air liquefaction achieved by a sequence of repeated compressions under constant temperature and adiabatic expansions, during which the air cools down. Since different gases present in air boil at slightly different

---

[19] See chapter 10, 260–262.
[20] Kapitza to Anna Kapitza, 16 February 1935 (IFP).
[21] Kapitza's diary, quoted in (Kapitza 1989e, 252–287).

temperatures, they can be separated during evaporation from the liquid. The conventional detanders—the machines for adiabatic expansion—of the 1930s operated on the basis of the Joule–Thomson effect, whereby the air expanded while passing through fine holes in a porous partition. Kapitza proposed using a turbine as a detander and designed one that operated efficiently at low temperatures. In spring 1938 he constructed a laboratory version of his turbodetander, sent letters to Molotov and Stalin about the accomplishment, and asked for permission to take out foreign patents on the invention.[22] He proceeded with developing an entire line of devices for the industrial production of oxygen, which included equipment for separation of liquid fractions and for producing gaseous oxygen. In the meantime, he also became involved with the industrial production of his machines.

Producing the industrial liquefier turned out to be more difficult than designing the prototype. In the Soviet system, scientific institutes were

Mechanic S. A. Mrisha near the apparatus for the production of liquid air, Institute of Physical Problems, Moscow 1941. That year Kapitza received his first Stalin Prize for the invention of a new method of gas liquefaction. Kaptiza's discoveries and apparatus revolutionized the field of low temperature physics and found many industrial applications. They also launched his career as an industrial manager in Stalinist Russia.
[*Courtesy*: RGAKFD.]

---

[22] Kapitza to Molotov, 20 April 1938 (Kapitza 1989a, 165–174), Kapitza to Stalin, 26 April 1938 (IFP); (Kapitza 1939a; 1939b).

also expected to take care of the *vnedrenie,* or the implantation of their discoveries, designs, and inventions into industrial and economic practice. This worked smoothly when a close link existed between the research institution and the industrial factory, as in the case of the State Optical Institute, or when a matter was high enough on the government's list of priorities to ensure direct orders to factories and continuous attention from above, as would later be the case with the atomic bomb project. Kapitza's case did not fit either of these categories, and he soon found out that it was actually very hard to expedite the production of a few industrial copies of his new machines.

His difficulties reflected the changing patterns of Soviet science policies and the relationship between science and industry. After the 1934 government decision to move the Academy of Sciences from Leningrad to Moscow, closer to the center of political power and to "the tasks of socialist production," the Academy quickly rose in status as the major agency controlling the nation's research in basic sciences. From the mid-1930s on, most of the newly established institutes engaged in fundamental research, including Kapitza's Institute of Physical Problems, would be organized within the Academy of Sciences. Many such institutes that had already existed, for example Joffe's Physico-Technical Institute in Leningrad, which had belonged to NKTP, the Commissariat of Heavy Industry, would be transferred to the Academy by the end of the decade. Although these academic institutes were still expected to conduct applied research as well, a clear bureaucratic gulf started emerging between them and the strictly applied institutes that remained within industrial commissariats. This effectively lessened the pressures on the Academy institutes to produce industrially applicable results, by the same token also leaving them with fewer means of applying whatever discoveries they had made. The plans for Kapitza's Institute of Physical Problems, for example, did not include any major practical goal or any prearranged link with industrial production or factory.

In the fall of 1938 the government ordered NKTP to build Kapitza's turbodetander. The Moscow plant Borets was to produce ten industrial copies by the end of 1939. As Kapitza soon found out, this relatively small and non-standard assignment was not as high on the plant management's priority list as the fulfillment of another government order, the annual plan for regular mass production. He tried to pressure the administration acting through the plant's communist party cell, arranging critical articles in newspapers, and sending monthly reports to the government about the progress of his work on the new inventions and the slowness of industrial *vnedrenie* (Kapitza 1989c). Through these combined pressures, Kapitza managed to have the first industrial turbodetander for the production of liquid air in operation by the summer of 1940.

Meanwhile, Kapitza designed a prototype of the additional device for separating oxygen from the turbodetander's liquid air. The Commissar of heavy industry, Viacheslav Malyshev, saw a possible military application and ordered the immediate production of small transportable oxygen plants for the use in military aviation. Most industrial oxygen was then produced for use in autogenous welding. The production of Kapitza's devices was also entrusted to the Welding Trust (*Glavavtogen*). Members of Kapitza's institute consulted the Trust's engineers, and in the summer of 1941, as the German armies invaded the territory of the Soviet Union, the production of transportable oxygen plants was already under way. In the autumn, most scientific institutes were evacuated from Moscow further east, to Kazan on the Volga River, where Kapitza continued his research, now entirely subordinated to the war needs. His oxygen devices found application during the war not only in military aviation but also in the production of explosives.

In 1942 Kaptiza was ready with a design of a larger machine capable of producing about 2000 kilograms of liquid oxygen per hour. The State Committee of Defense, the emergency government that ruled the country throughout the war, ordered the Welding Trust to build a large oxygen plant in Balashikha, near Moscow. Kapitza calculated that it would be more efficient to transport liquid oxygen to various destinations in specially constructed railway tanks than to produce it locally on a small scale. He complained, as earlier, about the slow pace of industrial *vnedrenie* and construction. In late 1942 and early 1943 he wrote several desperate letters to Molotov proposing a radical solution. He argued that in the four years since the first government order for oxygen machines the Welding Trust had proved unsuited to the task: "To solve successfully the job of developing our apparatus, it is necessary to set up a special organization, call it for instance the *Glavkislorod* [the Oxygen Trust], directly subordinate to Sovnarkom [the government] and independent of all other industrial Commissariats." In a less official note enclosed with the letter, he added: "all this time I have worked as a muledriver, and I have been denied both a stick and a switch. I think that in one way or another, I should be granted official power to direct all the process of industrial *vnedrenie*." On 19 April 1943 Kapitza wrote directly to Stalin with further criticism of the Welding Trust.[23]

---

[23] Kapitza to Molotov, 6 April 1943 (IFP); Kapitza to Molotov 19 October 1942, 6 April 1943; Kapitza to Stalin 19 April 1943 in (Kapitza 1989a, 194–201).

This time he won completely. Within a month, a new state trust was set up with Kapitza as official director and chairman of the technical council. The new organization was entrusted to complete the liquid oxygen plant in Balashikha and to produce the third and final device of Kapitza's series, the machine for producing gaseous oxygen. The Oxygen Trust was also responsible for developing further applications of oxygen in various branches of industry, including metallurgy. Kapitza's new position was primarily administrative rather than scientific or engineering. Reporting directly to the government, he commanded his own little ministry and collaborated successfully with Stalin's commissars. Moreover, through his personal letters to Soviet leaders, Kapitza possessed a channel of communication most other ministers could only dream of. He continued writing regularly, informing Stalin, Molotov, and his new direct supervisor, Politburo member Georgy Malenkov (1902–1988), about the course of his work. He complained in those letters when other commissars were slow in giving him appointments or when they skipped meetings of the technical council of his trust in order to address more pressing responsibilities.[24] When some of his complaints remained unanswered, Kapitza dared to reproach Stalin: "I did not receive any answer. What am I to do in this case? There is nobody above you to whom I can submit a complaint! And as I took the oxygen job, I just can't keep silent" (Kapitza to Stalin, 14 March 1945 in Kapitza 1989a, 226–228).

The technical council met every two weeks to discuss the scientific and engineering problems of the industrial production and application of oxygen. Its proceedings were published in a special bulletin, *Kislorod* (Oxygen). The new applications of oxygen included the enrichment of fuel and a new method for smelting steel, the oxygen blast, developed by academician Ivan Bardin. The plant in Balashikha started operating in the fall of 1944, its output rising to forty tons of liquid oxygen per day, or about one-sixth of the total Soviet production at the time. The State Commission inspected and approved the completion of its construction, as confirmed by a government order of 19 April 1945.

In the course of his work on oxygen production, Kapitza earned most of his state honors. He received his first Stalin prize in 1941 for the turbodetander, the second in 1943 for the discovery of superfluidity. In 1943 and again in 1944 he was awarded the Order of Lenin. After the approval of the Balashikha plant, on 30 April 1945, the Presidium of the Supreme

---

[24] Kapitza to Stalin, 10 May 1944 (IFP); 24 February and 13 October 1944 (Kapitza 1989a, 210, 221–222).

Soviet awarded him the country's highest civil title, "Hero of Socialist Labor," together with another Order of Lenin "for the scientific development of the new turbine method for producing oxygen and for the construction of the powerful oxygen apparatus" (*Pravda*, 1 May 1945). The Institute of Physical Problems also received the Order of the Red Banner of Labor, and more than a hundred of Kapitza's co-workers at the Institute and at the Oxygen Trust got various decorations.

That moment of Kapitza's greatest personal triumph as a scientist, and also as an industrial manager and a minister, coincided with the fall of Berlin and the great Soviet victory in the war against Nazi Germany. In the midst of the nation's triumphant celebrations, Kapitza must have felt that he had found a successful way to work within the Stalinist government and understood its modus operandi. Of all Soviet scientists, he apparently managed to establish the closest rapport with political leaders, including Stalin himself. Like many others, however, he would find out that in Stalin's times the transition from political success to disaster could be very quick. In less than a year, Kapitza would fall completely out of political grace and lose practically all of his official positions. The event that set the course of his life in a new direction was, as for many other Soviet physicists, the American bombing of Hiroshima in August 1945.

Chapter 6

"To Catch Up and To Surpass..."

In a landmark book published ten years ago, David Holloway (1994) summarized what was known about the history of the Soviet atomic bomb. By then, the veil of excessive secrecy surrounding the entire project had started lifting, people talked about it more freely, and many personal recollections, stories about particular episodes, and wild conspiracy theories circulated in the media. Carefully and critically juxtaposing those sources, Holloway drew together a detailed picture as reliably as was possible without access to the still classified main body of primary documents. Most of the nuclear-related archives remain closed to academic research, but the situation with available sources has changed qualitatively since 1994. In a flurry of archival publications, many crucial documents have been made generally available, most notably in a fundamental multi-volume collection *Atomnyi Proekt SSSR* (Atomnyi 1998–2002). The proceedings of several scholarly conferences and new historical monographs have provided revised accounts with much new information on particular aspects of the story.[1] Murky issues remain, as is to be expected when we deal with classified topics, but many questions that formerly could be discussed only hypothetically can now be answered with sufficient certainty. The sheer amount of new, sometimes contradictory information and sources suggests

---

[1] See especially (Kruglov 1994–1995; Larin 2001, [HISAP] 1997–1999, Gorelik 2000, Vizgin 1998–2002a, Mick 2000).

that the time for another historiographic synthesis is approaching. To help advance towards this goal, this chapter provides a short overview of the existing evidence and discusses some interpretative issues.

The best way to see the similarities and differences between the atomic bomb project in the Soviet Union and its counterparts in other countries is to consider how the work was organized politically and managed practically. Three stages of Soviet activity can be recognized from this perspective: academic research before 1941, small-scale, classified laboratory investigations of 1943–1945, and the all-out military-industrial effort of 1945–1955. Soviet nuclear physics built up momentum relatively late, during the 1930s, but it was flexible and attentive to the newest developments and achieved a mature state by the end of the decade. This happened just in time for Soviet physicists to respond to the discovery of uranium fission with a series of cutting-edge investigations, which included a realistic and dangerous calculation of the uranium bomb's critical mass. After the Nazi invasion in 1941, however, all Soviet work on fission was shelved in favor of military projects with immediate relevancy. The devastating war left the nation no resources to spare, and a long-term and uncertain atomic project would probably not have been pursued seriously until victory was won, had it not been for intelligence reports about intensified British attempts to build a uranium explosive.

Soviet research on fission resumed in 1943 on a small scale with rich intelligence information and intensive theoretical studies, but with practically no uranium. The effort was reorganized on an industrial scale only after the Hiroshima and Nagasaki bombings at the end of the war. The socialist economy, with its centralized military-style management and an existing tradition of big-science institutions, was perfectly suited for replicating the Manhattan Project. The greatest problem, holding back the entire undertaking, was the lack of uranium. The first Soviet bomb of 1949 was a close copy of the American plutonium bomb, while the first Soviet thermonuclear device of 1953 had an original design. By 1955, the problem of devising nuclear weapons could be regarded as solved, and many physicists started returning to fundamental research. However, the experience and legacy of the bomb project had transformed their entire field.

1. Before Fission

The most outspoken enthusiast for and public proponent of the new field of radioactivity in early 20th-century Russia was the geochemist Vladimir

Vernadsky. In his view, not only was scientific knowledge fundamentally changed by the discovery of radioactivity, humankind "entered a new era of radiating, or atomic energy.... [T]he possession of large resources of radium [could give] its owners the power which might become superior to that of the owners of gold, land, and capital" (Vernadsky 1910, 43–44). Vernadsky appealed to the Russian government and public to support the survey of radioactive ores in the territory of the Empire and establish a Radium Institute similar to those in Paris and Vienna (Vernadsky 1911). A modest exploration and research effort commenced in 1911 under his

The first Soviet cyclotron at the Radium Institute in Leningrad. At the start of its operation in 1937, it was also the first cyclotron in Europe. The decision to arrange the chamber vertically rather than horizontally was unfortunate: it made the cyclotron harder to operate, while the expected dual use for the study of cosmic rays did not materialize. Engineer B. D. Brizemeister is sitting next to the cyclotron in this photo of 1940.
[*Courtesy*: RGAKFD.]

directorship at the Radium Commission of the Imperial Academy of Sciences in St. Petersburg.

After the revolution, the Radium Commission became part of KEPS, the Commission for the Study of Natural Productive Forces, and received a subsidy from the Bolshevik government. Vernadsky's earlier collaborator, chemist Vitaly Khlopin (1890–1950), organized a small factory in the Ural Mountains and successfully extracted radium from uranium ores by the end of 1921. Meanwhile, Vernadsky returned to Petrograd after his Civil War odyssey and fulfilled his pre-revolutionary dream. The Radium Institute opened on 1 January 1922 with three departments: mineralogical, chemical, and physical, directed by Vernadsky, Khlopin, and Lev Mysovsky respectively. During the 1920s the Radium Institute was the main center of research on radioactivity in the country. The first Soviet (and European) cyclotron was built there by the spring of 1937, although problems with its stable operation continued well into 1939 (Alkhazov, Shilov, and Eismont 1982; Igonin 1975, 340–346; Holloway 1994, 40).

The second center developed at UFTI in Kharkov starting in 1930, with emphasis on nuclear physics and nuclear reactions. UFTI did not become the world's first laboratory to artificially split an atomic nucleus—scientists there reported a successful disintegration of lithium in October 1932, six months later than John Cockroft and E. T. S. Walton in Cambridge—but even second place in the race was hailed in the press as a great start for the young Soviet nuclear physics (*Pravda*, 22 October 1932; Sinelnikow *et al.* 1932). UFTI physicists dispatched an open telegram to Stalin, Molotov, and Ordzhonikidze (the Commissar of heavy industry) proudly reporting their success in having split the nucleus by the fifteenth anniversary of the Bolshevik revolution and promising to split even more nuclei. The imagery of the atom being smashed resonated well with revolutionary propaganda and the spirit of the time (Val'ter 1933; Pavlenko *et al.* 1998, 148–152).

At the Leningrad Physico-Technical Institute, Joffe was slower in reacting to the rise of the new field, but he changed his mind in 1932, the so-called nuclear *annus mirabilis* that saw the discoveries of the neutron and the positron. By the end of the year Joffe organized work on the atomic nucleus with two experimental laboratories headed by Igor Kurchatov (1903–1960) and Abram Alikhanov (1904–1970), with Dmitry Ivanenko as theorist. Once the discovery of the neutron was announced in 1932, Ivanenko quickly conceived and published the idea that nuclei can be built entirely of protons and neutrons, with no need for electrons, which helped resolve many existing difficulties in nuclear theory and quantum electrodynamics (Iwanenko 1932). Starting in 1934 Kurchatov conducted experimental research on neutrons, their interactions with atoms, and artificial radioactivity along the lines of that of the Enrico Fermi group in Rome (Igonin 1975, 266–306). The field's new prominence was on display during the first All-Union Conference on Nuclear Physics in Leningrad in September 1933, which was attended by foreign luminaries, including Frédéric Joliot, P. A. M. Dirac, and Franco Rasetti (Atomnoe 1934; Frenkel 1985).

LFTI's main problem was the lack of strong radioactive sources, either natural (radium) or artificial (accelerator). Kurchatov thus had to travel to Kharkov to perform some of his experiments at UFTI. In Leningrad, he helped Mysovsky at the Radium Institute with the completion of the cyclotron, and after Mysovsky's death in 1939 he took over the directorship of the cyclotron laboratory and used the machine mainly as a source of neutrons. Through Joffe, Kurchatov lobbied for the building of a larger cyclotron at LFTI and obtained funds from the Commissariat of Heavy Industry in 1937. A year later, however, the funding stopped owing to "administrative disorganization," in the term some documents of the time used to refer to the Commissariat's grave problems (Atomnyi I (1), 1998, 17–19, 54).

The Commissariat of Heavy Industry (NKTP) under Sergo Ordzhodikidze had been a powerful patron of science and provided the main support for large physico-technical institutes in Leningrad and Kharkov, among others. A member of the Communist Party Politburo and one of Stalin's most loyal, influential, and trusted comrades, Ordzhonikidze committed suicide in February 1937 in desperation over continuing purges of party officials and his inability to prevent arrests of some of his closest associates (Khlevniuk 1995). His death left the Commissariat even more deprived of protection, and most of its central staff perished during the next few years of Great Purges, after at least three consecutive waves of arrests. Eventually, as the purges wound down, the once powerful ministry was dismantled into seven (later more) smaller commissariats, each responsible for a particular branch of industry, such as electrical or chemical. The new commissariats inherited and distributed among themselves the vast research empire of the former Commissariat, but none of them had either its resources or its broad agenda.

It immediately became much more difficult for physico-technical institutes to have their research projects approved and funded. Nuclear physics was in a particularly vulnerable situation because it required large investments and at the time could promise almost nothing industrially useful in return. Elsewhere—in Berkeley, Cambridge, and Copenhagen—physicists were applying for funds to the Rockefeller Foundation for the construction of large cyclotrons as machines intended for essentially medical purposes (Heilbron 1986; Aaserud 1990). Even if Soviet physicists had been aware of this rhetoric, their supervisors, officials in heavy industry, would not have found it very appealing. In search of a useful application, one meeting at the Academy of Sciences in April 1938 discussed the possibility of using chain reactions as a path towards nuclear energy, but no concrete mechanism was seen at the time (Atomnyi I (1), 1998, 25–27). In their proposals to the Soviet government, physicists used the argument of international prestige and cited the fast rate of cyclotron construction in America, the leader in this new field.[2] Such reasoning had worked earlier with the old Commissariat of Heavy Industry, which could afford to invest in some non-profitable research fields for the sake of international image of Soviet science. The agendas of the new and smaller commissariats were much more narrow and utilitarian, and nuclear physics did not fit there.

Sergei Vavilov, director of the Physical Institute of the Academy of Sciences, proposed to take over the Kharkov and Leningrad nuclear laboratories and to concentrate the entire field under the Academy's administrative umbrella (Atomnyi I (1), 1998, 36–43). Directors of the physico-technical institutes were not excited about the idea, and after a year of bureaucratic deliberation, appeals, and proposals found another administrative solution. Rather than losing their important laboratories, the institutes asked to be transferred to the Academy of Sciences, which after the demise of the Commissariat of Heavy Industry became the most powerful sponsor of scientific research in the country. During 1939 and 1940 the Radium Institute and LFTI came under the auspices of the USSR Academy, while UFTI became part of the Ukrainian Academy of Sciences. In the meantime, with the discovery of uranium fission, nuclear physics acquired a realistic if remote possibility of useful application and a breathtaking if uncertain prospect of gaining control over nuclear energy.

2. As if it were not about the Bomb

By the time in late 1938 when Otto Hahn and Friedrich Strassmann made their discovery of nuclear fission in Berlin, most direct contact between Soviet physicists and their foreign colleagues, especially in Nazi

---

[2] By the mid-1930s, *Physical Review* was devoting about half of its space to nuclear physics. Meanwhile, other major European countries were only then building their first cyclotrons and in this respect were not further advanced than the Soviet Union (Atomnyi I (1), 1998, 38).

Germany, had been broken off. Still, news of the discovery arrived at Joffe's institute in Leningrad quickly, certainly not later than February 1939, via a letter from Joliot in Paris and his paper published in *Comptes rendus*. Vernadsky at the Radium Institute was alerted later, in June 1939, by a clipping from the *New York Times* sent to him by his son George, professor of Russian history at Yale (Holloway 1994, 59–60). In the Soviet Union, as elsewhere, physicists were reacting to the news with a mixture of enthusiasm and caution, and with a flurry of research.

Pre-war studies of fission in the Soviet Union produced three major accomplishments. In early 1940 two younger members of Kurchatov's group, Georgy Flerov (1913–1990) and Konstantin Petrzhak (1907–1998), discovered that uranium nuclei could also split spontaneously, without incoming neutrons. To verify that the effect was not caused by cosmic radiation, they reproduced their experiments deep underground in the Moscow metro. Anxious for international recognition, they announced the discovery of spontaneous fission in the *Physical Review* and were surprised at the lack of response. It would occur to them only later that the silence was probably due to secrecy imposed on uranium research in America (Flerov and Petrzhak 1940; Igonin 1975, 395–399). Yakov Frenkel had already done some work on nuclear theory earlier in 1936, when he developed Bohr's idea of the compound nucleus into a statistical theory of the nucleus as a many-body system. Having learned the news about fission, in March 1939 he formulated the main ideas of the liquid-droplet model of fission, which later that year was further developed in the famous paper by Bohr and John Wheeler (Frenkel 1939; Bohr and Wheeler 1939).

Even more important results appeared in a series of papers by Yakov Zeldovich (1914–1987) and Yuly Khariton (1904–1994). Both regularly took part in the nuclear colloquium at LFTI, but their main job was at the nearby Institute of Chemical Physics, where they studied chemical kinetics and chemical, i.e. non-nuclear, chain reactions with the discoverer of those reactions, Nikolai Semenov. Relying on that expertise, Zeldovich and Khariton developed in their three papers of 1939–1940 what was at that time the most advanced theory of the uranium chain reaction and the conditions under which it is possible. Among other important conclusions, they determined that uranium with the natural mixture of isotopes 238 and 235 was incapable of developing a chain reaction with fast neutrons—i.e. a bomb-type explosion—but that, in combination with a moderator or enriched with the isotope 235, it could sustain a chain reaction with slow neutrons and be used for energy production (Zeldovich 1993, 1–27; Igonin 1975, 416–430).

All-Union conferences on nuclear physics became almost annual events—Moscow (1936), Leningrad (1938), Kharkov (1939), and again Moscow (1940) (Igonin 1975, 28–40). The last two conferences featured active debate on whether and how atomic energy could be used for practical purposes, either military or civilian. The prevailing view, summarized by Kurchatov in 1940, was that application was a realistic possibility, but over a longer term: "although the question about the possibility of nuclear chain fission has been answered in the positive, its practical realization in systems researched so far still faces enormous difficulties" (Kurchatov 1941, 344).

Some apparent differences in opinion and institutional rivalry developed between physicists proper, such as Kurchatov, Joffe, and their LFTI colleagues, and scientists from the Radium Institute, primarily chemists, who prioritized the problem of raw materials. As a chemical element, uranium had little economic use prior to the discovery of fission, and only small amount was then available in the Soviet Union. In July 1940, Vernadsky, Fersman, and Khlopin submitted a memorandum to the government about a possible technical application of atomic energy through a chain reaction with slow neutrons. They requested funds for geological exploration for uranium ores and proposed the establishment of a State Uranium Reserve to accumulate and control the existing stock of radioactive materials. Upon their urging, the Academy of Sciences set up a special Uranium Commission under Khlopin to coordinate research conducted at various institutes with the stated goal of developing the mining of some 10 tons of uranium annually. Less than 1 ton was actually mined in 1940 (Komlev *et al.* 1982; Atomnyi I (1) 1998a, 200–202).

Although some Soviet scientists applied for patents and wrote letters to government offices about the military and other implications of their work, the community continued to publish openly about uranium fission and did not try to impose secrecy. In their treatment of uranium research as still a primarily academic field, Soviet scientists differed from English, American, and German colleagues but resembled the French (Weart 1979). Soviet standards actually required any science to have some important practical applications, or at least prospective ones. Previously, nuclear physics lacked such a prospect and was regarded deficient even as an academic field. With the discovery of fission, however remote and uncertain its potential uses could be, presented the field to Soviet officials as a respectable branch of science in its own right. Funding for it became available once again, after the hiatus of 1938, and the construction of the large cyclotron at LFTI resumed in 1939, its completion expected by the

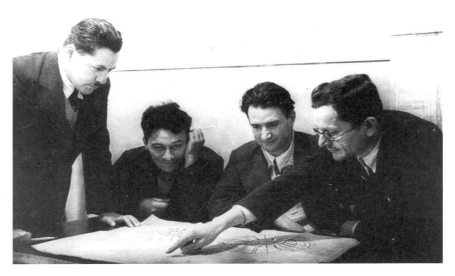

Designing a larger cyclotron for the Leningrad Physico-Technical Institute, 1940. Left to right: architect Ya. D. Galkin, physicists A. I. Alikhanov and I. V. Kurchatov, engineer A. F. Zhigulev. Construction did not finish in 1941 as planned due to the attack on the Soviet Union by Nazi Germany. Work resumed after the start of the atomic bomb project under Kurchatov's direction in 1943. Some of the prepared equipment survived the siege of Leningrad and was used for the 1944 cyclotron in Kurchatov's Moscow laboratory, while construction in Leningrad was completed only after the war's end.
[*Courtesy*: RGAKFD.]

end of 1941.[3] A Leningrad factory delivered a powerful magnet weighing 75 tons, and a special building for the cyclotron opened in June 1941, just before Hitler's armies attacked the Soviet Union. Most of the stored equipment survived the Leningrad blockade to be used in the cyclotron constructed after the war (Holloway 1994, 38–41; Aleksandrov 1988, 132–135).

One of the last pre-war accomplishments was also an extremely dangerous one. In the fourth paper in their series, which Zeldovich and Khariton wrote together with Isai Gurevich (1912–1992) from the Radium Institute, they explicitly stated that an explosion-type chain reaction with fast neutrons was possible in uranium-235 and estimated the critical mass of the separated isotope to be 10 kilograms (Atomnyi I (2), 2002, 469–495). This calculation made an atomic bomb look immediately more feasible, and a similar (somewhat more optimistic but less accurate) estimate of

---

[3] In April 1941 the government also approved funds for another large cyclotron in Moscow, at FIAN, which was supposed to open in 1943 (Atomnyi I (1), 1998, 227).

1 kilogram calculated in the famous secret memorandum by Otto Frisch and Rudolf Peierls gave the start to the British atomic bomb project in 1940. Had the paper by Gurevich, Zeldovich, and Khariton been published, and had it come to the attention of German physicists, it would have likely altered the course of the uranium project in Nazi Germany. But—as far as is known—German war-time uranium reports lacked this important calculation of the critical mass of pure uranium-235 (Bernstein 2003). The Soviet authors actually included their result in a paper submitted to a freely circulating scientific journal in the spring of 1941. The issue, however, did not come out, owing to the disruption of Soviet academic publishing after June 1941, as the Soviet Union struggled for her very survival in the Great Patriotic War against Nazi Germany.[4]

The catastrophic war ruled out even the hypothetical possibility of the start of a military-oriented Soviet atomic project. For Britain, the war had by then entered the attrition stage, promising to become protracted unless decided by some unexpected major development on either side. In such a situation, even a slight possibility that the Germans might become first to acquire a terrible new weapon inspired fear that effectively drove scientists and officials into a practical attempt at the atomic bomb. For Soviet scientists and officials, the threat from a hypothetical future German bomb was incomparably more vague and distant than that from Panzer divisions that quickly overran Kiev and Kharkov and were approaching the outskirts of Moscow and Leningrad. The mortal danger they posed required other actions, and it should not be surprising that Soviet scientists abandoned nuclear physics immediately and as a matter of course and switched to more clearly useful topics. Zeldovich and Khariton devoted themselves completely to practical research on chemical explosives. Kurchatov joined the work on protecting navy ships against magnetic mines, some of his associates helped evacuate their institutes to the east, and others volunteered for the military (Aleksandrov 1988, 159–174).

3. Atomic Secrets

Lieutenant Flerov's military service brought him to a training airfield near Voronezh to study the skills of an aviation technician. In late 1941 at a local university library he checked the latest available pre-war issues of

---

[4] The unpublished manuscript by Gurevich, Zeldovich, and Khariton would later become one of the first classified reports of the Soviet atomic bomb project. Their 1941 estimate of the critical mass would first become known in 1983 (Zeldovich and Khariton, 1941).

English-language physics journals, but found no references to his and Petrzhak's discovery of spontaneous fission. Moreover, papers on fission seemed to have disappeared altogether from the journal pages, which made Flerov suspicious that the allies had classified the problem as a military project (Smirnov 1996). In February 1942 Flerov wrote to Kurchatov urging a return to uranium research and proposing a crude sketch of an atomic explosive. He sent a similar proposal to the state plenipotentiary on science Sergei Kaftanov (1905–1978) and possibly also to Stalin (Atomnyi I (1), 1998, 251–252; II (2), 2000, 415–427). Flerov traveled to Kazan, to which most academic institutes had been evacuated, to discuss the matter with senior colleagues. They, in particular Joffe and Kapitza, were skeptical. A bomb, if possible at all, would require tremendous resources, an amount of uranium that simply was not available, and an estimated ten years of work. In the summer of 1942 the Soviet war against Hitler's Germany was at its most critical stage—the desperate defense of Stalingrad—and the situation could be saved only by mobilizing all existing efforts for the immediate fight.

With Joffe's support, however, Flerov was recalled from the army to concentrate on weapons research, and in a memorandum to the government Joffe and Kaftanov proposed to resume the work on uranium in order to determine whether it was realistic to achieve either a bomb or energy production. On 28 September, the State Committee of Defense—the five-man emergency committee under Stalin's chairmanship that governed the country throughout the war—approved the proposal. A week later, Stalin received a report from the Commissar of Internal Affairs, Lavrenty Beria (1899–1953), with a summary of intelligence information that had been arriving from Britain about the ally's intensive efforts to produce new uranium explosive (Atomnyi I (1), 1998, 263–274).

With the benefit of hindsight one can say that skeptics were right in their estimate that the atomic bomb would make no contribution to the war in Europe other then diverting already strained resources. The main importance of nuclear weapons would be for the post-war period, rather than for the outcome of World War II itself. Had the parties to the war known in advance how much time and effort the bomb would require, probably none of them—at least in Europe—would have decided to go ahead with the attempt. But British early estimates of the prospect of achieving isotope separation were too optimistic, resulting in the decision to start the Tube Alloys project in 1940. These calculations in the form of MAUD report of 1941 were shared with the Americans and stimulated the start of a large-scale work in the United States (Rhodes 1986, ch. 12).

The same estimates made their way to the Soviet Union by the way of spy reports.

Soviet intelligence knew as little as other allies about the situation with uranium work in Germany, but it had several highly placed spies in British offices (Modin 1994). Their first communications about the Tube Alloys project reached Moscow in October 1941 (Vizgin 1992). A draft report from Beria of March 1942 stated that the British had decided to go ahead with building a powerful uranium bomb and believed the project could be completed expeditiously. The information apparently was not delivered to Stalin until October, perhaps because of the catastrophic emergency on the southern fronts. Once the situation at Stalingrad stabilized and then delivered the great victory at the end of 1942, immediate survival became a less pressing imperative and some planning for the war's end and the post-war period could begin. Soviet scientists' independent estimates of the possibility of an atomic bomb were cautious, but Kurchatov was impressed by British self-confidence and determination when in November 1942 he was allowed to see some of the espionage materials. The available reports contained no information about developments in 1942, but they still allowed Kurchatov to conclude that the scale of allied efforts even back in 1941 had considerably exceeded what the Soviet decree envisioned for 1943 (Atomnyi I (1), 271–280). This awareness and complaints about the slow start contributed to a new round of bureaucratic proposals, resulting in another decree by the State Committee of Defense on 11 February 1943 that uranium work be intensified. Kurchatov's group was transferred from Kazan to Moscow, and the forty-year-old physicist was appointed head of the newly established secret Laboratory no. 2 of the Academy of Sciences (Atomnyi I (1), 1998, 297–307). With a handful of mainly pre-war collaborators, Kurchatov resumed research he had abandoned since June 1941.

David Holloway characterizes that undertaking as a "small hedge against future uncertainties" (Holloway 1994, 90). Indeed, for the duration of the war, Kurchatov's laboratory counted approximately 100 workers, about 25 of them scientists. Several other groups of Soviet workers pursued related tasks of pilot exploration of radioactive deposits and purification of uranium and graphite, but all the activities remained relatively small-scale. Concentration and overspending of resources and manpower on anything close to the scale of the hundred-thousand strong Manhattan Project were out of the question anywhere in war-devastated and bombed-out Europe. The overall dimensions of the Soviet wartime effort were comparable to, but apparently even smaller, than similar projects in Germany and in Britain proper. All these undertakings differed from the Manhattan Project in their limited size, their academic style of management with no military general at the top, and the absence of large-scale industrial development. What the Soviet Union lacked in comparison with other countries was uranium.

The lack of fissionable material would remain the main bottleneck of the Soviet drive for the atomic bomb until 1949. Before the war, only a few deposits of uranium were known worldwide, and even fewer were mined. The oldest and best-developed uranium mines in Czechoslovakia were under German control for the duration of the war, and the entire accumulated stock from the world's richest and best-quality mines, in the Belgian Congo, was appropriated by General Leslie Groves for the Manhattan Project. In the Soviet Union, the geological search had only started when the war broke out, and the only known possibility of mining existed at a relatively poor deposit in Tajikistan. The Soviet atomic bomb thus required not only researching and designing the bomb itself, as in the Manhattan Project, but also the additional task of building from scratch an entire industry of mining and enriching uranium. The resumption of geological exploration was ordered in November 1942, but it would take a full five years until the uranium industry was fully developed (Kruglov 1995, ch. 10). Kurchatov must have regretted his pre-war disagreements with Khlopin in the Uranium Commission, for until the end of the war he would not have sufficient uranium even for laboratory measurements.

Poverty required resourcefulness, and the coefficients of neutron diffusion were measured in so-called exponential experiments, whereby the scarce amount of uranium available was stretched out thinly in one dimension and the results of scattering and absorption measurements recalculated for the case of three dimensions. Some progress was also achieved in the technologies of uranium and graphite purification. In 1944 Kurchatov had a cyclotron running in his Moscow laboratory and used it to obtain his first miniscule quantities of plutonium for chemical analysis. Building a reactor, or a pile as it was then called, was still completely out of the question, since there was no uranium for it (Aleksandrov 1988, 276–277; Zhezherun 1978).

While the lack of raw materials hampered experimental progress, the theoretical work advanced rapidly. By 1944 Gurevich and Pomeranchuk developed a sophisticated theory of the uranium-graphite pile with optimal heterogeneous arrangement of the moderator. In the related non-classified field of cyclotron physics, Ivanenko and Pomeranchuk predicted the cyclotron radiation in 1944, and in the same year Vladimir Veksler (1907–1965) published his proposal of a new method in particle acceleration, automatic phase stabilization, which after the war allowed physicists to build an entire new generation of particle accelerators, synchrocyclotrons[5] (Veksler 1944; Ivanenko and Pomeranchuk 1944; Igonin 1975, 475–482).

The bulk of intelligence information about the Manhattan Project was also gathered before the war's end. The Soviet Union did not have a well-developed spy network in America, as it had in Britain, but assistance came from volunteers, Manhattan Project employees who contacted Soviet diplomats with offers of information. They were mostly anti-fascists who viewed their work on the bomb as a part of the common struggle against Nazism and thought that the information about the new weapon had to be shared with the ally who carried the main burden of fighting. Soviet intelligence officers code-named their file on the Manhattan Project ENORMOZ (in Cyrillic transliteration), which probably reflected their astonishment at the scale of the industrial enterprise (Atomnyi I (1), 1998, 345). The name also befitted the file itself, which grew to impressive dimensions owing to the quantity of smuggled reports. The quality of documents varied, but the most valuable information came from physicist Klaus Fuchs (1911–1988), a communist refugee from Nazi Germany who started to work on the atomic project in Britain and, while there, established contacts with diplomats at the Soviet embassy. In 1943 Fuchs arrived in Los Alamos as a member of the British team and resumed sending information to Soviet officials (Williams 1987).

During the war, the intelligence offices in Moscow were severely short of translators and could normally make only limited use of the materials in their possession. A dozen employees were reading random samples of the arriving English-language documents, guessing about the importance and usefulness of their contents (Modin 1994, 39). The priority of the atomic project reports was properly appreciated, but doubts persisted about the reliability of data. Secrecy ruled supreme, and intelligence officials were always mindful that the smuggled documents might contain wrong or outdated information (as, in all probability, some of them did) or simply disinformation. Beria was distrustful of both the providers and the potential users of the reports. Among physicists, only Kurchatov was allowed to see sizable selections of materials about the Manhattan Project, and on a few occasions he requested that particular information be shared

---

[5] Also suggested a year later by Edwin McMillan in Berkeley.

with one or two of his top associates. Other workers in the Soviet atomic effort were not supposed to know that any such documents even existed (Vizgin 1992; Atomnyi I (1), 1998, 278).

Many of the reports came directly from the Los Alamos laboratory, and Fuchs was not the only informant there. No such good source apparently existed at Chicago's Metallurgical Lab, and the information about Fermi's success in obtaining the first chain reaction in December 1942 would reach Kurchatov only eight months later, by the end of July 1943 (Vizgin 1992, 119; Atomnyi I (1), 1998, 375). Some indirect hints in smuggled reports suggested to Kurchatov the second possible route towards the atomic bomb—plutonium—two years before the release of the Smyth Report to the U.S. congress would make this information public. Soviet physicists had been working on the uranium-235 option, but the information about plutonium was new to them in 1943. Overall, the intelligence information suggested some key ideas to Soviet physicists, possibly saving them from a few bad mistakes and blind alleys, but it also apparently deflected their priorities in isotope separation from the original path of the ultracentrifuge—which years later they would find the most efficient method—to American-preferred gaseous diffusion and electromagnetic separation (Atomnyi I (1), 1998, 314).

Many additional details about know-how were learned from the documents, yet the single most important secret delivered by spies remained political rather than technological: it was the early warning in 1941 about the start of the Tube Alloys project in Britain. Without that information, the Soviet leadership would have most likely not appropriated any substantial amount of severely strained resources to uranium research and geological exploration while the war was still going on. Having resumed his research in 1943, Kurchatov was much better prepared when the time came for an all-out effort in 1945. No extra time or intelligence information, however, could compensate for the lack of uranium, which continued to block progress and prevent him from building an experimental reactor.

Soviet intelligence expected the first test of the American atomic bomb on 10 July 1945 (Atomnyi I (2), 2002, 335). It is not clear whether Stalin knew of these predictions when he departed for Potsdam to take part in the Big Three meeting. Neither espionage reports nor Truman's hint during the meeting—after the successful test on 16 July—that the United States had acquired a powerful new weapon led to a significant reorganization of the Soviet effort. A radical change was instituted a month later, after the American atomic bombing of Japanese cities, which provided the second crucial piece of foreign information that influenced the entire direction of the Soviet atomic bomb project. This second "atomic secret"—the definite proof of the bomb's feasibility and existence—was probably even more important than the first one, the start of the British project in 1941, but, having been so publicly demonstrated, it did not have to be delivered by spies.

4. Strategic Choices

On 20 August 1945, as the war against Japan drew to a close, the State Committee on Defense ordered the creation of a new Special Committee for all work related to uranium energy. After the destruction of Hiroshima and Nagasaki, the atomic bomb moved to the top of Stalin's list of state priorities. The American nuclear monopoly presented him not only with a possible military, but also more immediately with a geopolitical and diplomatic challenge. It threatened to deprive the country of the fruits of victory, which had been paid for so dearly, by undermining Soviet bargaining power in diplomatic negotiations with the Allies over the post-war world order. Stalin was determined not to lose the peace after winning the terrible war and was demonstratively refusing to let the atomic bomb argument play any role in the diplomatic peace settlement. In the meantime, he wanted the Soviet bomb to be produced as quickly as possible to ensure the country's protection from future military threats and to reaffirm its role as one of the three superpowers (he did not anticipate Britain's losing this role so quickly) (Holloway 1994, 150–171).

The appointment of Beria as head of the Special Committee reflected the ultimate urgency of the project. As a candidate member of the Communist Party Politburo, Beria participated in decision making on most important political decisions; as a member of the State Committee of Defense and subsequently first deputy prime minister, he commanded necessary branches of industry, and as former chief of state security he continued to coordinate the work of intelligence and could mobilize the forced labor of thousands of prisoners. Coordination of all these resources in a gigantic military-industrial project was crucial for eventual success, and Marshal Beria would prove himself through this work to be a ruthless, cynical, and very efficient administrator.

Together with seven politicians and industrial ministers, two physicists—Kapitza and Kurchatov—were made members of the Special Committee, and five more were appointed to the affiliated Technical

Council to discuss scientific and engineering aspects of the project (Atomnyi II (1), 1999, 11–14). Kapitza was not an expert on nuclear physics and initially was rather skeptical about the feasibility of the bomb, but he was arguably the most politically connected and most decorated scientist in the country at the time. Just several months after May 1945, when he received the highest honors for his wartime work on oxygen production, Kapitza continued to expand his administrative influence. In a letter to Stalin he proposed to concentrate all industries that dealt with low temperatures and gas liquefaction in his Oxygen Trust, which meant absorbing a rival Welding Trust. The head of the latter, M. K. Sukov, objected to the idea and complained to Stalin with personal accusations against Kapitza.[6]

The government approved the reorganization Kapitza was asking for in September 1945, but he still felt that Beria, who presided over the meeting, gave too much attention to Sukov's complaint by quoting from it and by suggesting Sukov's appointment as Kapitza's deputy in the enlarged Oxygen Trust. Feeling offended, Kapitza protested to Malenkov and also to Stalin.[7] As was typical of his letters, he started by discussing a specific problem but proceeded to larger issues, describing his serious disagreements with Beria about ways to organize the atomic bomb project. Tensions between them developed quickly. Beria was notoriously rude while Kapitza had a long record of protesting what he regarded as inconsiderate treatment by officials. This time he raised the level of his criticism, not only complaining of Beria's personal character but also suggesting that the new increased importance of scientists needed to be acknowledged by politicians and calling for "men like Comrade Beria to learn more respect for scientists."[8] The first letter to Stalin produced no noticeable effect, and a month and a half later Kapitza wrote a lengthier and more elaborate one:

> In my opinion there is much that is abnormal in the organization of the work on the A[tomic] B[omb]... Although it will be difficult, we must try to produce an A.B. quickly and cheaply, but not by the path we are following at present, which is completely unmethodical and without any plan. The main deficiencies of our present approach are that it fails to

---
[6] M. K. Sukov to Stalin, 22 August 1945 (IFP).
[7] Kapitza to Malenkov, 27 September 1945, Kapitza to Stalin, 3 October 1945 (Kapitza 1989a, 231–235).
[8] Kapitza to Stalin, 3 October 1945 (Kapitza 1989a, 234; 1990, 370). See Chapter 11, 292–293 for an additional analysis of this letter.

make use of our organizational possibilities and that it is unoriginal. We are trying to repeat everything done by the Americans rather than trying to find our own path. We forget that to follow the American path is not within our means and would take too long...

The problem is like that of a Commander-in-Chief who has been offered several proposals for taking a fortress... Invariably only a single plan is chosen and a single general to direct it. That is how we should proceed in science... If... we strive for quick success, it will inevitably involve risk and concentration of the main forces along a well-defined path in a very limited and wisely chosen direction. On these questions I do not agree with the other comrades. The next question, the selection of leaders to direct the work, is also a big problem. I take the view that the basis of choice should not be what a person promises to do, but what he had already achieved... The correct organization of this work is possible only subject to one condition... that great trust must be established between scientists and the statesmen... My turbine oxygen-producing installation... only got going when I, quite abnormally for a scientist, became the head of the Oxygen Trust...

Comrades Beria, Malenkov, and Voznesensky behave in the Special Committee as if they were supermen, particularly comrade Beria. It is true he has the conductor's baton in his hand. This is fine, but after him, a scientist should play the first violin, for it is the violin that sets the tone of the whole orchestra. Comrade Beria's basic weakness is that as a conductor, he should not only wave the baton, but also understand the score... I can't get anywhere with Beria—as I have already said, I don't like at all his attitude towards scientists... We must choose for leading roles wisely selected scientists and trust them completely... for example by requiring the signature of the scientist in order to approve the protocols of the Special Committee and orders of other administrators. We should then establish scientific commissars like political ones in the military...

With current conditions of work I don't see any use in my presence at the meetings of the Special Committee and the Technical Council... Since I am involved in this project, I naturally feel my responsibility for it, but I have no power to turn it my way. This is impossible because comrade Beria and most of the other comrades do not agree with my objections... So I ask you once more and very insistently to release me from my participation in the Special Committee and the Technical Council.[9]

Kapitza's letter was first published with excerpts in (Kapitza 1989a, 237–248), but rumors about it had long circulated among physicists. Some

---

[9] Kapitza to Stalin, 25 November 1945 (Atomnyi II (1), 1999, 613–620). Partial English translation in (Kapitza 1990, 372–377; Holloway 1990, 25).

commentators interpreted it as a refusal to take part in bomb construction, others praised his bravery in opposing Stalinism and its evil incarnate Beria, whose name had long been a symbol of the repressive state security apparatus. Yet behind the rhetoric of offering his resignation, the letter reveals Kapitza's longing for more power within the atomic bomb project and his disagreement not only with Beria but also with Kurchatov about the main strategy.

The principal dilemma was whether to imitate the Manhattan Project by working along a number of parallel routes that could lead to a bomb (plutonium and uranium-235, reactors with graphite and with heavy water, different methods of isotope separation, etc.) or to try to narrow the choice to a cheaper and possibly more original solution. The first option was more secure but guaranteed huge overspending of resources, while the second promised to be more economical but entailed a larger risk of mistakes. For Stalin at that particular juncture, as for General Groves earlier, the chief priority must have been time and security, rather than saving resources or satisfying scientists' egos. He thus clearly preferred the first option, siding with Beria and Kurchatov and going against Kapitza's advice. In December 1945 Stalin complied with Kapitza's "request" and released him from membership in the Special Committee and the Technical Council. In a public speech several months later, Stalin offered scientists a slogan, *"Dognat' i peregnat'!"* ("To catch up and to surpass!") which indirectly referred to the atomic bomb and reflected the strategic choice of following the American path (Stalin 1946).

For the time being, Kapitza remained at the head of the Oxygen Trust, but not for long. Although direct documentary proof that Beria engineered Kapitza's bureaucratic downfall is lacking, available circumstantial evidence suggests this scenario. In March 1946 the Soviet government underwent its major post-war reorganization. Its name changed from the revolutionary Soviet of Peoples Commissars, or Sovnarkom, to Council of Ministers, which in the perception of old communists sounded reactionary. In the reshuffling of ministries, Beria, who by that time had become a full member of the Politburo, extended his responsibilities. Among other agencies, he now also oversaw Kapitza's Oxygen Trust and in this official capacity received Kapitza's request to create a state commission to evaluate the last in the series of industrial oxygen devices, the one that produced gaseous oxygen for blast furnaces in metallurgy.[10]

---

[10] Kapitza to Beria, 2 April 1946 (Kapitza 1989a, 254–256).

The commission, headed by the engineer and high party official Maksim Saburov (1900–1977), initially drafted a generally positive report mentioning some minor defects of the machine, but the Council of Ministers ordered another, broader review of the Oxygen Trust's entire performance and added five new members to the commission. Three of them were engineers whom Kapitza knew as opponents of his inventions, but his protest against their appointment was in vain.[11] The expanded commission produced a much more critical report which argued that the German-made machines by Linde and Fränkl—based on the standard method of liquefaction using the Joule-Thomson effect—were smaller but more efficient than Kapitza's turbodetanders. The experts thus recommended continuing work to improve Kapitza's devices, while also using German machines brought to the Soviet Union as war reparations.[12] Kapitza argued that the report was prejudiced and wrote a lengthy refutation, but his actions only made matters worse. As the authorities continued to examine the case, accusations against him were becoming more and more serious.[13]

The final conclusion of the Saburov commission suggested releasing Kapitza from his position as head of the Oxygen Trust.[14] The Council of Ministers decided in addition to fire Kapitza from the directorship of the Institute of Physical Problems. Stalin as prime minister signed the official decree stating that Kapitza had failed to fulfill government orders for the construction of more efficient oxygen machines, that he deliberately ignored foreign innovations and stubbornly rejected the suggestions of Soviet specialists. Finally, Kapitza was accused of being more occupied with his scientific experiments than with industrial applications. Sukov made good his bid to replace Kapitza as head of the Oxygen Trust. Physicist Anatoly Aleksandrov (1903–1994), a student of Joffe's and a corresponding member of the Academy of Sciences, was appointed the new director of the Institute with the commission of reorienting its research towards topics directly relevant to the atomic bomb project. In line with

---

[11] Decree of the Council of Ministers, no. 1034, 14 May 1946 (IFP); Kapitza to Stalin, 19 May 1946, two letters, one official and one personal (Kapitza 1989a, 259–265).
[12] Committee of experts under the chairmanship of V. A. Malyshev, "Conclusion," 14 June 1946 (IFP).
[13] Kapitza to Stalin, 2 June and 16 July 1946; Kapitza to Malenkov, 25 June 1946 (Kapitza 1989a, 265–272); Kapitza's response to the conclusion by experts (IFP).
[14] One member, academician and engineer-metallurgist Ivan Bardin, disagreed and wrote a separate opinion in favor of Kapitza.

the government decree, the Academy of Sciences also issued a similar decision about Kapitza.[15]

In vain, Aleksandrov pleaded with Beria to allow Kapitza continue his personal research in the Institute (Atomnyi II (3), 2002, 486–487). Having lost almost all his official positions except the Academy membership, Kapitza retreated from his Moscow house, which was part of the Institute, to a dacha outside Moscow. He arranged there a garage laboratory nicknamed the "Hut for Physical Problems" and conducted some research, assisted by his two sons. He did small hydrodynamic experiments and theoretical work that did not require elaborate equipment. Several research papers by Kapitza were published in scholarly journals during those years, and he continued writing occasional letters to high officials, including Malenkov and Stalin, although fewer than before. In them, he tried to exonerate himself from the accusations, pointing out that his methods of oxygen production were gaining international recognition, and attempted to gain the interest of politicians in his new scientific proposals with possible military use.[16] With the exception of few academic publications, Kapitza's name practically disappeared from public view until 1953, when Beria lost his bid for power to Khrushchev in the political reshuffling after Stalin's death. In the intermediate seven years, Kapitza's Moscow institute under the directorship of Aleksandrov worked on the problems of isotope separation, the production of deuterium, and theoretical calculations of the nuclear explosive yield (Atomnyi II (3), 2002, 76–83).

5. Socialist Management at Work

General Groves' immediate post-war optimism about maintaining the U.S. nuclear monopoly for an extended period of time rested mainly on his estimates of the availability of uranium. Unlike most journalists, politicians, and many later historians, Groves understood that control over the world production of fissionable materials was more important than something called "the atomic secret," a concept that was easy to invoke but hard to define. One of Groves' first actions as head of the Manhattan Engineer

---

[15] The Saburov commission to Stalin (IFP); Decree of the Council of Ministers, no. 1815–782, 17 August 1946 (Atomnyi 2002, 7–12); the Presidium of the Academy of Sciences, "On the leadership of the Institute of Physical Problems," 20 September 1946 (ARAN, Kapitza's personal dossier, 411-3-445, p. 291).
[16] Kapitza to Stalin, 6 August 1948, 30 December 1950; Kapitza to Malenkov, 25 June 1950 (Kapitza 1989a, 281–297).

District had been to acquire the entire stock of uranium mined in the world's richest deposit in the Belgian Congo. He subsequently managed to ensure American control of or exclusive rights to most of the world's identified resources of uranium. It remained unknown to Groves how much uranium could exist in the vast territory of the Soviet Union, but neither did the Soviet authorities and experts know this at the time (Norris 2002, 178–179, 475; Holloway 1994, 174; Atomnyi I (1), 2002, 450–462; II (2), 2000, 313–316).

The fist sizable amount of uranium that became available for the Soviet bomb was found in May 1945 in occupied Germany. After the end of fighting, Soviet ministries dispatched several groups of engineers and scientists disguised in military uniforms to look for valuable industrial equipment and information. In the strategically important fields of atomic and missile technologies they did not find much: most of the hardware, documentation, and personnel had already fled or moved to Western occupation zones and came under the control of analogous programs on the American and British sides. Some German experts, however, remained in the Soviet occupation zone, and several groups of them agreed to move to the Soviet Union to work on military programs.[17] The most important find for the Soviet atomic project was the remnant of the German stock of uranium oxide—about 100 tons—serendipitously discovered at a factory near Berlin, where it was disguised as an ordinary chemical (Drovenikov and Romanov 1998). After extraction and purification, this amount allowed the construction of an experimental nuclear pile capable of sustaining a chain reaction. Kurchatov's first "boiler" F-1 used almost 35 tons of pure metallic uranium, 13 tons of uranium oxide, and 420 tons of pure graphite. It went operational at Laboratory no. 2 in Moscow on 25 December 1946 (Zhezherun 1978; Atomnyi II (1), 1999, 631–632).

Subsequent stages of work, as in the United States, involved industrial development more than scientific research. The Soviet economic system was well suited to replicating the Manhattan Project, sharing as it did the basic organizational principles of military-style management, hierarchical order of command and centralization of resources. Big science projects had been a familiar feature of the Soviet system long before the American developments in World War II, but now the dimensions of the enterprise

---

[17] (Mick 2000; Atomnyi II (2), 2000, 60–61, 319–321). By the end of 1946, 257 German specialists worked in the Soviet atomic project (64 scientists, 48 engineers, 145 technicians and qualified workers). Of them 122 were hired from occupied Germany; the rest were recruited from among war prisoners (Atomnyi II (3), 2002, 594).

had to be increased an order of magnitude over what had been typical. Kapitza's advice to look for more direct and economical ways was brushed aside. Stalin met with Kurchatov for the first time in February 1946, shortly after releasing Kapitza from the Special Committee, and urged him not to spare resources but to pursue the work on a "broad Russian scale" (Atomnyi II (2), 2000, 428–436). Like General Groves before him, Marshal Beria ordered the work to be conducted in parallel, often mutually duplicating and superfluous directions, and launched the construction of industrial objects well before the completion of necessary research. The result, in both cases, was gross overspending of resources, but also an increased probability of overall success and some reduction in total time, which was the main priority.

The difference, and the main difficulty in the Soviet case, remained the need to find and mine uranium. During the initial post-war years, most of the uranium was coming from Eastern Europe, from the already severely depleted mines in Czechoslovakia and some deposits in East Germany and Bulgaria. Altogether, they provided almost 80 percent of the stock of uranium acquired by the Soviet project through 1949, the year the first bomb was tested (Kruglov 1995, 263). Meanwhile, in 1946 geologists discovered new uranium deposits in Soviet Central Asia and the Ukraine, and the balance started shifting gradually from importation of uranium towards internal production (Atomnyi II (3), 2002, 523, 540, 686–688). As it turned out eventually, uranium was altogether not that rare on Earth, but Soviet ores were of a considerably poor quality and required the invention of new methods of enrichment and extraction. Economically, the development of the mining industry and the supply of uranium were the most challenging part of the entire Soviet atomic project, consuming the lion's share of total expenditures and manpower and determining the ultimate time schedule for the acquiring of a bomb.

Most of the work on construction and mining in the Soviet atomic industry until Stalin's death was done by convicts from labor camps, who worked in dismal conditions and suffered large numbers of casualties (Medvedev 2001). The atomic project was a terrible burden on the nation's economy, devastated as it was by the four years of war, with half the country's cities lying in ruins and peasants starving in villages, stripped once again of all their resources so the needs of speedy post-war industrial reconstruction could be met. The total human toll of the Soviet atomic bomb is hard to estimate, especially considering indirect casualties caused by the diversion of resources from human consumption into the defense projects of the escalating Cold War. Yet these sacrifices seemed justified to contemporaries, who had just barely survived the most destructive war in modern Russian history and did not need further proof of the consequences of a lack of military and technological preparedness in the face of a stronger adversary.

The flow of intelligence from overseas dried up after 1945, since "friendly espionage" was no longer an option. It was one thing for volunteer spies to help the chief ally in the common fight against Hitler and quite another to provide classified information to a potential enemy after the start of the Cold War. Possibly in an attempt to find fresh sources, in November 1945 Beria sent a young Moscow University physicist, Yakov Terletsky (1912–1993), to Copenhagen to meet with Niels Bohr. After taking some minor part in the Manhattan Project, Bohr had recently returned to liberated Denmark and resumed correspondence with Kapitza. Terletsky arrived in Copenhagen with a letter of introduction from Kapitza and during their discussion asked Bohr a series of questions about the physics of chain reactions and explosions (Andreev et al. 1994).

Bohr was eager to restore contact with Soviet colleagues and tried to answer in a friendly yet noncommittal manner. He presented Terletsky a copy of the recently released Smyth Report to the U.S. Congress and generally kept his responses within the limits of the officially declassified information.[18] Despite the meager results of the Copenhagen mission, Beria reported it to Stalin as a success. What the episode does illustrate well is the degree of compartmentalization that existed and Beria's distrust of scientists. Terletsky—a talented theoretical physicist who prior to the war had developed a theory of a relativistic accelerator, the betatron—was not exactly an expert on fission. He was entrusted with this delicate mission precisely because he did not actually work on the Soviet bomb, knew next to nothing about the real state of affairs, and could not possibly reveal anything about it. His role in the bomb project was that of an expert hired by intelligence to help sort out and translate the goldmine of technical reports about the Manhattan Project. Apparently Kurchatov was

---

[18] One of Bohr's answers is hard to decipher, most probably as a result of miscommunication rather than mistaken information. Bohr's language, especially his pronunciation, was often obscure, even when he tried to be crystal clear, even more so when he was speaking on a delicate subject. The entire interview process was somewhat comical, because Terletsky could not really speak foreign languages, while the interpreter who accompanied him knew no physics. Both thus tried to memorize what parts of Bohr's answers they could understand and put them together on paper upon returning to their hotel. Interestingly, the document so produced still for the most part makes sense (Andreev et al. 1994).

no longer allowed to see those documents himself because Beria wanted to use the reports as a check on his performance. Terletsky was suitable for the job because he did not belong to Joffe's Leningrad school, from which Kurchatov and most leading scientific cadres of the atomic bomb project had been recruited, but represented a rival group of Moscow university physics (Andreev et al. 1994).

The reliance on existing academic conflicts reflected Beria's distrust of Kurchatov, but he also distrusted intelligence reports, suspecting not merely mistakes but deliberate disinformation. Typically, security apparatus used espionage data versus the research by Soviet scientists as a system of checks and counterchecks. A Russian-born German chemist, Nikolaus Riehl (1901–1990), reported one characteristic exchange of compartmentalized information. Riehl, his team of a dozen German specialists, and a group of Soviet engineers worked at a small factory outside Moscow on chemical purification and production of uranium metal required for nuclear reactors (Atomnyi II (3), 2002, 595). As they struggled to overcome serious difficulties during the initial stage of work, the authorities offered "help" in the form of a sophisticated lesson in Russian obscenities and public dressing down. In one later episode, as they appeared to be stalled for too long with a bottleneck at the final stage of production, an official arrived from Moscow with a hint about a sophisticated chemical process, which helped solve the difficulty (Riehl and Seitz 1996, 100, 107). Riehl's group eventually managed to organize the production of pure metallic uranium in blocks at a rate of about one ton per day. He was allowed to keep a shortwave radio and listened clandestinely to BBC, to be surprised one night in 1949 with the news that the Soviet Union had exploded an atomic device. Among with other key participants in the project, in October 1949 Riehl was awarded the title of Hero of Socialist Labor, a Stalin prize, a car, and a house in appreciation of his work, apparently the most important contribution made by a German specialist to the Soviet atomic bomb (Atomnyi II (1), 1999, 556). After ten years of working in the USSR, he was allowed to return to East Germany and later moved to the West (Riehl and Seitz 1996).

Riehl believed that the information delivered to him in the episode described above was of transatlantic origin, but it also might have originated from another Soviet group whose work he did not know. Many concrete research and design tasks in the Soviet project were duplicated, with independent teams of scientists and engineers working in an indirect race against each other. Often they became aware of each other's existence only when, at the final stage, they had to meet face to face and argue whose product or solution was to be applied industrially. Recollections by participants in the Soviet project mention such surprise confrontations. In their own recollections, German specialists express astonishment at the directness and vicious level of criticism to which the Russians were accustomed when conducting their scientific disputes, without respect for formalities and academic hierarchy (Mick 2000, 183–185).

In one such debate, the chemical engineer Nikolai Dollezhal' successfully defended his group's design of an industrial uranium-graphite pile, in which the chain reactions would slowly produce plutonium atoms (Dollezhal' 1999). The reactor was built and started operating in June 1948 in a secret township near Cheliabinsk in the southern Urals. The plutonium industrial complex also included radiochemical and metallurgical plants for the extraction of plutonium. The parallel construction of the isotope separation complex—gaseous diffusion and electromagnetic separation plants, producing uranium-235—proceeded meanwhile at another secret location near Sverdlovsk in the central Urals (Atomnyi II (3), 2002, 830–831; Kruglov 1995). At both sites, the industrial production of fissionable material for the bombs ran into unforeseen difficulties.

The large uranium reactor had to keep the chain reaction going for about a year to accumulate the necessary amount of plutonium. Roughly in the middle, on 20 January 1949, a serious accident occurred, typical of the cases when industrial construction is rushed ahead without waiting for the completion of research and testing. A previously unknown type of corrosion developed and threatened to destroy the reactor's cooling system. To make the needed repair, it was necessary to shut down the reactor and let it cool, which would have interrupted the production of plutonium for another half-year. Such a delay was apparently politically unacceptable, and Kurchatov decided to risk making necessary replacements while the reactor was still relatively hot and highly radioactive. He personally commanded the emergency work, subjecting himself and other workers to high doses of radiation. In the immediate post-war period, scientists, administrators, military officials, and many ordinary workers judged the acceptability of personal risks by the standards of wartime rather than peace and often blatantly ignored basic safety regulations (Atomnyi II (3), 2002, 836–837). With sleepless nights, round-the-clock work, and exposure to radiation, Kurchatov managed to put the reactor back into operation on 26 March 1949—in two months instead of six—and by the summer produced enough plutonium for the first Soviet bomb. This and other accidents and breeches of workplace safety during the first, most strenuous years of work up to 1951 caused over 2000 cases of radiation sickness and some fatalities among ordinary workers as well as top personnel (Kruglov 1994, 105; 1998; Larin 2001).

6. The Bomb and the Fallout

Yuly Khariton directed the final construction of bombs at a secret installation deep in the forests surrounding the town of Arzamas in the Volga region. The laboratory used the buildings of a wartime munitions factory on the premises of what had been, before the revolution, one of the holiest Orthodox shrines, the monastery of Sarov. The name Sarov disappeared from official maps and other Soviet documents. For the next forty years, internal documents referred to the place as "the Volga branch" or P.O.B. Arzamas-16. Unofficially, the inhabitants nicknamed their secret city

The Volga branch, Arzamas-16, Los Arzamas, Sarov—all these names were used to refer in different times and contexts to one of the most secret locations in the post-war Soviet Union. The laboratory for the production of nuclear weapons was established on this site in 1947. The photo was taken during the early years, probably before 1950, when the laboratory occupied some of the buildings of a former orthodox monastery, *Sarova pustyn'*.
[*Source*: http://nuclearweaponarchive.org/]

Los Arzamas, since it was a close functional equivalent of the Los Alamos laboratory in New Mexico (Negin 2000).

Problems with the industrial process of isotope separation lasted well into 1950, delaying the uranium-235 bomb (Kruglov 1994, ch. 8). The plutonium produced and extracted by the summer of 1949 allowed Khariton to go ahead with the construction of the first plutonium bomb, which was code-named for secrecy RDS or "rocket engine S." This part of the project

Yu. B. Khariton, the scientific director of the Arzamas-16 laboratory since its establishment and until the end of the Soviet Union. Together with Ya. B. Zeldovich, Khariton developed in 1939–1941 what at the time was the most advanced theory of chain reactions in uranium, which correctly estimated the critical mass of a uranium-235 bomb. In this official photo Khariton is shown wearing his government awards, three stars of the Hero of Socialist Labor, and medals for three Stalin Prizes and one Lenin Prize.
[*Courtesy*: IPPI.]

was well supplied with intelligence information, at least with respect to the crucial technology of implosion. In order to explode efficiently rather than just "fizzle," plutonium needs to be compressed almost instantly into a state of high density. To achieve this, the metallic plutonium core is surrounded by ordinary explosives and the shock waves from the blast are focused inward. The main difficulty is that of ensuring a simultaneous and even pressure on all sides of the spherical core. Klaus Fuchs was among the key authors of this technology in Los Alamos, and he was simultaneously sharing information on his own design with Soviet agents. The quality and reliability of Fuchs' data reportedly allowed Khariton to skip some dangerous tests of the critical mass of plutonium and thus save some time during the final stages of the preparations (Holloway 1994, 106–107; Vizgin 1992; Kruglov 1994, 325).

In order to minimize time and risks, the first Soviet bomb was built as the closest possible replica of the first American plutonium bomb tested at the Trinity site (Khariton and Smirnov 1994). Implementation of further improvements and more original ideas was postponed for the second device, RDS-2, two years later. Beria personally attended the first test,

The first Soviet atomic test, RDS-1. Semipalatinsk testing grounds, Kazakhstan, 29 August 1949. Often, the event is mistakenly illustrated with photos of some later explosions. This photo, at least, comes from the most direct and authoritative source: (Negin 2000).

which took place in the early morning of 29 August 1949 at the Semipalatinsk testing ground in Kazakhstan. About a month later, results from the monitoring of radioactive debris convinced President Truman that the United States no longer possessed a nuclear monopoly (Machta 1992; Friedman, Lockhart, and Blifford 1996). Truman's public announcement that the Soviet Union had successfully tested an atomic device came as a shock to the American public and officials and intensified the purges of communists, sympathizers, and American leftists, who were made scapegoats for the loss of the "atomic secret" (Wang 1999). In an attempt to maintain the upper hand in the arms race, Truman authorized an all-out effort to build a hydrogen bomb. Alarmed by earlier American mentions of the "Super" and communications from Fuchs, the Soviets were already studying whether this was a bluff or a real threat.

Two theoretical groups, under Zeldovich at the Institute of Chemical Physics and under Tamm at FIAN, the Physical Institute of the Academy of Sciences, investigated the model they called *Truba* (the Tube), which was most likely analogous to early Los Alamos proposals of 1946. A young member from the Tamm group, Andrei Sakharov (1921–1989), proposed an alternative design nicknamed *Sloika* (the Layer Cake), which his colleague Vitaly Ginzburg (b. 1916) supplemented with the second crucial idea, that of using Li^6D as material. The Truba design eventually proved a dead end: it could be made to produce a thermonuclear explosion, as the American 1952 test showed, but only as a huge device, far too heavy for a bomb. The much more compact Sloika was successfully exploded by the Soviets on 12 August 1953 (RDS-6s) and was hailed by them as the first hydrogen bomb. Unwilling to admit the Soviet accomplishment, some American experts disputed the claim by pointing out that the larger portion of the Sloika's total yield came from fission, rather than from hydrogen fusion. A subsequent powerful American thermonuclear test in March 1954 (*Bravo*) demonstrated to Soviet scientists that a more efficient configuration was feasible, which led them to what they called "the third idea" (or the Teller-Ulam scheme, in U.S. terminology), that of a two-stage weapon. In this improved combination, a fission explosion helped compress and ignite the thermonuclear Sloika bomb, which in principle allowed building hydrogen bombs of practically unlimited yields. A Soviet bomb with the "third idea" was tested by airdrop on 22 November 1955 (RDS-37). The two Cold War adversaries thus followed different paths in their race for the hydrogen bomb, but their designs converged at the final stage, perhaps as a result of the mutual monitoring of radioactive debris (Goncharov 1996; 2002; Gorelik 2000).

To complete the work on Sloika, Tamm and Sakharov moved to "the Volga branch" in 1950, where they joined Khariton and Zeldovich. Tamm returned to Moscow and to his work on fundamental particle physics after the successful test in 1953, but the other three theoreticians remained in Arzamas-16 for many years, working on further improvement of nuclear weapons. The Stalinist system invented a special kind of research institution, *sharashka*, where prisoners with high qualifications worked on technological projects (Kerber 1996; Kopelev 1981). Though technically not prisoners, nuclear engineers and researchers worked in conditions approaching those of *sharashka*: confined within remote closed towns, bound by the highest degree of secrecy and discipline, watching columns of regular prisoners marching every day to the construction site, and aware of the possibility of either high rewards or severe punishment in the end. Psychologically, they saw the work as their personal contribution to the Great Patriotic War that had just ended and a direct continuation of the war effort of defending the country against a superior threat. Hardly anyone doubted the moral imperative of nuclear deterrence on ethical, political, or professional grounds. As in the Manhattan Project, more critical attitudes developed among nuclear specialists later, after the main job had been accomplished and some tensions over the question of who should control the product had arisen. For many American scientists this process started in 1945, while Soviet scientists could feel their main patriotic duty fulfilled and entered a similar stage of awakening doubts and changing priorities approximately ten years later, around 1955.

The experience of the nuclear project transformed physics and science in general. Although the project itself, as in the United States, was more an industrial than a scientific enterprise (Hughes 1989), nuclear physicists took most of the public credit for it. After the success of the bomb work and Stalin's death, the Soviet cult of science rose to an even higher level, considerably surpassing the analogous cult in the Cold War United States. In their new celebrity status, physicists and their profession captured the public imagination. Nuclear physicists got used to working with industrial-scale resources and institutions and started demanding a similar level of support for their civilian projects. In part to accommodate these expectations, some of the formerly secret townships were declassified and converted into "scientific cities," places where civilian scientists lived and worked in their own social milieu, outside of major metropolitan centers. The popularity of this social innovation ensured that more such scientific cities, and not just for physicists, were eventually built in new locations, to continue the institutional expansion of Soviet science. The scientific

"To Catch Up and To Surpass..."

elite used its new authority in dealing with politicians to shape the profession and institutions of science, if not Soviet society in general, in a way more to their liking.[19]

These important changes in the role and status of science happened mostly during Khrushchev's period in the second half of the 1950s. The years before 1953, however, while Stalin was still alive and the work on the atomic bomb was still in progress, are on the contrary usually characterized as one of the darkest and most difficult periods in the history of Russian science. It was, after all, the time when the infamous Lysenko affair destroyed Soviet genetics and when science as a whole—despite being held in extremely high regard—was subjected to intensified political pressure and ideological censorship. The next three chapters seek to sort out this contradiction and analyze the public life of Soviet science in the political context of late Stalinism. With this goal in mind, we shall start with following the life and career of Sergei Vavilov, the first post-war president of the Soviet Academy of Sciences, who was mainly responsible for representing science to Stalinist politicians and to the public in general during those difficult times.

---

[19] On the Soviet cult of atomic science see (Josephson 2000; Vizgin 2002b); for a history of Akademgorodok, a Siberian city of science, see (Josephson 1997); and for post-Stalin reforms in science see (Ivanov 2002).

Chapter 7

President of Stalin's Academy: The Mask and Responsibility of Sergei Vavilov

It has become a cliché, when speaking about Sergei Ivanovich Vavilov (1891–1951), to start with a dramatic comparison of two fates: that of Sergei himself—a high-ranking official, president of the Soviet Academy of Sciences at the apogee of Stalin's rule, 1945–1951—and that of his brother Nikolai (1887–1943)—a famous geneticist who became a victim of the purges and starved in prison in 1943. Although the general reader may find it paradoxical, such a pattern was not at all unusual for that epoch. Participants in political games accepted the cruelty and dangers of that life as inevitable, and even Politburo members Viacheslav Molotov and Mikhail Kalinin had convicts among their closest relatives. Sergei Vavilov was also a high-level politician who both administered the scientific Academy and delivered the required political speeches during a time characterized by the Cold War, Stalin's dictatorship, nationalism (including anti-Semitism), and ideological purges in culture and science. He managed to perform this job without committing any major political mistakes and died in 1951 in full official recognition. The Politburo member Nikita Khrushchev speaking at his funeral called Vavilov a "non-party Bolshevik," a laudatory term for one who was not formally a member of the Communist party (Poslednii 1951).

However, Vavilov did not fit any of the stereotypical images of Soviet officials, such as the energetic and rude revolutionary or the gloomy apparatchik. He was a phlegmatic and solitary intellectual whose hobby was searching for rare books among the heaps in secondhand bookstores and who translated Isaac Newton's *Lectiones Opticae* from Latin into Russian. He is also known to have acted in some cases quietly and unofficially against decisions that he himself promoted on the official level. Later changes in politics destroyed the reputations of many of Stalin's heroes. In the years of Gorbachev's perestroika it would not have been difficult for a zealous journalist to glean from Vavilov's writings enough quotations to declare him a Stalinist. But there was a conscious effort on the part of the scientific community, especially among physicists, to describe both Vavilov brothers as heroes, emphasizing Sergei's efforts to protect science in those difficult times (Vavilov 1991).

These contradictions should serve as a warning that simple black-and-white pictures are inadequate representations of Stalinist society's *tempora* and *mores*. The complexity of moral challenges of life under a dictatorial regime offered a sharper form of the universal and everlasting question: What does it mean to be moral in the immoral world? The choice then was often between open protest, which would also harm innocents, and compromise with what one considered as a lesser evil. Vavilov's case may be seen as an attempt to live in awareness of this moral dilemma.

Those highly disciplined times, when the rules of public speech and behavior allowed few variations and even privately expressed heterodoxy was extremely dangerous, have left behind few direct sources for the social historian. To understand Vavilov's roles and what brought such an unlikely individual to a high political position in Stalinist society will require a look at some general features of the Soviet culture: its ritualistic pattern of career-building, the influence of the cultural revolution on the academic community, and the aesthetic image of the scientist. Though not giving a full explanation of Vavilov's career, this general analysis will provide a context that helps bring its meaning into focus. By the same token, of course, the individual biography can, in return, serve as a mirror that facilitates a better understanding of Stalinist society in general.

A public official in Stalin's era had to keep many of his thoughts secret, not trusting even private conversation, correspondence, or a diary. Vavilov's real views thus have to be reconstructed through a hermeneutic reading of the highly ritualized Soviet political literature of the 1940s, which at first sight seems undifferentiated in language and meaning. The authors of such literature were able, however, to communicate different, even opposing, messages through inventive play with and around politically acceptable expressions. As one of the key interpretative tools, I will use Vavilov's seemingly apolitical biography of Isaac Newton, which will be shown to reflect the author's self-image, his political views and life strategy. Textual criticism can also reveal some of the hidden subtexts of his official publications, as well as better interpret Vavilov's actions and the difficult choices he made.

1. Early Career

"For the first twelve years of revolution it looked as if Nikolai would become the chief administrator of science, while Sergei seemed destined to remain a modest teacher and researcher," remarks David Joravsky (1965, 382). Indeed, Sergei Vavilov, unlike his elder brother, did not appear to be a genius and a leader. His strengths lay not so much in making contacts with others and influencing people as in hard and tedious work. His other characteristics included unhurried manners, politeness and a quiet voice, broad interests, scientific ideas that were solid rather than highly original, a gift for writing but not for speaking, modesty, and restraint from public activity. Altogether, he was a scholar whose ideal career would be that of a professor and qualified specialist in a narrow field of research, a quiet worker who shuns publicity.[1]

Indeed, as of 1929 when he turned thirty-eight, Sergei was working as a rank-and-file member at the Institute of Physics and Biophysics in Moscow and teaching in a position similar to *Privatdozent* at Moscow University. Like many of his generation, Vavilov had had an opportunity to travel abroad, and he worked for several months in 1926 with Peter Pringsheim in Berlin. He was interested in the microscopic structure of light: initially he concurred in Planck's criticism of the Einsteinian light quanta but accepted the concept later, around 1923, and continued to use quantum ideas in the experimental study of the phenomena of luminescence, or the emission of light by some substances in the absence of high temperatures. He produced work of good professional quality, but nothing very exceptional grew of it for the time being.

Less than three years later, before the end of 1932, we find Vavilov at the age of forty-one a full member of the Academy of Sciences and directing two research institutes. Such rapid advancement cannot be easily

---

[1] This psychological type shows through all, even rather official, reminiscences. For the most complete collection of them see (Frank 1991a). Hardly anyone could avoid mentioning 'hard-working' as Vavilov's most characteristic feature. While reading *Faust* in the trenches of World War I, Vavilov remarked in a notebook that Wagner, not Faust, represents the model scientist. He had just graduated from Moscow University and was thinking about science as a calling and his own future (Keler 1961, 54–56).

explained, and much of its inner mechanics remains hidden. Neither Vavilov's scientific accomplishments nor his political writings, which began appearing in this period, can sufficiently account for his stellar rise. The interpretative key can be found, however, in the phenomenon of the cultural revolution and what it meant for the Soviet academic community. As described in chapter 4, the cultural revolution undermined the positions of many senior academic leaders with pre-revolutionary backgrounds, or "bourgeois specialists," advancing instead younger radicals in both science and politics. Vavilov did not belong to this latter group: far from being a rebellious reformer or a militant critic, he does not seem to have been much involved in the struggle.[2] Still, he profited from defeats of some of his senior colleagues.

Between 1928 and 1932 the communist government forced several successive elections to the Soviet Academy of Sciences in order to increase its membership, diminish the influence of the old guard, and, finally, establish control over the organization (Graham 1967; Perchenok 1991). These elections took place under strong political pressure, and one might expect political criteria to have influenced the choice of candidates. This was indeed the case in the social sciences, where an open struggle broke out over several communist candidates. In the physical sciences, however, there was no open polemic between politicians and members of the Academy; rather, a quiet tug-of-war ensued. The internal politics of Academy elections usually remain closed, but if one compiles and classifies the results, it is difficult to avoid the conclusion that the elections reflected power relations within the physics community rather than external political influences.

In physics, four out of the five full members—Piotr Lazarev, Abram Joffe, Dmitry Rozhdestvensky, and Leonid Mandelstam—represented major research institutions, two in Leningrad and two in Moscow. Once elected, each of the four managed to secure three corresponding member positions for his closest collaborators and pupils. Such a careful balance must have been achieved by quiet agreements among the senior

---

[2] However, in 1928 Vavilov joined other Moscow University physicists in an attempt to overthrow Arkady Timiriazev, whose power in the Physics Department rested mainly on his communist party membership and a responsible post in the Commissariat of Enlightenment (ARAN, 641-3-79). Vavilov also took an active part in the radical educational reform at the Physics Department during the cultural revolution. See "Luchshie udarniki fizicheskogo otdeleniia," *Za proletarskie kadry*, 30 September 1931, and "Pervyi professor-udarnik," ibid., 19 October 1931. He continued to express support for the pedagogical experiment even when the political mood was becoming more conservative (Vavilov 1932).

academicians. The Academy had a long tradition of choosing candidates by such agreements; usually in the formal balloting procedure only one person was proposed for each vacancy. If that was the calculation, Vavilov became the third of the disciples of Lazarev to be elected corresponding member on 31 January 1931.

Lazarev (1878–1942), a physicist and biophysicist, was accused of "wrecking," arrested in March 1931, and exiled. Restoring the balance of academic representation by promoting a member of his school is the most plausible interpretation of Vavilov's election to full membership on 29 March 1932. This promotion did not, however, save Lazarev's Institute of Physics and Biophysics from being closed (Frank 1991b, 11).[3] Soon after Vavilov's election, another senior physicist in the Academy, Rozhdestvensky, got into trouble with industrial Commissariat officials and with the young communist members of his Optical Institute over the proper proportions of applied and fundamental research conducted there. Rozhdestvensky resigned as director in March 1932 and proposed the newly elected academician Vavilov as his successor, because the latter was an expert in optics and, as an outsider, was not involved in the institute's conflict (Frish 1992, 190–195). However, in accordance with a common pattern of the time, the Commissariat appointed a communist functionary as the official administrative director, while the specialist Vavilov became deputy director for scientific matters.

Before 1934, the Academy did not have a research institute in physics but only a rather nominal department in the Physico-Mathematical Institute. When in the spring of 1932 Joffe, Rozhdestvensky, and Mitkevich blocked the attempt by young theoreticians to transform that department into a separate institute for theoretical physics, they promoted instead the idea of a full-size physics institute at the Academy to be directed by an experimental physicist. Each of the senior physicists at the Academy already supervised a big research institute elsewhere. During one of the discussions in April 1932, Rozhdestvensky proposed Vavilov as director of the new institute, and the meeting decided to ask him whether he was willing to take the job (Gorelik and Savina 1993). In September 1932 Vavilov moved from Moscow to Leningrad and took up his two new posts—the Optical Institute and the department at the Physico-Mathematical Institute, which in 1934

---

[3] In 1932, after the attitude towards old specialists had changed, Lazarev returned to the Academy, worked quietly in the Institute of Experimental Medicine and the Institute of Theoretical Geophysics, and died a respected scholar in 1942. His file is preserved in the KGB Archives.

Full members	Corresponding members	Institute
Lazarev (1917) Vavilov (1932) ←	Shchodro (1929) Shuleikin (1929) ──Vavilov (1931)	Institute of Physics and Biophysics, Moscow
Joffe (1920)	Semenov (1929) Kapitza (1929) Frenkel (1929)	State Physico-Technical Institute, Leningrad
Mitkevich (1927, **1929**)		Special Bureau on Military Technology, Leningrad
Rozhdestvensky (1927, **1929**)	Ignatovsky (1932) Terenin (1932) Fock (1932)	State Optical Institute, Leningrad
Mandelstam (1928, **1929**)	Arkadiev (1927) Papaleksi (1931) Landsberg (1932)	Research Institute of Physics at the Moscow University
	Gamow (1932)	Physico-Mathematical Institute of the Academy of Sciences, Leningrad

Physicist members of the Academy of Sciences during the early 1930s. The country's four main physics institutes had equal representation by one full member and three corresponding members. There are a couple of exceptions to this general pattern. Vladimir Mitkevich was elected as corresponding member in physics, but as a full member he filled a slot in engineering sciences. Gamow's elections were orchestrated by young theoretical physicists and supported by the applied mathematician Aleksei Krylov, director of the Academy's Physico-Mathematical Institute. Most of the biographical data for the table comes from (Frenkel 1990).

would be officially transformed into a separate Physical Institute of the Academy of Sciences (FIAN). Vavilov thus became the first of the younger generation of physicists to rise to the top of academic hierarchy, and he remained one of only two full members until the next major elections in 1939. In promoting him, the Academy paid tribute to current politics; still, a conservative body, it preferred the moderate Vavilov to even younger and more radical fellows.

The following year Vavilov took part in a discovery that eventually received a Nobel Prize. As part of his attempt to develop FIAN into a major research institute in physics, he encouraged the development of newer fields, such as nuclear physics. He asked an *aspirant* of his, Pavel Cherenkov, to study whether luminescence in some solutions, ordinarily induced by incoming light, can also be caused by gamma rays from a radioactive source. A clever experimental method developed by Vavilov in the 1920s, when equipment was scarce, allowed observation of faint luminescence, almost at the threshold of single quanta, by the naked eye after two hours of adaptation in complete darkness. Following this technique, Cherenkov noticed, in addition to expected luminescence, that passing gamma rays produced a faint background glow that remained present in pure liquids like water, not only in luminescent solutions (Cherenkov 1999).

Vavilov's expertise helped recognize this phenomenon as a novel, heretofore unknown kind of radiation. The discovery of blue light was published in two related papers—an experimental report by Cherenkov and a theoretical discussion by Vavilov—in the same issue of *Doklady Akademii Nauk* (Cherenkov 1934; Vavilov 1934). In the preceding decades physics had seen several cases of widely publicized discoveries of spurious rays (Kragh 1999, 34–38). Some Soviet colleagues and foreign visitors saw FIAN's search for an almost invisible radiation in the dark as more spiritualism and ghost-hunting. In the face of widespread skepticism, Cherenkov stubbornly persisted and always had Vavilov's support. He defended his first thesis (*kandidat nauk*) in 1935, and in a series of subsequent studies that became the basis of his second thesis (*doktor nauk*, defended in 1940) discovered unexpected properties of the new radiation, including its distinctive directedness. This feature helped FIAN theoreticians Igor Tamm and Ilia Frank explain the phenomenon in 1937 (Tamm and Frank 1937; Frank 1984; Izluchenie 1984).

The Cherenkov or Vavilov–Cherenkov radiation is produced when electrons travel through a substance faster than light does but more slowly than light travels in a vacuum so that no contradiction with Einstein's theory of relativity arises. (It is the optical analog of shock waves in acoustics, the sound produced in the air by an ultrasonic projectile or jet.) The phenomenon of Vavilov–Cherenkov radiation can thus be used for registering fast-traveling subatomic particles. Such new detectors, Cherenkov counters, were designed and became widely used in high-energy accelerators after World War II, when they helped physicists discover new elementary particles, for example the antiproton. The 1958 Nobel Prize in physics was awarded to Cherenkov, Tamm, and Frank for the discovery and the interpretation of the Cherenkov effect. Vavilov was no longer alive and thus could not be nominated. In the Soviet Union the discovery received the Stalin Prize in 1946.

However important this discovery proved later, it could hardly have changed Vavilov's fortunes much in the 1930s, for he was already recognized by political officials as a leading scientific manager and was spending most of his time doing administrative work.

2. Rise with and within the Academy

> "Besides scientific aspirations and absorption in thoughts, a talent for administration was also characteristic of Newton.... In a time when bribery was more than a usual thing, Newton, according to the available information, fulfilled his duties honestly and severely.... A great deal of administrative and practical activity had torn him away from his scientific work" (Vavilov 1945, 188, 170).

When Vavilov was elected to full membership in the Academy in 1932, the cultural revolution was already coming to an end. The reversal of revolutionary social and cultural policies was not as sudden and outspoken as their outbreak in 1928 but came under the heading of 'correcting excesses' and took several years of quiet but consistent changes in one field after another. The emerging period of 'high Stalinism' was characterized by the purge of revolutionaries, traditionalism and even conservatism, the restoration of social hierarchy, the withering away of class rhetoric, and a shift from internationalism to nationalism.[4]

In the field of science policy this resulted in the decline of the superpower of the Commissariat of Heavy Industry (NKTP) and in the rise of the Academy of Sciences. The most important steps in this process were the subordination of the Academy under the auspices of the *Sovnarkom* (Council of Peoples' Commissars) and its transfer from Leningrad to Moscow in 1934; a series of purges in NKTP and its dissolution in 1937 into a number of smaller, specialized ministries; and the subsequent incorporation of many research institutes into the Academy in 1938–1940. As a result, the Academy succeeded in its long-standing desire to dominate national science, having become much more privileged than the universities, and gathered under its administrative auspices most of the

---

[4] The process is sometimes called "The Great Retreat" after (Timacheff 1946), but the term is somewhat misleading. What was happening was the emergence of a new order out of revolutionary chaos rather than a return to the old order. A synthetic study of this process comparable to the volume (Fitzpatrick 1978) is still lacking.

country's leading research institutions in fundamental science. Its role changed correspondingly from an assembly of a few respected scholars to a functional substitute for the otherwise lacking ministry of science, and it received an increasing share of the rapidly expanding government support for scientific research (Kozhevnikov 1990).

Along with the 1934 decision to move the Academy to Moscow, the government officially approved the proposal to split its Physico-Mathematical Institute in two. Vavilov's job as director of a ghostlike physics department thus turned into a very important post when he got the commission and resources to develop it into the nation's largest physics institute, FIAN.[5] Physicists from FIAN always described Vavilov as an ideal administrator who protected them from the thunderstorms of the larger society, took on the job of securing funds and equipment for them, and did not interfere in the research of the laboratories.[6] Three main collaborators of Mandelstam—Grigory Landsberg, Nikolai Papaleksi, and Igor Tamm—came from Moscow University to head laboratories of optics, oscillations, and theoretical physics, respectively. Vavilov directed a small laboratory for the study of luminescence. Gamow, the institute's only expert in nuclear physics, remained in the West in late 1933, so Vavilov commissioned several graduate students to start research on their own and appointed himself the formal head of the nuclear laboratory to ensure its administrative protection (Frank 1981, 307–315).

The years after 1934 marked an enormous increase in Vavilov's administrative obligations. He continued to direct the Optical Institute, commuting between Moscow and Leningrad. From 1935 to 1938 he served in the Academy's ruling body, the Presidium, and had a dozen other smaller commissions. Archival documents reflect his futile attempts to reduce his duties, in particular to avoid being included in the Presidium and to give up the directorship of FIAN, while retaining responsibility only for the Optical Institute. In a letter to the Presidium, Vavilov observed that even though he worked twelve hours a day his administrative chores took so much time that he had almost none left over for his own experimental

---

[5] Vavilov took the job of organizing a new institute in Moscow as a personal tribute to his teachers Piotr Lebedev and Lazarev. Indeed, FIAN, or the Lebedev Physical Institute of the Academy of Sciences, occupied the building of Lazarev's former Institute of Physics and Biophysics in Moscow (Vavilov 1937; Novik unpublished).

[6] Even the emigré physicist Yakov Alpert, whose status as a refugee required him to talk about anti-Semitism in the Soviet Union, drew a totally idyllic picture of FIAN under Vavilov (oral history interview with Yakov Alpert, National Air and Space Museum, Smithsonian Institution).

S. I. Vavilov, one of the youngest members of the USSR Academy of Sciences and its Presidium in the 1930s. Vavilov's involvement in optical research, in particular in the studies of the Cherenkov–Vavilov radiation he helped discover in 1934, decreased as a result of his enormous administrative chores. Vavilov mastered the art of perfectly correct behavior in complicated social and political circumstances, but his true thoughts remained an enigma for many an outside observer.
[*Courtesy*: IPPI.]

work and reading. Indeed, since the mid 1930s he had been able to make only sporadic and desperate attempts to perform experiments with his own hands.[7]

---

[7] Vavilov's request to remain outside the Presidium is in Vavilov's letter to G. M. Krzhizhanovsky (Vice-President of the Academy) 17 November 1935 (Vavilov's personal dossier, ARAN, 411-3-124); His proposal to give up the directorship of FIAN and the calculation of his work schedule are in a letter to the Presidium of 1936 or 1937 (published in Frenkel 1990, 118–120); Novik (unpublished) quotes other similar documents (ARAN, 596-2-1a; 596-2-2); On Vavilov's futile attempts to continue his experimental work see (Frank 1981, 235, 253).

Vavilov's role at the Optical Institute is not described in such unreservedly glowing terms as his tenure at FIAN. Sergei Frish, speaking in his later recollections for the pupils of Rozhdestvensky, criticized Vavilov for compromising too much with communist officials. In particular, Vavilov had failed to defend Rozhdestvensky, whom the Commissariat of arms production continued to criticize for doing too much fundamental rather than applied research. After some of his research proposals were declined, Rozhdestvensky moved with his laboratory to Leningrad University. According to Frish, Rozhdestvensky characterized Vavilov's behavior in this conflict as follows: "He carries in his pocket a letter of resignation, already signed but without a date on it. When the moment comes and he is forced to agree with something totally unacceptable, he is prepared to take this letter out of his pocket. But he can never decide whether this moment has already come" (Frish 1992, 240).

Party membership became quite common among science administrators after about 1940, when even some older scholars such as Joffe joined the party. Vavilov, however, remained unaffiliated and was the last noncommunist president of the Soviet Academy. In response to a question by a puzzled junior colleague, Ilia Frank, he alluded, without elaborating, to the case of his arrested brother Nikolai as the reason. Nikolai Vavilov was the leading administrator in agricultural research in the years of the cultural revolution and collectivization. After the revolution was over he suffered critical attacks, some from the growing Lysenkoist movement, and in 1935 had to resign the presidency of the Academy of Agricultural Sciences. In 1940 Nikolai was arrested, convicted of "wrecking," and died in prison in 1943 (Popovsky 1984; Boiko and Vilensky 1987). The implicit meaning of Sergei's reply to Frank's naïve curiosity was, "The party would not trust a convict's relative and accept me as a member"[8] True or not, this excuse effectively masked Vavilov's apparent reluctance to join the party.

However, in all his administrative positions, Vavilov worked in close and apparently harmonious contact with communists. In compliance with the usual standard, he always appointed one of FIAN's few communists as either Deputy Director or Academic Secretary of the Institute. Until his

---

[8] It is ironic—and a telling example of how meanings can reverse along with cultural values—that in 1991 Frank interpreted the answer as: "I don't want to, because they arrested my brother" (Frank 1991b, 16). In the 1940s the party was synonymous with the very best, while Nikolai Vavilov was officially a criminal. By 1990 Nikolai was publicly perceived as hero and innocent victim, while the Party had become synonymous with evil. The dialogue between Sergei Vavilov and Frank probably took place before late 1943, when Sergei learned about his brother's death (Vavilov 1991, 33).

President of Stalin's Academy

arrest in 1936, the Marxist philosopher of science Boris Hessen worked as Deputy Director of FIAN. At an institute meeting in April 1937, Vavilov admitted that party membership was the reason for appointing Hessen (Gorelik 1992, 17). Vavilov's ability to avoid conflicts must be due to an important character trait, which was grasped by a somewhat jealous colleague, Piotr Kapitza:

> Vavilov... is young, only 45. I doubt if you know him by name, his work was in the fluorescence of liquids. You know the sort of work when you pass a beam of light through a vessel filled with liquid and observe the light perpendicularly. Once installed, you can play with the apparatus for all your life, changing the liquids, the number of which is immense, and you can also vary the spectra of the primary beam. And thus you have such a number of combinations that will keep a research student busy all his life and give him the feeling of satisfaction that he is doing scientific work. He never did anything else. I was always surprised why Vavilov got into the Academy when even with our poor stock of physicists we have such people as Skobeltsyn, Fock and others, who are miles better than Vavilov. I think you will find the secret in that Vavilov is a very polished man, who knows what to say and when to say it so as to please everybody (Kapitza 1990, 281).

The quote is from Kapitza's 1936 letter to Rutherford with a caricature of the members of the Academy's Presidium. Having spent thirteen years in Cambridge, Kapitza probably was not aware of the politics in and around physics that had brought Vavilov into the Presidium. He must have known about the discovery of Cherenkov radiation but underestimated its importance. It came about almost exactly as Kapitza's rather disparaging description suggests: as a result of replacing the incoming beam of ordinary light in the experimental set-up with the gamma beam from a radioactive source. Kapitza apparently was one of the skeptics who doubted the validity of FIAN's discovery.

On the surface, it looks as if Vavilov never sought administrative positions but simply accepted offers and followed orders from above. This should not be surprising, for the Soviet cultural pattern of career building required just such self-effacing patience. Public rituals and moral norms strongly prohibited open demonstration of interest in higher positions and overt self-promotion; such behavior could only backfire. Officially acceptable was a fatalistic stance based on the assumption that the genuine accomplishments of someone who did his best in his present position would not escape the attention of the authorities and coworkers. In due time—in politics just as in the military—an offer of a more responsible post would be forthcoming. A candidate then had to express modest doubts about his abilities but not demur too long, because the offer from above was usually not much different from an order.

In actual practice, of course, careerists could maintain a public stance of ritual indifference while unofficially complementing it with active self-promotion and lobbying. Regarding Vavilov, however, nothing of this sort has been found in archives or recollections. He seems to have taken the cultural norm seriously and followed it, being a man of duty and responsibility rather than one who made demands. Apparently, he did not want a greater responsibility than directing a research institute and did not strive for a higher position. In 1945, however, he found himself in charge of the whole Academy.

3. The Call

> "The position of the Warden and later the Master of Mint was gradually transforming a modest Cambridge professor into a courtier and grandèe ... Worldly honors rained down on Newton as the court turned its attention towards him ... In 1703 Newton became the President of the Royal Society, a position he held until his death. In 1705 he was knighted by Queen Anne; he was called "Sir Isaac," took part in various parliamentary and ministerial commissions, visited the court, and became a salon philosopher"(Vavilov 1945, 188, 170).

The May 1945 victory over Germany was followed in the Soviet Union by a series of grandiose celebrations. The Academy shared the general mood as it convened in June to mark its 220th anniversary. The seventy-five-year-old president, Vladimir Komarov, presided over the jubilee meetings. Just a fortnight after the end of the celebration, on 14 July, he suddenly submitted to the Presidium a formal letter of resignation and proposed Vavilov as his successor. Though Komarov's health was poor, this was a forced and unexpected resignation, and it is not clear exactly what triggered it. Perhaps it was Komarov's 7 July letter to the Party Central Committee denouncing the academician-secretary N. G. Bruevich. A direct complaint against a subordinate was against the rules of Soviet bureaucratic intrigue and must have been interpreted by the authorities as a sign of Komarov's inability to control his organization.[9]

---
[9] Komarov to Secretary Georgy Malenkov, 7 July 1945, (CPA, 17-125-359, p. 86). For Komarov's resignation and recommendation see "Protokol obshchego sobraniia Akademii Nauk SSSR, 17 iiulia 1945 g." in Vavilov's personal dossier (ARAN, 411-3-123). For discussion of other possible reasons for Komarov's resignation see (Esakov 1991).

President of Stalin's Academy

The Academy convened urgently on 17 July to elect a new president. There was only one candidate; everyone knew—but no one mentioned—that he was approved by the highest political authority. Elections were a play with a rather strict script and rules: though the politicians directed the action from behind the scenes, all the public performance was done by the academicians. Komarov was not present; a member of the Presidium informed the meeting of his resignation and nominated Vavilov for the post, while several other academicians supported the proposal in their speeches. Vavilov did not have to do or say anything before the vote, in accordance with the fatalistic model of career building. Only once elected would the new president express his gratitude in ritualistically modest words (Obshchee 1945).

Despite silent but definite political approval and unanimously positive discussion, Vavilov received only 92 of the 94 votes cast. According to folklore, Kapitza lobbied against Vavilov's candidacy; he must have been one of the two who voted negatively. One document related to the search for a new president has surfaced from the formerly classified archives. It contains a list of short personal characterizations provided to the Politburo from the security police files and pays more attention to the morals of academicians and their reputation among colleagues than to their political orthodoxy, as this sample suggests:

> Academician B[ardin I. P.]—prominent specialist in the field of metallurgy. Seldom meets colleagues because of the excessive stinginess of his wife;
> Academician Vavilov S. I.—physicist. At the peak of his abilities. Brother—Vavilov N. I.—geneticist, arrested in 1940 for wrecking in the field of agriculture, sentenced to 15 years, died in Saratov prison;
> Academician V[inogradov I. M.]—is respected only among mathematicians. Bachelor. Drinks alcohol in considerable quantities;
> Academician Lysenko T. D.—not a party-member, director of the Institute of Genetics. President of the Agricultural Academy, was awarded Stalin Prize two times. Academician Lysenko does not enjoy the respect of others, including the president Komarov. Everybody believes that Vavilov N. I. was arrested because of him.[10]

Nikolai's tragic fate certainly received attention when Sergei was considered for the Academy presidency. Mid-level authorities would not have

---

[10] (Volkogonov 1989, 132). In September 1964 KGB conducted a special investigation of its files and reported to the party leadership that Lysenko was not responsible for Nikolai's arrest (Rossiiskii Gosudarstvennyi Arkhiv Noveishei Istorii, formerly Archive of the Central Committee, hereafter CCA 89-65-12).

The official photo released on the occasion of Vavilov's election to the presidency of the Academy of Sciences in July 1945. The photo actually dates from 1941, when Vavilov served as the State Plenipotentiary for Optical Industry. In that position he was responsible for the entire production and supply of optical equipment and devices for the needs of the fighting Red Army and answered directly to the State Committee of Defense, the war-time emergency government.
[Source: *Vestnik AN SSSR* no. 7–8 (1945).]

taken responsibility for promoting a close relative of a convict, but Stalin and the Politburo could do as they pleased; indeed, it could be argued that loyalty that had survived a relative's arrest had thus been proved.

Quite another set of sources also helps to explain Vavilov's perceived suitability for the role. In the culture of high Stalinism there was a canonical aesthetic image of the great scientist, best represented in the cinematography. In most respects this figure resembled a survivor of the "old regime" whom the cultural revolution of the late 1920s had tried to eliminate: he was bearded, old-fashioned in clothes and manners, involved in somewhat sacred activities, and not very well connected with everyday practical life (in comedy often absentminded). The great movie scientist of Stalin's era differed from the 19th-century type or from the "bourgeois professor" in just one crucial detail: he used convenient occasions to deliver laudatory remarks about Soviet rule and life, present and future. Since Stalin watched and approved Soviet movies before they started running publicly, this image can be taken as matching his own perception of the prominent scientist. Vavilov fits the type quite well, especially in his polite, old-fashioned manners and glorifying speeches:

> A black suit, a tie of the same color and a white shirt with a straight collar was the clothing of Sergei Ivanovich. He was always dressed in this manner. Even on vacations in the Crimea he retained his usual tightness and a starched collar. Only at his dacha could he allow an exception, and a light silk shirt came to replace the suit (Keler 1961, 115).

The differences between Vavilov and the movie hero scientist were that he was younger, only fifty-five, wore a mustache instead of a beard, and was an efficient administrator. These deviations could be excused: after all, the president of the Academy was traditionally elected for a term of many years and Vavilov could age into this role. He was also expected to perform administrative as well as ceremonial tasks.

There is a story developed by his physicist colleagues about why Vavilov accepted the presidency:

> Recently Yakov Alpert...told me the following story (which he had learned from Leontovich, who supposedly heard it directly from Vavilov). Vavilov had been informed...that there were two acceptable candidates for the post of president of the Academy: Vavilov was the first choice; if he didn't accept, it would be Lysenko. Vavilov sat up the entire night pondering his reply, smoking through several packs of cigarettes, and decided to accept the post—thereby saving the Academy and Soviet science from the devastation Lysenko's election would inevitably have caused.... According to Evgeny Feinberg, the other candidate was not Lysenko but Andrei Vyshinsky, the former Procurator General—which sounds somehow more likely, and even more frightening (Sakharov 1990, 77).

To let a candidate think overnight before accepting a job could well be part of the accepted ritual, but blackmailing as a style of job offering certainly was not. The legend too obviously reflects physicists' fears of what could have happened had Vavilov not taken the job, their view of Vavilov and Lysenko as opposing figures, and their desire to find an excuse for

Vavilov's subsequent compromises. We should not expect any of these thoughts to be shared by the communist leaders who made the offer. Even physicists must have developed them only later, after 1948, when Lysenko became a powerful figure and Vavilov was making his most shameful speeches and appearances. The mood and political situation in 1945, when Vavilov took on the presidency, were very different.

Amid the loud celebrations at war's end, the public silently hoped that the country's easier situation would make the regime looser. Signs of that optimism can be found scattered through various documents, including those related to physics. In May 1946 theoretical physicists from FIAN submitted an official proposal to Vavilov, in his capacity as president, arguing for the development of international exchanges, conferences and foreign travel.[11] There were encouraging signs: after 1943 Soviet papers started appearing again in British and American journals, and the 1945

Vavilov inspecting the ruins of Pulkovo Astronomical Observatory near Leningrad, 1946. During the siege of Leningrad, the battlefront ran through the site of the legendary observatory, which was completely destroyed by continuous shelling. Its reconstruction became the chief priority for the post-war Academy of Sciences.
[*Courtesy*: RGAKFD.]

---

[11] I. E. Tamm, D. I. Blokhintsev, and V. L. Ginzburg, "O meropriiatiiakh po razvitiiu teoreticheskoi fiziki v SSSR" (ARAN, 596-2-156).

jubilee meeting of the Academy was the first Soviet conference since 1937 to which foreign scientists were invited. Similar hopes show through Vavilov's writings in the first year of his presidency and even into 1947 and must have shaped his interpretation of the offer in the summer of 1945. But the optimistic signs were short-lived. It was soon Vavilov's role to follow the ever-stricter standards of Cold War politics, while still trying to calm militant tendencies and to moderate ideological discussions in science.

4. Political Profile

"Prominent physicist. Many of his works are known abroad. Right-leaning (*pravo nastroen*), but lately strives to work with us. Gave talks at a meeting of physicists–materialists and at the courses of the Communist Academy. Readily contributes to our press" so wrote the militant communist Aleksandr Maksimov about Sergei Vavilov in October 1929 in a confidential report to the party Central Committee that detailed the political, scientific, and personal characteristics of the Moscow University physicists (Andreev 1993). Maksimov certainly had good instincts for class distinctions; however, his term 'right-leaning' could represent quite a broad spectrum of political positions—anything to the right of the official party line as it was at the time. Given the lack of any document that can directly attest to Vavilov's political views (unless one wishes to take literally what he says in his ritualistic publications), interpretation of his writings must involve reading between the lines and literary criticism. After all, Vavilov was a fine stylist, and even his deliberate omissions are themselves eloquent.

In the late Soviet years colleagues spoke of Vavilov only as a scientist and administrator; they did not want to talk about his political writings. An earlier posthumous biography of 1954, however, gives a full appraisal of Vavilov's political activity, describing him as a patriot and a convinced and outspoken dialectical materialist. In addition to presenting quotations from his official publications, the biographer—a close friend and lifetime collaborator—mentioned one fact about Vavilov's early, pre-revolutionary years: in the upper class of the Gymnasium he read Lenin's major philosophical book *Materialism and Empiriocriticism* (Levshin 1954, 9). The apparent source of this information is an autobiographical manuscript that Vavilov began in 1950, in the last months of his life, after a serious illness. Frank avowals were dangerous in 1950, even in private diaries or confidential talks. For officials of higher rank, especially, almost no private sphere was left. Hence Vavilov's formulations are careful; still, he is not as cautious here as in the papers he wrote for publication:

> We were reading—or pretending to read—brochures of Marx and Engels, Bebel, Dietzgen, empiriocritical essays of Carstanjen, Lunacharsky. In 1909 I bought *Materialism and Empiriocriticism* by V. Il'in [one of Lenin's pseudonyms—AK], there are still some comments of mine in the margins of the book. I had no idea of who Il'in was, but of the ideological brawl between materialists and empiriocritics, I had a complete notion, much better than people usually have now (Vavilov 1981, 93).

What is striking is the seeming lack of judgment. Vavilov includes no standard enthusiastic appraisal of Lenin's book, but instead indicates indifference with respect to what were then regarded as right and wrong philosophies. Searching for further details, we find meaningful word choices: the Russian *potasovka* (brawl) has a connotation of disdain. In everyday language, it can be used to refer to the romping of children or drunks. In Soviet political slang, it could be applied to a conflict between aliens or enemies, but not to a serious struggle between good and evil. To publish such a comment during Vavilov's time would have been quite risky. Even the posthumous biography of 1954 proclaims Vavilov's early interest in Marxist philosophy but does not provide the quotation (Levshin 1954, 9). The sensitivity to nuances and the attentiveness of critical readers, however, decreased with the years. A 1961 biography provides a partial quotations, without its dubious second half (Keler 1961, 33). A 1981 memorial volume publishes the entire quotation as if there were nothing suspicious in it (Vavilov 1981, 93).

In the 1922 preface and commentary to a popular book on Einstein's relativity theory, Vavilov expressed some of his genuine views on the relationship between physics and philosophy. In physics he seems a radical empiricist, accepting the equations of relativity as straightforward generalizations of experimental facts and downplaying its philosophy of space and time as preliminary and speculative. Vavilov follows Ernst Mach in his skepticism towards theoretical conceptions in science and in paying attention to direct sensory experience and measurements, but not in disregarding philosophy per se. "Philosophical space and time are not subjected to physicists' 'abuses,'" he writes, thus securing an independent domain for philosophy (Auerbach 1922, 148). The little he said is not enough to identify his philosophy precisely, but it is safe to conclude that he was probably neither an empirical positivist nor a dialectical materialist.

The autobiographical notes hint at some of Vavilov's political values, too. These emerge throughout his description of events in pre-revolutionary Russia, which continued to have direct political meaning for the people of 1950. Vavilov's father was born to a peasant family and grew to be a rich merchant.[12] The family lived in their own house in a district of Moscow called Presnia, which became the center of violent fighting during the uprising of Moscow workers in December 1905. Almost every pupil in the upper high school classes read forbidden socialist literature. Vavilov's recurrent trope is represented by phrases like his "pretending to read" above: "My own attitude was not clear yet. Left, took part in the construction of a barricade, had torn to pieces a portrait of tsar, hid leaflets, but all this was a children's game"; "[in discussions with my father] I was a defender of socialists (but understood very little, like him)"; "I was writing a charter for a sort of a *kruzhok* (circle) and was concocting, without understanding anything, a paper on socialism" (Vavilov 1981, 100, 102).

What an official commentator would have emphasized as early manifestations of true political beliefs Vavilov disavowed by the refrain "I did not understand anything." His clearest passage states:

> "As far back as I remember myself..., I always felt 'left,' 'democrat,' 'for the people.' ... But my leftism and democratism never went into politics, its harshness and even cruelty (I understood their objective necessity but could not proceed from thoughts to deeds). At present this is usually called *'miagkotelost'*. My organic *bespartiinost'* originates here. The revolution of 1905 had frightened me. I threw myself into science, philosophy, the arts. In this state I arrived in 1917" (Vavilov 1981, 102).

The two words left without translation in the quotation are labels from Soviet political slang. The first can be approximately translated as 'flabbiness'; it denoted inability to make difficult and responsible decisions. The second, in this context, meant not only "not being a party member" but in literal sense psychological or ideological inability to meet the high standards of the party and to internalize its spirit. Either handicap was a permissible failing for a lay person, but they could serve as accusations against officials. Though not as serious as the crimes of 'political mistakes' or 'wrecking,' they could be used as a basis for demoting an administrator. It was not uncommon for Vavilov to express a milder self-critique of this sort in public.

---

[12] Ivan Vavilov left the country in 1918, settled in Berlin, and died during a visit to Leningrad in 1927. As his social origin (a very important point in Soviet questionnaires), Sergei Vavilov usually indicated *sluzhashchii* (white-collar employee), which represented a middle stage in his father's career. Apparently this created for him neither privileges nor serious problems.

"The revolution of 1905 had frightened me" is a typical Soviet reference to a widespread mood among Russian intelligentsia, the symbol of which was the 1909 volume *Vekhi* (Landmarks). After the failed revolution, many of those who had sympathized with revolutionary parties began to view radicalism as destructive and the prospect of revolution as tragic. Vavilov's nostalgic description of life in Presnia suggests that even in 1950 he continued to see revolution this way. He comments that [church in the neighborhood] "is ravaged now and turned, probably, into a club." The word *razorena* (ravaged) implies regret and moral reproach, in contrast to the neutral *razrushena* (destroyed) (Vavilov 1981, 82). Such a sentence would be typical of post-communist, not loyal Soviet, rhetoric.

In 1921, after three years of civil war, a new generation of political writers argued from abroad in *Smena Vekh* (Changing Landmarks) that since the tragedy had actually happened and the Bolsheviks had proved to constitute the only possible stable government, their regime should be accepted as a fait accompli and destructive opposition abandoned in favor of constructive collaboration. There were many among the "bourgeois specialists" who openly supported this position, but the attitude of political compromise for the sake of the country lasted only until 1928. After that point, collaboration was possible only on the basis of a declaration of full and sincere support for the regime. These were well-known dilemmas, which Vavilov certainly faced, but he was never able to describe his post-revolutionary development. The last of his published autobiographical notes is dated 11 January 1951—two weeks before his death—and it still deals with his school years. But even the available text reflects Vavilov's uneasiness with the matter of revolution, and we can infer that he had difficulties in accepting one of the most basic of Soviet political values. Though he was president of the Soviet Academy of Sciences, his political position seems to have been one of conscious mimicry rather than unconscious adaptation and acceptance of the values of the society of the time. If it was mimicry, it was perfectly done, for throughout his life nobody dared openly question his orthodoxy.

Additional support for this conclusion comes from Vavilov's biography of Newton. What is of interest here is not the validity of Vavilov's picture of Sir Isaac but how he wanted to portray his hero: what questions he was interested in and what features he sympathized with. The opening statement reads:

> "Newton was born in the year of the great Civil War..., witnessed ... the execution of Charles I, the rule of Cromwell, the restoration of the

Stuarts, the second...revolution, and died under stable constitutional rule...But political storms apparently could not influence his life greatly. He remained, at least on the surface, an apolitical 'philosopher'" (Vavilov 1945, 9–10). A few pages later, describing Newton's life strategy, Vavilov quotes in Russian translation his letter to Francis Aston, which reads "When you come into a fresh company, 1, observe their humours; 2, suit your own carriage thereto...3, let your discours bee more in Quaerys & doubtings yn peremptory assertions or disputings...You will find little or noe advantage in seeming wiser or much more ignorant yn your company. 4, seldom discommend any thing though never so bad...Tis safer to commend any thing more then it deserves yn to discommend a thing so much as it deserves" (Newton 1959, 9–10).

He goes on with a commentary: "Some biographers describe this letter as a naive piece or even a joke on the part of the young Newton...This conclusion is unjust. Tactical rules...may seem naive because of their old-fashioned expression, but they are in fact very far-sighted and have not lost their importance and usefulness even now." Vavilov then quotes, for the second time, Newton's third maxim. Both times, his translation contains an almost Freudian mistake: I shall provide the back translation from Russian, with the mistake emphasized, instead of the original Newton's text: "You will find little or no advantage in seeming wiser or *less* ignorant than the society where you find yourself" (Vavilov 1945, 27–28).

5. Political Writings and the Ideological Image of Soviet Science

"Newton's theological and historical studies can be viewed as an inevitable tribute to his time....Newton's theological production...did not differ much from what was done by others...and did not bear a sign of individuality" (Vavilov 1945, 10).

Although Vavilov might have thought the same about his own philosophical and historical studies, I would not join him in that view. Of course, once one had taken the job of writing on politically charged matters one had to pay tribute to the existing canons. Soviet political texts of the Stalin era consisted in large part of stereotypical statements. The most important thing for interpreting those texts is to understand where the ritualistic part ends and improvisation begins. Contemporaries were, in one sense, better prepared for the task of interpretative reading than we are, for they used to take the daily *Pravda* as a cryptogram and try to decode the shifts in politically correct language. Although we lack their feel for nuances, we do have a broader historical perspective. In Vavilov's political writings, one can find him following the rules, playing around with them, delivering a personal message, and—since he himself was a political authority—setting new models.

Political publications were not so much a prerequisite for advancing one's career (they were this too, but not so decisively as is often thought) as the necessary consequence of holding an administrative position above a certain level. One could try to avoid both political career and writings, but Vavilov chose another course, and the quantity of his political publications far exceeds the sufficient minimum appropriate for his status. Apparently he did not hesitate to write what was required at a given moment, whether or not he believed it to be true. Still, his writings deviate considerably from the 'average,' with respect to both the topics he preferred to handle and how he dealt with them.

As an administrator, Vavilov did much for military and industrial applications of science, especially in his role as scientific director of the Optical Institute. During the war, he served as state plenipotentiary for the optical industry, coordinating the production of military optics for the Red Army. In peacetime, he pioneered the development of fluorescent lamp technology. As the Academy's spokesman, however, he argued to secure a niche for the 'theoretical' or 'abstract' sciences (the term 'pure science' was mainly reserved for pejorative use in Soviet discourse), stressing their relationship to important philosophical questions and their contribution to the international prestige of the country (Vavilov 1946, 38). Most of Vavilov's political writings are devoted to the philosophy and the history of science.

The most important characteristic of Vavilov's personal style can be described by one of Newton's maxims: "Seldom discommend any thing though never so bad ... [It] is safer to commend any thing more than it deserves [than] to discommend a thing so much as it deserves." Here was Vavilov's true reservation when compromising with political realities; he was ready to make any positive assertion (whether it originated from political authorities or expressed his own opinion), but he tried to avoid making critical judgments. We may call this 'politeness' or, following Kapitza, 'polish.' The neoclassicist aesthetics of late Stalinism gave rise to a political rhetoric of elaborate laudations and severe denunciations. Vavilov specialized in the former genre: he was quite inventive in the laudatory discourse while being unoriginal in or toning down the militant one. Because of his reputation as a lavish orator, he is sometimes credited with inventing one of Stalin's semiofficial titles, "coryphaeus of science." This particular attribution must be wrong, however, because the term was in use already before Vavilov became the chief rhetorician of the Academy.[13]

The history of science was considered an ideological discipline within the Soviet system at least until the 1960s. Vavilov's personal interest in this field coincided with the desires of politicians. In the 1920s he translated Newton's optical works and wrote several papers on the history of optics. To write popular, historical, and philosophical essays on science was the scientist's public duty, amounting to a formal obligation. FIAN covered this activity in a separate section of the institute's annual report to the authorities, consisting to a large extent of lists of Vavilov's works—as usual, he was taking on the role of a protective mediator between physicists and the demands of the outer society. During the war FIAN and the Optical Institute were evacuated to two cities in the Volga region, Kazan and Yoshkar-Ola. Vavilov commuted between the two, organizing the production of military optical equipment. In the midst of all this, at the most critical moment in World War II—the battle of Stalingrad in the fall of 1942—he wrote... a full-length biography of Newton. This was a political commission, too—the celebration of Newton's 300th birthday was a friendly gesture towards the British ally—and the necessary books were flown to Vavilov from Moscow and even from besieged Leningrad.

During the last years of Stalin's rule and after the Soviet military victory over Nazi Germany, nationalism became a political priority. Vavilov reflected this trend in his typical constructive way, writing about the history of the St. Petersburg Academy of Sciences and Mikhail Lomonosov in the 18th century and appraising the pre-revolutionary achievements of Russian science (something that had been almost forbidden during the 1920s). The history of Russian science became a major ideological preoccupation of the Academy and the topic of a special session in January 1949 (Voprosy 1949). Vavilov's appraisals of Lomonosov followed the rise of the nationalist tide, culminating in 1949 when he accepted the name "The Lomonosov Law" for the title of an article he published in *Pravda*. In the article itself, however, he masked behind baroque phraseology the absence of a definite claim for Lomonosov's priority in discovering the laws of matter and energy conservation. Not making such a claim—which, in reality, was based on an extremely free interpretation of a letter

---

[13] The 1939 official protocol of Stalin's election to honorary membership in the Soviet Academy of Sciences had proclaimed him "the greatest thinker of our time and the coryphaeus of vanguard science" (Obshchee 1939, 2–3).

from Lomonosov to Leonhard Euler—was then regarded as utterly unpatriotic.[14]

In July 1945 Vavilov assumed the presidency of the Academy and with it the responsibility of representing the whole of Soviet science to the political authorities. One of his public duties was developing a laudatory, protective image of Soviet science in a rapidly worsening ideological climate. His public addresses of 1946 clearly reflect those profound changes.

Until the summer of 1946, despite diplomatic tensions and Stalin's angry reply to Churchill's speech, in which the phrase "Iron Curtain" was coined, Soviet propaganda continued working for the preservation of the Grand Alliance. It restrained from criticizing the Western allies explicitly, and representations for both internal and external consumption pictured the Soviet Union and its allies as democratic countries, except that Soviet democracy was of a truer, better kind. Democracy was the main slogan of the elections to the Supreme Soviet in February, and Vavilov's pre-election speech, given in January, sounded all the right notes. He proclaimed that the joint democratic forces—of which the Soviet democracy was the most consistent—had achieved a joint victory over the fascist tyranny. A new era of "strengthening the role of science and democracy in the life of peoples" was beginning. "Science serves progress only when combined with democracy."[15]

In July, Vavilov mentioned dialectical materialism as the basis of Soviet science, and in November he had to invent a substitute for the already risky term 'international science,' writing instead that "knowledge is indivisible," and that this unity of science is one of the criteria for distinguishing between truth and falsehood. However, "Soviet science is not just a part of world science, but a science with a particular order and character." Its four main specificities, in Vavilov's list, were: democratism (open to the people and serving society and the state); "deep ties between abstract thought, theory and practice"; collectivism; and "full clarity regarding its philosophical world view."[16]

---

[14] Vavilov used the following elaborate praise in lieu of an explicit priority claim: "As if Lomonosov confined in common brackets for all centuries ahead all types of the conservation of matter properties" ("Lomonosov na vse veka kak by vzial v obshchie skobki vse vidy sokhraneniia svoistv materii") (Vavilov 1949a). See an appraisal of this Aesopic language by a contemporary in (Frish 1992, 338–339).

[15] Vavilov, "Soviet science in the service of the Motherland" (January 1946), in the collection of Vavilov's political speeches of 1946 (Vavilov 1946, 7–34, on 32–34).

[16] Vavilov, "Scientific problems in the next five-year plan" (July 1946), and "Specificities and prospects of Soviet science" (November 1946), both in (Vavilov 1946, 71–90, 35–51).

The party decision to change the tone of propaganda and increase the role of ideology in response to the intensifying Cold War was made public in June–July 1946. The transformation of the allies into the most probable enemies and the transition from celebration to confrontation had to be explained to Soviet audiences through a number of model publications orchestrated by Agitprop—the Department for Propaganda and Agitation of the Party Central Committee. The new emphasis on ideology became widely known under the name *Zhdanovschina* (after Andrei Zhdanov (1896–1948), the Secretary of the Central Committee and Stalin's newest favorite). The highly publicized start of the campaign came with the August–September 1946 party decisions on literary journals, theater, and movies. But even before that, in July and August, newspapers had started criticizing the Academy's institutes in economics and law for a tendency to view favorably the economics and politics of Britain and the United States.[17] Apparently because of this critique, issue 8–9 of the Academy's official journal *Vestnik Akademii Nauk* was delayed in publication and, when it finally came out, reflected tightening ideological requirements. This marked the shift from the celebratory mood of 1945 to a critical one that was becoming increasingly militant in tone.

The international confrontation grew faster than the audience learned the lessons. In early 1947 a political showcase was made out of an attempt to arrange the American publication of a book by two Soviet medical researchers. The so-called case of Kliueva and Roskin was described as trading state secrets. The Central Committee ordered the Ministry of Public Health to elect a special "court of honor" (*sud chesti*) aimed at the moral and administrative punishment of unpatriotic behavior. In October 1947 a similar court was established in the Academy of Sciences. On this occasion Vavilov delivered one of his most shameful talks, "On the Dignity of the Soviet Scholar," in which he stated that it was unpatriotic to provide a possible enemy with any valuable scientific information and that the political campaign was prompted by "a possible attack on Soviet territory."[18]

With the strengthening of ideological demands, Vavilov's description of Soviet science changed also. He presented a modified list of its main characteristics in the address "On Stalin's Scientific Genius," given at a

---

[17] *Kul'tura i zhizn'*, 20 July 1946, 10 August 1946; *Pravda*, 12 August 1946.
[18] On the Kliueva-Roskin case see (Esakov and Levina 1994; Krementsov 2002). The complete 'internal' version of Vavilov's talk was published only recently (Prostovolosova 1991). The publisher was not aware that a shorter 'civilized' version had been published long before as (Vavilov 1950b).

public celebration of Stalin's seventieth birthday in December 1949. Democratism was replaced with *narodnost'* (folkishness)—that is, science of the people and for the people. Collectivism gave way to *partiinost'* (party-mindedness)—that is, science that serves the right political and social interests, which are formulated by the party (Vavilov 1949b).[19] It is important that there was no mention of either "bourgeois" or "proletarian" science among these theses. The class issue, which had played the central role in Soviet ideology during the cultural revolution, had become a rather marginal, pro-forma topic by the late 1940s. Vavilov, though much less a sincere Marxist, was a more sophisticated ideologist than Trofim Lysenko. Lysenko's 1948 manuscript "The Situation in Biological Science" contained a strong portion of class rhetoric, so that its political line had to be corrected by Stalin's personal editing (Rossianov 1993a; 1993b).

Vavilov was a non-communist, most probably not even a sympathizer, who happened to become an exemplary Stalinist politician. To make sense of such a case requires a non-standard combination of explanatory resources. The preceding story shows how the cultural revolution, though designed to promote communists into scholarly ranks, also advanced Vavilov into the Academy; how the incorporation of the conservative tendency into the modernizing regime transformed the Academy into something like a ministry of science and Vavilov into a high-level administrator; and how moral and aesthetic criteria, more than political considerations, were important for his appointment as president in 1945. Vavilov's case, though unique, is also revealing, as it helps to understand Stalinism not simply as a totalitarian dictatorship but as a society and a culture with specific rituals, mores, and styles.

The careers of the two Vavilov brothers are often used to symbolize the different fates of biology and physics under Stalin's rule. In fact, they better represent two different historical periods. Nikolai even in his childhood was a street fighter and an atheist rebel (Frank 1981, 86). He sympathized with Bolshevik goals, and he rose to power in the revolutionary 1920s, putting forward grand and radical tasks and pursuing them with energy and vigor. He also made compromises, but he was also accustomed to taking huge risks. Nikolai's moment passed with the cultural revolution, and some change in leadership had to occur. His counterpart in physics, Joffe, also lost his dominating role. But in physics, instead of a

---

[19] *Partiinost'* was already mentioned in the November 1946 speech, but not among the four main specificities. It determined then "the size, the direction of growth of the Soviet science and, first of all, its deep, organic *democratism*" (Vavilov 1946, 37).

figure like Lysenko, Sergei Vavilov and Kapitza—each in his own way—took over the job of representing the field to politicians.

Sergei, at least in his early years, was religious (Frank 1991, 19) and politically far from revolutionary, and his career developed in the more conservative late 1930s and '40s. He did not strive for great ends and high positions, but he accepted responsibilities dutifully and fulfilled them—as he wrote that Newton did—"honestly and severely." He learned how to become a perfect Stalinist politician by mastering the language and games of this culture. On the surface, his political writings appear entirely conformist in language and meaning, but when approached as texts conforming to specific genre rules they reveal other hidden messages as well. Hardly anyone could have performed his role better than Vavilov did, although personally he would have probably preferred sitting in his dark optical laboratory during the day and reading poetry before bed.

Wearing a mask was not an easy duty for Vavilov. Both political conformism and taking care of science were among his duties. Sometimes these duties came into conflict, and in these cases Vavilov faced hard decisions and moral compromises. To this extent, then, he also chose to become a Stalinist politician. Some of the most difficult choices he had to make came with the series of fierce ideological discussions in Soviet science in the late 1940s, which resulted, in particular, in an ideological ban on Mendelian genetics as supposedly "idealist," "formalist," and simply wrong science. As the next chapters will show, those events required from Vavilov a mobilization of all his skills as a master of grand philosophical rhetoric, as well as as a specialist in quiet bureaucratic intrigue.

Chapter 8

Games of Soviet Democracy

"It is generally recognized that no science can develop and flourish without a battle of opinions, without freedom of criticism."
(Stalin 1950, 427)

Following the end of the World War II, science moved to the top of the list of state priorities in the Soviet Union. This new importance was not limited to nuclear physics and other fields related to military projects but embraced all fields of scholarship, or *nauki* in the Russian sense. *Uchënye* (scholars in this wider sense) came to form an elite social group next to party apparatchiks, industrial administrators, and the military and became more privileged than engineers. In material terms, this change of status was decreed by the Council of Ministers in March 1946 (O povyshenii 1946). Not only resources for research but also individual salaries were raised higher than at any other time in Soviet history. As in the case of other elites in Stalinist society, with increased privileges came increased dangers, and with attentive care tighter control.[1] As an elite group, scientists came into a closer dialogue with politicians and accepted some of their values, language, and games.

---

[1] On the notion of elite group extended to intelligentsia see "Introduction: On Power and Culture" in (Fitzpatrick 1992, 1–15).

One result of the increased concern with science was a series of ideological discussions in various scholarly fields that took place during the last years of Stalin's life. In particular one of these events—the denunciation of genetics by Lysenko in 1948—has become universally known as a symbol of the ideological dictate in science and its damaging consequences. In a standard textbook on the history of Russian science, Loren Graham (1993, 121–122) summarizes the existing consensus as follows:

> "Under Stalin, dialectical materialism was used to terrorize scientists. If a certain scientific theory was branded by Stalin's ideologists as "idealistic" or "bourgeois," the scientist who continued to defend that theory immediately came under the suspicion of political disloyalty. The scientist would be subjected to humiliating criticism at political meetings in his or her institute or laboratory. Demotion or imprisonment (often the same as death sentence) were clear possibilities... The episode of this period best known in the West is Lysenkoism in genetics. Lysenkoism was only the most extreme of many manifestations of philosophical dogmatism and political oppression under Stalin."

Elsewhere, Graham demonstrates that the relationship between science and ideology should not be reduced to simple oppression (Graham 1987; 1993, ch. 5). Earlier chapters of this book also discussed several cases that necessitate a more complex approach to the problem. Collectivist ideology provided useful heuristic metaphors for Frenkel and other theorists in their search for better mathematical models for the interactions between atoms and electrons in dense bodies. Revolutionary allusions helped stir an enthusiastic reception for the radical new theories of relativity and the quanta in the Soviet society of the 1920s. Boris Hessen, Sergei Vavilov, and others used philosophical arguments from Marxist dialectics to build justification and support for novel developments in 20th-century science.

This chapter reexamines the infamous Lysenko affair and concludes that the simple model of an ideological purge does not work even for this primary example. Typical generalizations about the suppression of science by Soviet ideology equally do not hold up against a wider empirical background if one brings into consideration a variety of ideological discussions in Soviet sciences in the 1940s. A more flexible interpretation will be developed that relies on the analysis of specific social rules that governed debates in sciences and the behavior of discussants. This model will account for different ways in which ideology was used in scientific discussions and for the actual diversity of results achieved.

1. The Campaign of Ideological Discussions in Sciences, 1947–1952

Science and ideology, especially Soviet ideology, are typically seen in a conflict mode. It is often explained that in the years following World War II the Stalinist leadership launched an ideological and nationalistic campaign aimed at the creation of a Marxist-Leninist and/or distinctively Russian, non-Western science. Concepts and theories that were found idealistic or bourgeois were banned and their supporters silenced. In no other science was this process completed to the same degree as in biology after the August 1948 Session of the Soviet Academy of Agricultural Sciences, at which Trofim Lysenko (1898–1976) declared the victory of his "Michurinist biology" over presumably idealist "formal" genetics. This resulted in an effective ban on research on Mendelian inheritance and chromosome genetics in the Soviet Union, which officially lasted sixteen years. The August Session, in turn, served as the model for a number of other ideological discussions in various scholarly disciplines (Krementsov 1997, 193).[2]

This widely accepted interpretation, however, encounters two serious difficulties. The first arises from a selective focus on one particular debate which best fits the stereotype. It was the critics of the Stalinist system who singled out the Lysenko case as the most important and symbolic example of the application of Soviet ideology to science. The Soviet communist party viewed things differently. It did regard the 1948 event as a great achievement of the party's ideological work and a major contribution to the progress of science (until 1964, when the mistake was quietly acknowledged). But what is more interesting, and less expected, is that communists claimed not just one but five, major ideological successes of this sort: discussions in philosophy (1947), biology (1948), linguistics (1950), physiology (1950), and political economy (1951) (Istoriia 1959, 606).[3] The additional four cases did not become as widely known outside the Soviet Union as the biological one, apparently because they did not fit as well the standard picture of the campaign as an ideological purge. Their effects on scholarship were not obviously damaging, their patterns and outcomes were much more confusing than those of the "clear" Lysenko case, and they did not

---

[2] David Joravsky (1989, 400–405) emphasizes the ideology of nationalism and anti-cosmopolitan policies, rather than dialectical materialism, but he also agrees that the aim of the campaign was the introduction of ideology into science and the drive towards a monolith.
[3] Later editions dropped the biological discussion as damaging to the party reputation. The same list of five main discussions can be found in other communist publications, for example (Bol'shevistskaia 1950; Zhdanov 1951; Iovchuk 1952).

present the critics of communism with such a perfect example of scandalous failure for use in Cold War propaganda.

The second difficulty concerns the apparent incoherence of events. No straightforward generalization based on the single case of Lysenko could effectively be sustained against a wider factual background. Those who assume that the goal of the campaign was to subordinate science to ideology disagree considerably on what constituted the ideology that had to be applied in the sciences. Indeed, many different and often contradictory ideological principles were pronounced, and none was consistently carried through the entire campaign. Dialectical materialist and Cold War slogans suffused the rhetoric, calling for unity in the struggle against idealism, cosmopolitanism, and obsequiousness before the West. At the same time, however, one also frequently encounters attacks on monopolism in science and encouragement of creativity and free criticism. David Joravsky has characterized this ideological mess as a "bizarre mixture of elements," "obvious self-contradiction" for "the outsider," and the "most astonishing incongruity in the Stalinist drive for monolithic unity" (Joravsky 1989, 405–406). At the same time, he notes that for Stalin there was no self-contradiction here.

These five particular ideological cases acquired the importance of a general political event and had to be publicized far beyond the circle of directly concerned scholars because they attracted the attention of Stalin, who participated in them either openly or behind the scenes. But even having been approved by the same authority, they still form a rather chaotic set in light of their varied conflicts, contents, and outcomes. Philosophers met in June 1947 to criticize a book by Georgy Aleksandrov, a high party official who, although demoted, was subsequently appointed to supervise the work of his critics (Diskussiia 1947; Esakov 1993). The August 1948 Session, as mentioned above, led to the banning of the internationally accepted approach in genetics in favor of an idiosyncratic and specifically Soviet version (Situation 1949; Joravsky 1970; Krementsov 1997). The linguistics controversy presents quite a contrast. In June 1950, after a series of polemical publications in *Pravda*, the candidate for Lysenkoism in linguistics— revolutionary and anti-Western Nikolai Marr's "new doctrine on language"— was silenced in favor of a quite traditional and internationally recognized comparative approach (Soviet 1951; Alpatov 1991). Conceptual disagreements in physiology were not so pronounced when in July 1950 representatives of this field gathered at a joint session of Academies of Sciences and of Medical Sciences. Nevertheless, the disciples of Ivan Pavlov fought a serious battle over which of them followed the orthodoxy of their deceased teacher more closely and should therefore direct his physiological institutes

(Nauchnaia 1950; Joravsky 1989). Finally, in November 1951 a closed panel of economists and politicians at the party's Central Committee discussed the project of a new textbook on political economy. This meeting apparently did not end in any definitive resolution, but it provided a pretext and inspiration for Stalin's last major theoretical opus, *Economical Problems of Socialism in the USSR* (Stalin 1952; Openkin 1991).

The variety already displayed in these most-controlled cases increases considerably when one takes into account dozens of other critical discussions reported in the press between 1947 and 1953. They could be as large as an All-Union conference and as small as an institute's meeting of an institute to review a monograph or a textbook. Political authorities at some level were occasionally involved, but most of the meetings were organized solely by academics. Ideological argumentation and accusations sometimes were used very heavily, whereas in other cases the discourse was almost scholarly in style and paid only lip service to political rhetoric. In the majority of episodes it is difficult if not impossible to classify the participants into two categories, such as "Lysenkos" and "true scientists." Disputes could reflect serious conceptual disagreements, but also institutional conflicts or merely personal animosities. Some critical discussions led to serious changes in the academic hierarchy, others only confirmed existing power relations. Their general effect on scholarship can be described as mixed: sometimes negative, sometimes, as in linguistics, more positive, and in many other cases largely irrelevant.[4]

Diverse patterns and results notwithstanding, these discussions taken together constituted a political campaign in the Soviet sense: several highly publicized model events and a number of local reactions and imitations. The very fact of holding a discussion already had a political meaning prior to what its particular outcome would be. The goal of this chapter is to understand what in this campaign made it look coherent to insiders, communist practitioners, even though it appears irregular and chaotic to us, cultural outsiders. It will be shown that while insiders generally misunderstood the consequences of their actions, outsiders usually have misconstrued the insiders' goals and intentions.

---

[4] The discussion on cosmogony was almost academic in style and did not involve politicians (Levin 1991). Changing the institute's director was apparently the main goal of a discussion in literary criticism (Protiv 1948). A major discussion in which the existing academic hierarchy did not allow a revolt to happen took place in chemistry (Graham 1987; Pechenkin 1994). The institutional conflict between Moscow University and the Academy of Sciences was the driving force behind the discussion in physics (Gorelik 1991).

Understanding the logic of a different culture—Stalinist culture in this case—asks for an anthropological approach. A regularity, indeed, can be found, but on the level of formal rules and rites of public behavior rather than in the contents and results of disputes. The analysis of the 1947 Philosophical Dispute—chronologically the first, and also the purest, performance staged by politicians themselves—will reveal the rules of the Communist games of *diskussiia* (disputation) and *kritika i samokritika* (criticism and self-criticism). An inquiry into the rituals of Stalinist political culture and its special domain called "intraparty democracy" will then be needed to understand both the ascribed functions of these games and the possible motivations of politicians who proffered them to scholars as methods for handling scientific disputes. Provoked from above, scholars engaged in a variety of academic conflicts while pursuing their own agendas and inventively using available cultural resources in dialogues with politicians. An important feature of these games was that, in theory and often also in practice, their outcomes were not predetermined but depended upon the play. How scholars interpreted and exploited this particular feature will be shown by a new analysis of the textbook Lysenko case of 1948.

The campaign of ideological discussions in sciences will thus be reinterpreted as the transfer of the rites of intraparty democracy from communist political culture into academic life. In the process, the rules of public behavior and, to some degree, rhetorical vocabulary were relatively stable, but they left sufficient room for the unpredictability and diversity that actual events displayed.

2. Exercises on the Philosophical Front

> "In Marxism perfectly / he could express himself and write, / admitted mistakes easily, / and repented elegantly." Soviet folkloric play on a line from Pushkin's *Eugene Onegin*[5]

Even in dictatorial and hierarchical Stalinist Russia, authorities were not exempt from grassroots criticism. On special occasions such criticism was not only possible but welcomed and even required. Soviet philosophers knew this when the Central Committee summoned them for a

---

[5] "On po-marksistski sovershenno/Mog iz"iasniat'sia, i pisal/Legko oshibki priznaval/i kaialsia neprinuzhdenno." Pushkin's original: "On po-frantsuzski sovershenno/Mog iz"iasniat'sia, i pisal/Legko mazurku tantseval/i klanialsia neprinuzhdenno."

representative gathering on 16 June 1947. Andrei Zhdanov, the Politburo member responsible for ideology and Stalin's current favorite, presided over the meeting and, in a few introductory words, informed the participants that their task was to discuss the book *The History of West-European Philosophy* by Georgy Aleksandrov (1908–1961). Having expressed the hope that "the comrades invited to the discussion will take an active part in it and will freely voice all critical remarks and suggestions," but stopping short of providing any more detailed instructions, Zhdanov opened the meeting and let the panel go (Diskussiia 1947, 6).

To understand the humor of the situation, one has to imagine oneself in the shoes of a rank-and-file philosopher who also had to be a party member and for whom Aleksandrov was the official authority, within both the profession and the party. Having not yet turned forty, Aleksandrov had accomplished an extraordinary career within the party apparatus. Zhdanov's protégé, he was appointed in 1940 director of the Department

Secretary of the Central Committee A. A. Zhdanov at the podium. Between 1946 and 1948, Zhdanov occupied the second position in the Communist Party hierarchy and presided over the meetings of the Secretariat. He emphasized the strengthening of the party's ideological work during the time of intensifying Cold War and international tensions. The resulting period of increased ideological activities, campaigns, and censorship in the fields of culture and science became infamous under the name *Zhdanovshchina*. Zhdanov's downfall was apparently caused by his, or rather his son's, ill-prepared attempt to intervene on behalf of genetics against the powerful biologist T. D. Lysenko.
[*Courtesy*: RGAKFD.]

of Propaganda and Agitation (Agitprop), which together with the Department of Cadres was the most important department in the Central Committee apparatus. The following year he was elected candidate member of the Central Committee and member of its Orgburo. Aleksandrov's philosophical publications were devoted to topics more scholarly than one would have expected from a party bureaucrat: Aristotle and pre-Marxist philosophy. In the fall of 1946 he reached the apex of his political career and added to it signs of academic recognition by receiving a Stalin prize for his textbook *The History of West-European Philosophy* and by becoming a full member of the Academy of Sciences. Zhdanov's rise to favor in 1946 and the renewed stress on ideological work put Agitprop, and Aleksandrov as its head, into the center of the party's political activity. In normal circumstances, he would be the one who would call in philosophers, scold them for mistakes, and deliver instructions on their job, while they would have considered it a great honor to be invited to publish a laudatory review of his book.

G. F. Aleksandrov, newly elected member of the Academy of Sciences in the field of philosophy, 1946. The future seemed bright for the up-and-coming party bureaucrat, but troubles were brewing in the following year.
[*Courtesy*: RGAKFD.]

At the June 1947 Philosophical Dispute, however, the roles were reversed, and philosophers were encouraged to develop a principled critique of Aleksandrov's book and its highly placed author. What sort of criticism was exactly expected was not obvious: the first attempt to engage in a serious discussion had already been made in January 1947 at the Academy's Institute of Philosophy. It had been prepared by Alexandrov's colleague from Agitprop Piotr Fedoseev, but the level of criticism failed to satisfy the Central Committee.[6] In Zhdanov's words, the first discussion was "pale (*blednaia*), skimpy (*kutsaia*), and ineffective." For the second try, Zhdanov himself presided over the meeting, and more participants, in particular from outside Moscow, were invited and encouraged to freely express their disagreements.

The audience fulfilled Zhdanov's hopes and demonstrated a great deal of activity. For more than a week, almost fifty speakers presented their critical comments on the book, and twenty more who had not been allotted time to speak insisted on including their texts as an addendum to the published minutes. Several remarks made it clear that the event was taking place because Stalin had expressed dissatisfaction with the book (Diskussiia 1947, 267, 289). Historian Vladimir Esakov has suggested that the entire chain of events was started by a letter of criticism, or denunciation, from one of Aleksandrov's foes, Moscow University philosopher Zinovy Beletsky. Beletsky's letter, dated November 1946 and addressed to Stalin, was discussed at the Central Committee Secretariat and prompted the decision to organize the critical discussion (Esakov 1993, 87).[7]

The philosophers did not know the particulars of Stalin's and the Central Committee's criticism, if indeed there were any, so they had to develop critiques of their own, guessing about the essence and seriousness of Aleksandrov's mistakes. Within certain limits, the gathering produced a variety of conflicting views on the book's scholarly and political shortcomings. Mark Mitin and Pavel Yudin, the "old guard" of communist philosophy and Aleksandrov's personal foes, apparently hoped that the event would shake up the Young Turk's career and restore their own importance in the field.[8] Supported by Beletsky and Aleksandr Maksimov, they spoke

---

[6] For a report on this meeting see the archive of the Communist Party: *Rossiiskii Gosudarstvennyi Arkhiv Sotsial'no-politicheskoi Istorii* (RGASPI, hereafter-CPA), 17-125-477.
[7] A few others also claimed to have signaled to the Central Committee about mistakes in the book (CPA, 17-125-477, pp. 111; 17-125-527, pp. 9–37).
[8] On the feud between Mitin and Aleksandrov and groupings among philosophers see D. Chesnokov's and Agitprop's 1949 report to Georgy Malenkov and Mikhail Suslov (CPA, 17-132-155, pp. 20–26; 17-132-161, pp. 8–36).

against "conciliatory attitudes" displayed during the previous discussion of the book and called for a "principled criticism" and for "militant struggle" with bourgeois ideology. More moderate critics included a group of up-and-coming young philosophers like Bonifaty Kedrov and Mikhail Iovchuk, who proposed such slogans as "creative criticism" and "further creative elaboration of Marxist philosophy."[9] Many who did not belong to either "militant" or "creative" camps and had no personal reason to be for or against Aleksandrov used the opportunity to speak before Zhdanov, demonstrating their talents, loyalty, and activity while not forgetting to mention various personal agendas.

Only after listening to the others did Zhdanov deliver his talk, in which he summarized the results of the discussion and drew further conclusions. In his opinion, although the book deserved encouragement as the first attempt to write a Marxist textbook on the history of philosophy, it had in general failed to meet its goals. Zhdanov criticized several examples of bad style and unclear definitions and accused Aleksandrov of committing not only factual mistakes but also such political ones as "objectivism"—insufficient criticism of pre-Marxist bourgeois philosophy. According to Zhdanov, the textbook's deficiencies reflected the generally unsatisfactory situation "on the philosophical front." The uncritical reception and laudatory reviews of the book, until Stalin intervened, had demonstrated "the absence of Bolshevik criticism and self-criticism among Soviet philosophers." Combining the slogans of rival philosophical camps, Zhdanov said that Soviet philosophical publications were often scholastic and conciliatory rather than creative and militant; they stopped short of developing Marxist doctrine further and of fighting against idealistic perversions. Aleksandrov failed to ensure good leadership in the field; "moreover, he relied in his work too much on a narrow circle of his closest collaborators and admirers,"—at this point Zhdanov was interrupted by the applause and shouts of "Right!"—and "philosophical work had thus been monopolized by a small group of philosophers" (Diskussiia 1947, 269).

At the end of the meeting, Aleksandrov was given an opportunity to engage in self-criticism. His role was technically the most difficult one: on the one hand, the ritual strictly forbade the use of a defensive tone; on the other, his career would not benefit were he to accept the most serious

---

[9] For a comparison of the slogans of the "creative" and "militant" parties, see the editorials "Za tvorcheskuiu razrabotku marksistskoi filosofii" and "Za boevoi filosofskii zhurnal" in *Voprosy Filosofii* 1948, no. 1, and 1949, no. 1, respectively.

accusations. For the game to be performed and resolved successfully, and to convince the spectators that his repentance was sincere, Aleksandrov had to estimate correctly the mood of the audience and higher referees and find the right tone of self-accusation. Having done this in the first part of his speech, as he thanked everybody for exposing his mistakes and summarized them once again, Aleksandrov towards the end shifted his tone to that of a philosophers' instructor and urged everybody to learn from his case and to improve work on the philosophical front (Diskussiia 1947, 188–199).

The Stalinist system preferred distinct black-and-white colors over shadings and had difficulty drawing an intermediate line between unequivocal political praise and complete political denigration. In Aleksandrov's case, however, the discussion did not destroy him either as a politician or as a person, but it did constitute a turning point in what had been an extraordinary rapid and successful career. Although Aleksandrov survived for another three months as director of Agitprop and even submitted a proposal for further work, his career was in danger. In September 1947, the Central Committee Secretariat reviewed the results of the philosophical discussion and decided to remove him from his influential party post.[10] Demoted, he was appointed director of the Institute of Philosophy, in which position, presumably, he had to supervise in person how his critics were learning from his mistakes.

Stigmatized by the event, Aleksandrov was repeatedly criticized within the party apparatus, especially after the death of his patron Zhdanov in August 1948.[11] In July 1949, he was accused of political mistakes, fired from the editorial board of the party's main theoretical journal, *Bolshevik*, and disappeared for a while from the public political arena. He managed to return to it in 1950 and even came back into favor during the flurry of political changes that followed Stalin's death. In 1954, Aleksandrov was appointed minister of culture, only to be removed the following year in a sex scandal. He was transferred to Minsk and died there in 1961 as a rank-and-file member of the Belorussian Institute of Philosophy. Such was the end of this turbulent and unusual career for a Soviet bureaucrat (Medvedev 1972a, 215, 221; Kedrov and Gurgenidze 1955).

---

[10] "Stenogramma soveshchaniia Upravleniia propagandy i agitatsii TsK VKP(b) ot 19 sentiabria 1947 g. 'O sostoianii raboty v oblasti propagandy i agitatsii'" (CPA, 17-125-493). Piotr Fedoseev, who had organized the earlier discussion of January 1947, was also criticized for a lack of principles.

[11] CPA, 17-132-160, pp. 91–98; 17-132-155, pp. 1–51.

3. Games of Intra-Party Democracy

> "We cannot do without self-criticism, Aleksei Maksimovich. Without it, stagnation, corruption of the *apparat*, and an increase of bureaucratization would be inevitable. Of course, self-criticism provides arguments for our enemies, you are completely right here. But it also gives arguments (and a push) for our own progressive movement" Stalin to Maxim Gorky (Stalin [1930] 1949, 173).

The ritualistic performance described in the preceding section may seem strange to a modern reader, but for Soviet audiences it was an example of the familiar cultural games *diskussiia* (disputation) and *kritika i samokritika* (criticism and self-criticism). These games originated and were usually played within party structures and belonged to the repertoire called "intraparty democracy."

Soviet and more narrowly intraparty democracy is a controversial topic. Some Sovietologists referred to it as mere propaganda and a "verbal masquerade" (Fainsod 1953, 180–196), while others took it seriously as an element of true democracy and argued against violations of its principles in Soviet life (Medvedev 1972b, 124–156). More recently, historian Arch Getty called attention to its function of controlling local party bosses with the help of rank-and-file members and argued that under certain conditions the process could get out of control and produce a massive purge (Getty 1985, 97–98, 141–142). Communists themselves, in public and in private, viewed intraparty democracy as a mechanism for making officials accountable to the party masses and as the main tool in the struggle against bureaucratism and corruption in the party apparatus. Although openly preferring administrative centralization and hierarchical discipline as the organizing principles of social life, they were also aware that local bosses were in a position to abuse their power and to prevent higher authorities from receiving objective reports about local conditions. The Stalinist leadership tried to establish a system of counterbalances designed to provide feedback as well as to define situations and limits within which grassroots control of the apparatus was possible. In combination with the principle of administrative hierarchy, this system was called by the idiosyncratic term "democratic centralism," and, as we shall see shortly, it could lead to idiosyncratic results.

Intraparty democracy could perform all of the aforementioned functions—propagandistic, democratic or populistic, controlling, and purging—but it would be a simplification to reduce it to any particular one of these or to define it simply by function. The phenomenon is more complex and might be better understood as a system of cultural rituals specific, and of central importance, to Stalinist society. For members of that culture it had a high ideal value in its own right, not only because of its presumed practical goals. It also had sufficient power to ensure the public compliance of even the highest officials, such as Zhdanov. In modern anthropological studies, rituals are no longer described as rigid, strictly repetitive, and noncreative activities but are seen as forms of life. They are formalized collective performances, a unity of spatial movement and verbal discourse, which constitute the core of social identity in all communities and have both sacred and practical meanings. Although rule-governed, ritual activity is not petrified or simply symbolic: rituals "are not just expressive or abstract ideas but do things, have effects on the world, and are work that is carried out." "[Ritual] is an arena of contradictory and contestable perspectives—participants having their own reasons, viewpoints, and motives—and in fact is made up as it goes along" (de Coppert 1992, 1–4, 13–18).

Social life under Stalinism was ritualized in this sense to a very high degree. In its political sphere, the most typical space of formalized

This meeting is neither *diskussiia*, nor *kritika i samoskritika*, but a festive gathering (*torzheztvennoe zasedanie*) convened by the Academy of Sciences to celebrate Stalin's 70th birthday, Moscow, 23 December 1949. Vavilov is on the podium delivering his laudation "On Stalin's scientific genius," in which he elaborated on the specific features of Soviet science. [*Courtesy*: RGAKFD.]

collective action and discourse was a local meeting of a party organization or an institution. The repertoire of distinctive types of meetings, with their specific genres of discourse, was quite rich, and there were many words for "meeting with discussion" in the political language: *sobranie, soveshchanie, zasedanie, vstrecha, obsuzhdenie, priëm, sessiia,* and others. Some correspondence, although not one-to-one, between genres and names can be established. The English word "discussion" is too general and too neutral to convey that diversity. In the following I will use "discussion" as a generic term and more specific words to stress when necessary the differences in genres.

For instance, a local meeting *(sobranie)* which invited participants to discuss and draw conclusions from an authoritative decision or decree would typically be called *obsuzhdenie* (consideration). When a meeting was announced as a *diskussiia* (disputation), this was a sign that participants were invited to demonstrate polemical skills in a theoretical matter that had not been decided by the authorities. A *diskussiia* allowed for temporary, public disagreement over important political questions. It was often used for, or followed by, resolution of the controversy and formulation of a decision. The decision was sometimes taken by participants' voting and sometimes by authorities who either observed the meeting in person or reviewed its minutes later. In the most serious *diskussii* that threatened to split the party several times during the 1920s, it was the party congress, or *s"ezd*, that resolved the controversy. Officially, a *s"ezd* was the highest party authority. By voting, it settled the disputes once and for all, and the opposition, or the losing party, had to stop any further polemics with the majority.

In addition to *diskussiia, kritika i samokritika* (criticism and self-criticism) also belonged to the repertoire of intraparty democracy, but it usually dealt with personal rather than theoretical matters. Berthold Unfried (1994) has characterized it as a ritual central to the culture of the party and as a dialectical combination of two functions: initiation (educating and enculturating party cadres) and terror (exposing and destroying enemies). Standing the trial of *kritika i samokritika* was a necessary part of the training of new party members and officials. Subordinating one's personal views and ego to those of the collective and accepting criticism and delivering self-criticism in the proper way were the proof of successfully internalized cultural values and of one's status as an insider. The same ritual could also be used as a mechanism for purging, for revealing and accusing internal (but not external) enemies and outsiders. Its cultural force was so strong that even communist oppositionists who faced the death penalty were still proving their insider status by admitting imaginary crimes and accusing themselves in the public performances of Moscow trials, while denying their guilt in their last private letters to Stalin or to the party.

Another role of *kritika i samokritika*, identified by Arch Getty (1985, 50, 67, 134–5, 145, 224), allowed and provided an institutional framework for grassroots criticism of local bosses. Party secretaries normally would rule in an authoritarian way, exempt from criticism from below, but within ritualistic space-time constraints the usual hierarchy could be temporarily reversed and horizontal or upward critique welcomed. The requirement of self-criticism forbade the local authority under fire from using his power to suppress criticism. In communists' self-descriptions, this democratic institution supplemented the hierarchical structure of the party and was steadily at work revealing and repairing shortcomings and local abuses of power, "however unpleasant it might be for the leaders." In practice, *kritika i samokritika* was performed mainly on special occasions and usually required permission or initiative from above. It could be applied when higher authorities wanted popular justification for their desire to remove a local functionary, when they were not sure about denunciations against him and wanted to test him publicly, during elections to party posts, or simply as a substitute for the Christian ritual of penance for the regular cleansing of the system.

Analyzing the Philosophical Dispute of 1947 as a combination of two rituals, *diskussiia* and *kritika i samokritika*, reveals some of their rules. Rule-governing in the ritual does not necessarily imply the existence of an explicit code but may only indicate a shared perception that there are some rules: "Even when neither observers nor participants can agree on, understand, or even perceive ritual regulations, they are united by a sense of the occasion as being in some way rule-governed and as necessarily so in order [for a public ritual] to be complete, efficacious, and proper" (de Coppert 1992, 15). Party members learned most of their cultural rules not from such texts as party statutes but from watching and participating in actual performances; their behavior and discourse at a meeting depended in the most critical way on the announced type of ritual. A sense of definitive rules permeated the entire procedure of the Philosophical Dispute, as participants watched each other's behavior and often criticized perceived violations. They protested when, in their opinion, speakers were expressing personal animosity instead of principled criticism, and especially strongly when self-defense was being offered in place of self-criticism. Aleksandrov provided a good example of playing according to the rules and thus proved his loyalty and his status as an insider. But at the 1950 physiological discussion, when

Leon Orbeli protested against the accusatory style of criticism, the audience got more infuriated at this "violation of rules" than at his other alleged mistakes, and at the end of the meeting Orbeli had to deliver another, much humbler talk (Nauchnaia 1950, 94, 187, 501).

The rules of both *diskussiia* and *kritika i samokritika* assumed the openness of the result, which was not supposed to be determined prior to public performances. Although the structure of the discourse was quite rigid, its critical content and outcome depended very much upon the activity of the players. On the theoretical side, Aleksandrov's mistakes were not exactly known to participants but had to be found out during *diskussiia*. On the administrative side, the ritual of *kritika i samokritika*, like the ritual of confession, could be constructive as well as destructive. In the regular training and elections of party cadres, self-criticism could often be followed by a promotion. At a trial of an official, such as Aleksandrov's case, the procedure was certainly a purgatory for him, but it could still end up anywhere between purge and practical acquittal. Public contestations which, like *diskussiia* and *kritika i samokritika*, had more or less fixed rules but open results would be more appropriately termed "games."[12]

The Philosophical Dispute can also illustrate the characteristic role structure of both games. Since both constituted a temporary challenge to the normal order—conceptual or hierarchical—the play often required permission or encouragement from a higher authority, either in a concrete form or as an announcement of a general campaign of, say, *samokritika*. Typically, a meeting was moderated by a representative from an agency further up the administrative hierarchy, who was neither openly partisan nor completely impartial. He often had his own agenda but was not supposed to be associated directly with actively contesting parties—he played above them. Occasionally the meeting could get out of his control.[13] Zhdanov's role as moderator at the Philosophical Dispute was to announce the type of game to be played and the topic, to suppress by his

---

[12] The metaphor is used here not in the narrow sense of game theory but in the wider, Wittgensteinian sense. See examples of games in Wittgenstein's *Philosophical Investigations* (Wittgenstein 1958, § 66–71). An analogy with a game such as amateur soccer, where players follow certain models, but can also argue about and improvise rules, suffices for the purpose of this study.

[13] Getty describes a *kritika i samokritika* meeting during which angry communist insurgents ousted a district party secretary despite the protective attempts of the higher representative. Although Getty does not assign a special role to the moderator as the third party (besides the mob and the local boss) in his examples of *kritika i samokritika*, this figure appears in his narratives whenever he describes a meeting in detail (Getty 1985, 72, 151–153).

aura of power the usual hierarchy between Aleksandrov and his subordinates, and to enforce procedures and rules. Various agencies could fulfill the role of referee. Many participants in the Philosophical Dispute included indirect appeals to the Central Committee in their speeches. As it turned out, the Central Committee Secretariat played referee with regard to *kritika i samokritika* by deciding about Aleksandrov's career after reviewing the minutes of the dispute. Zhdanov himself refereed the *diskussiia*, when at the end of the meeting he summarized its results and fixed the theoretical consensus.

The roots of these games are not to be found in Marxist doctrine, either in its original form or as it was developed by Lenin. Apparently they were first established in communist practice and only later in theory. *Diskussiia*, as a way of sorting out and resolving factional disagreements within the party, existed in some form before the Revolution and in a fully developed version certainly by 1920. Within its space-time constraints, the opposition was then arguing for and partially achieving the freedom to criticize party authorities. *Samokritika* as a political slogan and campaign first appeared in 1928 and meant "the purge of the party from below," which allowed young radicals to criticize authorities and do away with the moderate New Economic Policy (NEP) (Spravochnik 1930; Alikhanov 1928; O samokritike 1928). By 1935 the game had changed its name to *kritika i samokritika* and was playing an important role in the party purges. Among Soviet leaders, Zhdanov always appeared as its chief promoter and propagandist. It was considered one of the main principles of party life, known to all members and applied on various occasions within the party and Soviet structures. But by the time of the 1947 Philosophical Dispute, it had not yet received a higher justification from Marxist theory.

In his talk at the dispute, Zhdanov presented the first outline of such a justification. In the draft of the talk he attempted to ascribe authorship of the concept to Stalin, but Stalin crossed this out from the manuscript (Esakov 1993, 92). Thus, Zhdanov said,

> "the party has long ago found and put into the service of socialism this particular form of exposing and overcoming contradictions in socialist society (these contradictions exist, although philosophers are reluctant to write about them), this particular form of the struggle between old and new, between withering away and emerging in our Soviet society, which is called *kritika i samokritika*.... Development in our society occurs in the form of *kritika i samokritika*, which is the true moving force of our progressive development, a powerful tool in the party's hands" (Diskussiia 1947, 270).

Further theoretical rationalizations for the existing practice explained that criticism and self-criticism did for socialist society what "bourgeois democracy" did for capitalism—provided mechanisms for change. In the one-party system, so the argument ran, when no competing political party was providing external criticism, the communist party had to carry the burden of self-criticism to reveal and repair its own defects if it was to cleanse and improve itself. Such was the communist interpretation of the democratic idea as applied to the party itself (Bol'shevistskaia 1950).[14]

4. Opening Pandora's Box

> "The great and serious tasks arising before Soviet science can be solved successfully only through the wider development of *kritika i samokritika*— 'one of the most serious forces, that pushes forward our development' (Pervye 1948, 13, quoting Stalin)."

In the official judgment, the Philosophical Dispute "enlivened work on the philosophical front and stimulated further progress in it." The immediate consequence was the establishment of the professional philosophical journal *Voprosy Filosofii*. Bonifaty Kedrov (1903–1985), who during the meeting had argued in favor of such a journal and managed to pass a note to Zhdanov asking for a personal appointment, became editor-in-chief (Esakov 1993; Kedrov 1988). The entire first issue of the journal was devoted to the minutes of the discussion. The theory of *kritika i samokritika* as developed by Zhdanov and sanctioned by Stalin was thus introduced to wider audiences as an important new contribution to Marxism–Leninism. It offered a basis and inspiration for lower-level politicians to develop derivatives and applications. A demonstration of zeal by initiating and carrying out a successful interpretation of a general slogan could certainly bring rewards and push one's career ahead. At the same time, risks could never be eliminated entirely. We shall see shortly that no matter how correct an official might try to be in his actions, the chance always remained that he might come under fire for real or assumed mistakes.

Although the minutes of the Philosophical Dispute did not yet suggest that the method ought to be applied within other academic disciplines,

---

[14] See also "Samokritika—ispytannoe oruzhie bol'shevizma," *Pravda*, 24 August 1946; Stepanian, "O protivorechiiakh pri sotsializme," *Pravda*, 20 August 1947; "Pod znamenem bol'shevistskoi kritiki i samokritiki," *Pravda*, 15 March 1948.

the slogan "*kritika i samokritika* in science" soon became one of the policies of Agitprop under its new leadership, the official director and Central Committee secretary, Mikhail Suslov (1902–1982), and the acting director, Dmitry Shepilov (1905–1995). However, it was mainly lower-level politicians whose names became directly associated with the initiative. Kedrov was apparently the first to publish, in February 1948, a theoretical paper on the topic (Kedrov 1948). Later the entire campaign was reviewed and praised by a former Agitprop officer, Mikhail Iovchuk, and by Yury Zhdanov (1919–), Andrei Zhdanov's son, a young Moscow University graduate in chemistry who came to Agitprop in late 1947 to head the Sector of Science (Zhdanov 1951; Iovchuk 1952; Za svobodnuiu 1950). Extending *kritika i samokritika* to the sciences could well have seemed a safe bet. The word for "sciences" in Russian, *nauki*, embraces not only the natural and social sciences but also humanities and ideological scholarship. The Dispute of 1947 was performed by party members who just happened to be philosophers, but since philosophy was also one of the *nauki* it was just as natural to apply the same, presumably so effective method to other fields as well. The double status of philosophy as both a party business and an academic field made it easier for the games *diskussiia* and *kritika i samokritika* to be transferred from party culture to academia.

Vestnik Akademii Nauk was the official monthly journal of the Soviet Academy of Sciences covering the life of academic community: events, appointments, meetings, and political issues related to science. Kedrov's article (1948), "The meaning of *kritika i samokritika* for the progress of science (To the problem of the role of negation in dialectics and metaphysics)," in its February issue could still be viewed as the author's personal opinion. The appearance of the editorial "The First Results of Creative Disputations" in the subsequent issue, however, signified to readers the existence of an ongoing political campaign. Unsigned editorials in newspapers were the usual means for delivering messages from authorities regarding sanctioned opinions and policies. The March 1948 editorial reviewed several early examples of "creative disputations": the Philosophical Dispute, the discussions of E. S. Varga's book on world economics, of textbooks in linguistics and law and on the history of the Soviet Union, discussion at Moscow University and the Academy of Sciences of intraspecies competition, and a few others. The editors mentioned that the initiative had come from the party press and appealed to scientists to follow these examples. Methods of creative *diskussii* and *kritika i samokritika* had to be applied in the work of scientists in order to "reveal our own mistakes and to overcome them" (Pervye 1948, 15).

The choice of games characterized the special place of science in Stalinist society. In an earlier ideological initiative—known as Zhdanovshchina—that hit mainly literary journals, films, theater, and music but also some academic institutes in law and economics, the emphasis was also on an increased level of criticism in various cultural fields. In his talk of August 1946, Andrei Zhdanov pronounced: "Where there is no criticism, there solidifies stagnation and stinking, there is no room there for progressive movement" (Zhdanov 1946). However, the judgement was expected then to come from the party. When the first plans for such an extension of ideological work were discussed at a closed meeting of Agitprop on 18 April 1946, Zhdanov was particularly concerned about the weakness of internal criticism in such hierarchically governed organizations as the Writer's Union and the State Committee for Cultural Affairs: "Who can correct these departments' attitude which spoils the work and contradicts the interests of the people? Of course, only the involvement of the party ... through the organization of party criticism in order to counterbalance the department's own criticism."[15] Open party involvement in cultural affairs followed. Politicians apparently considered themselves competent enough in literature and film to make expert judgments and to issue them publicly in the name of party bodies. Writers and film directors convened afterwards and held *obsuzhdenie* (consideration) of authoritative decisions.

In contrast, when it came to scholarly disputes in the fall of 1947, politicians preferred to act behind the scenes, left most action and public performances to scholars, and let decisions be issued in the name of a representative academic meeting. In sciences, politicians did not possess the knowledge and authority to define agendas and outcomes entirely on their own but required the active participation of, and dialogue with, experts. They therefore recommended games—*diskussiia* (with a special adjective, *tvorcheskaia*, or "creative") and *kritika i samokritika*—which implied open results and grassroots initiative and criticism.

Scientists were thus invited to play, within their own ranks, the games of intraparty democracy, and they could respond in a number of different ways. A sufficient demonstration of loyalty would be to hold an *obsuzhdenie* of the Philosophical Dispute at a local meeting and send a report with assurances that creative disputations and criticism had always been, and

---

[15] "Stenogramma soveshchaniia v TSK po voprosam propagandy, o rabote tsentral'nykh gazet i izdatel'stv, 18 aprelia 1946 g. pod predsedatel'stvom A. A. Zhdanova" (CPA 77-1-976, p. 40).

continued to be, crucial for scientific work. Some interpreted the invitation as permission for more freedom in academic discourse.[16] Many reacted with discussions imitating the Philosophical Dispute. Since the model event was a debate over a textbook, most of the early imitations also took the form of a discussion of a certain textbook or other book (Obsuzhdenie 1948a; Rasshirennoe 1948a). Being the best informed about the rules of the game, philosophers staged one more smooth performance. In January 1948 a *diskussiia* was organized in the Institute of Philosophy, and it became a minor remake of the 1947 Dispute. The cast of characters included Aleksandrov, who had become director of the Institute, presiding over the meeting as mini-Zhdanov; Kedrov, with his book *Engels and the Natural Sciences*, played mini-Aleksandrov. Both were apparently in control of the situation, and the meeting only confirmed the existing hierarchy. While presenting a mixture of moderate praise and criticism of Kedrov's book, the audience turned largely against his main opponent, Aleksandr Maksimov, blaming him for unfair and dogmatic use of criticism (Obsuzhdenie 1948b).

While agendas and outcomes were not predetermined, the rhetorical and cultural resources, in a certain sense, were. Rival groups of scholars were already used to including political argumentation in academic discourse and often sent political authorities letters of denunciation and complaints against colleagues. Agitprop files are filled with such letters, though only relatively few of them could receive any serious attention. With the new agenda of critical discussions, there emerged tempting opportunities for scholars to proceed with existing academic conflicts in more open and politically sanctioned ways. The campaign stimulated public as well as unofficial dialogue between scholars and politicians, wherein the common language was mainly that of current politics and ideology. By appealing to politicians as referees and striving for their support, scholars competed in translating scientific concepts and agendas into that language. Conflicting academic parties were developing ideological pictures of their field in ways that would support their positions in the controversies.

In these scenarios, politicians could fulfill different roles. That *"kritika i samokritika* is the law of the development in science" quickly became a commonplace slogan for them.[17] In fields such as philosophy, political

---

[16] See the next chapter about Vavilov's attempt to arrange for the publication and discussion of a paper by M. A. Markov.

[17] See, for instance, G. F. Aleksandrov, "Ob oshibochnykh vzgliadakh B. M. Kedrova v oblasti filosofii i estestvoznaniia," 23 February 1949 (CPA, 17-132-180, pp. 48–97).

economy, and law, Agitprop could initiate and set the direction of some discussions. In most other fields, it did not have its own agenda but welcomed scholars' initiatives and was more interested in the fact that a discussion came about than in its particular result. In these cases, disputes were performed within the academic hierarchy and depended largely upon internal conflicts and power relations. In some situations, politicians listened to appeals for support by rival groups of scientists and, if convinced by the rhetoric, could accept the role of referee. In biology, this resulted in the biggest setback for Soviet science, the ideological ban on genetics pronounced at the 1948 August Session of the Soviet Academy of Agricultural Sciences.

5. Resolving the Controversy and Defining Consensus

"In science as in politics, contradictions are resolved not through reconciliation, but through an open struggle." Andrei Zhdanov and Georgy Malenkov, July 1948.

The conflict in biology had ripened long before 1948. Lysenko, a poorly educated but charismatic agronomist, achieved his fame in the early 1930s, with newspapers praising his supposedly very effective new methods of improving crops. Scientists, including geneticist Nikolai Vavilov, initially helped him achieve academic recognition and positions, but Lysenko soon turned against some of the basic concepts of the new science of genetics, proposing instead his own, idiosyncratic theories of plant heredity. Early debates already used some ideological argumentation, but geneticists suffered particularly serious losses during the Stalinist purges of the late 1930s, when Vavilov and several other prominent figures perished in the carnage. Lysenko filled the vacated positions and rose to head the Academy of Agricultural Sciences and the Institute of Genetics in the Academy of Sciences. After World War II geneticists tried to regain some ground and to undermine Lysenko's position. Anton Zhebrak (1901–1965), a geneticist, and in 1945–1946 an Agitprop officer, wrote letters to the Central Committee arguing that Lysenko's monopoly was damaging the reputation of Soviet science among the Western allies and lobbied for opening another institute of genetics in the Academy, with himself as its director (Esakov and Levina 1991, Levina 1995).[18] The Academy's

---

[18] See also CPA, 17-125-359, pp. 5–8; 17-125-449, pp. 48–49, 108–111; S. I. Alikhanov to Stalin, 6 May 1948 (CPA, 17-132-71, pp. 4–41).

vacant position of a corresponding member in biology, however, did not go to Zhebrak in major elections during the fall of 1946—possibly because of complaints that too many Agitprop officers were seeking membership—but to another geneticist, Nikolai Dubinin, who was elected despite Lysenko's opposition. The Academy proceeded with preparations to organize a new institute in genetics. Soon after Yury Zhdanov became the head of the Science Sector in Agitprop on 1 December 1947, several of Lysenko's opponents made appointments with him and complained about the unsatisfactory situation in biology (Zhdanov 1993, 74).

With the start of the new campaign of creative disputations, biologists felt encouraged to criticize Lysenko more openly. A new dispute about the Darwinian theory and the problem of intraspecies competition erupted between Lysenko and his opponents in the pages of *Literaturnaia Gazeta*. With silent permission from Agitprop—and in line with the new policies—biologists organized conferences at Moscow University (November 1947 and February 1948) and at the Biology Division of the Academy of Sciences (December 1947), where they criticized some of Lysenko's views (O vnutrividovoi 1948; Nauchnye 1947; Moskovskaia 1948). On 10 April 1948, Yury Zhdanov entered the discussion with a lecture at a meeting of party propagandists on "Controversial Questions of Contemporary Darwinism," in which he partly sided with Lysenko's critics. According to Zhdanov, the struggle was between two schools of Soviet biology, rather than between Soviet and non-Soviet sciences. Both Neo-Darwinians (geneticists) and Neo-Lamarckists (Lysenkoists) had accomplishments, and both had succumbed to an undesirable radicalism during the struggle. Lysenko, in particular, should not claim to be the only follower of the great Russian plant breeder Michurin. Having started as an innovator, Lysenko later lost his self-critical attitude and, by suppressing other approaches, he had brought about direct damage. Monopolies in every field of scientific research have to be liquidated: creative disputations, developing *kritika i samokritika* in science, and cultivating a variety of research approaches would help achieve this (Zhdanov 1993, 74, 81, 85–86).

In trying to referee the biological controversy Yury Zhdanov, young and inexperienced apparatchik, acted prematurely. Although he had consulted with his boss, Shepilov, he had secured neither definitive approval from higher authorities nor the means to drive Lysenko towards *samokritika*. Zhdanov made it clear to the audience that he was delivering a personal rather than an official opinion. Lysenko, who was not invited to the lecture, he managed to hear it secretly and became intimidated, for he had apparently almost lost this round of *kritika*. Cleverly enough, he started a new one. Since Lysenko was the authority in the field of biology, he would have committed a breach of the rules had he decided to complain about the criticism from below. Instead, he built a new triangle of *kritika i samokritika* by complaining against the actions of Zhdanov Jr., who was the party authority for scientists, and appealing to Stalin as referee. In his letter of 17 April to Stalin and Andrei Zhdanov, Lysenko pictured himself as a non-party but loyal scientist who was offended by the attack from a high party official and worried whether the party had lost trust in him. In case this was true, Lysenko offered in another letter his resignation as president of the Agricultural Academy (Soyfer 1994, 172–177; Krementsov 1997, 164–165).

Lysenko's complaint impressed Stalin. At a Politburo meeting in June, Stalin expressed his dissatisfaction with Yury Zhdanov's talk. In later interviews with Valery Soyfer, Zhdanov Jr. and Shepilov made contradictory and obscure remarks about who in the ideological hierarchy, and in what form, admitted responsibility for the mistake. A committee was established to investigate the case. Following the unwritten rules of the bureaucratic *modus operandi*, Shepilov advised the younger Zhdanov to write a letter of self-criticism. According to Yury, rivals of Zhdanov Sr. among the upper level of Soviet leadership used the occasion to criticize the youngster for "insufficient disarmament" and the father for protecting his son (Zhdanov 1993, 87). Whether Yury's precipitate action contributed to his father's fall or whether it was Andrei Zhdanov's loss of power that helped the agricultural bureaucracy to prevail over the ideological one is still difficult to tell with certainty.[19] But some connection apparently existed, for Politburo decisions on the Lysenko case and on the Central Committee apparatus coincided.

Andrei Zhdanov became the main victim of the enacted changes, while most other concerned party officials managed to improve their positions. Malenkov, Zhdanov's chief rival, was added to the Secretariat on 1 July and took over the chairmanship there a week later when Zhdanov took two months' vacation, which resembled exile, during which he would die of heart failure. On 10 July the Politburo effected a major reorganization of the Central Committee apparatus, shifting its emphasis in work from Zhdanov's propaganda to cadres, in which Malenkov was the recognized expert. Suslov took charge of international relations, Shepilov was promoted to the official directorship of Agitprop, and Malenkov, in addition

---

[19] Shepilov's memoirs support the former option (Shepilov 1998, 4).

to cadres, oversaw the reestablished Agricultural Department. The younger Zhdanov received a severe moral reprimand, but Stalin spared him from any more serious punishment. He remained in his position at Agitprop, but only as long as Stalin was alive. Learning the rules of apparatus intrigue required years of experience; a hasty and amateur involvement in high politics could be very dangerous.[20]

On 15 July the Politburo met to discuss questions presented by the agricultural establishment—the Academy, ministries, and the new Agricultural Department of the Central Committee—and to repair the damage caused by the "incorrect report of Yu. Zhdanov on matters of Soviet biology, which did not reflect the position of the Central Committee." Stalin's expression of sympathy for Lysenko could possibly suffice to ruin the career of a Politburo member, but not to close the scientific dispute. On behalf of the committee investigating the case, Andrei Zhdanov had written, and Malenkov cosigned, a draft resolution on the situation in biological science and the mistakes of Yury Zhdanov, but the party once again stopped short of issuing a decision on a scientific topic in its own name. Instead, the Politburo approved the agricultural lobby's proposal to appoint a number of Michurinists as new members of the Academy and decided to compensate Lysenko for moral damage by allowing him to organize a discussion on his terms, and to present to the Academy and publish a report "On the Situation in Soviet Biology."[21]

The session of the Lenin Academy of Agricultural Sciences of the USSR. opened on 31 July with a major presentation by Lysenko. Although Stalin had edited the manuscript and corrected its ideological message, the party's support was not announced at first. Lysenko's task was to prove that he could control the field, mobilize enough grassroots support, and stage a smooth performance. Only after he had passed this test, on 7 August, the last day of the meeting, was Yury Zhdanov's repentant letter published in *Pravda*, and Lysenko was allowed to say that he had the backing of the Central Committee.[22] Having been sanctioned both by

---

[20] On the conflicting principles of the Central Committee organization and the Zhdanov-Malenkov rivalry see (Fainsod 1953, 172–177; Hahn 1982). Several months later, in November 1948, Kedrov also lost his job as the editor of *Voprosy Filosofii* as an indirect consequence of the August Session.

[21] A. Zhdanov and G. Malenkov, "O polozhenii v sovetskoi biologicheskoi nauke" (CPA, 17-77-991), and the protocols of Politburo meetings on 10 and 15 July 1948 (CPA, 17-3-1071).

[22] (Rossianov 1993a, 1993b; Situation 1949). On 6 August the Agricultural Department sent a long report to Malenkov detailing that the majority at the ongoing session was supporting Lysenko (CPA, 17-138-30, pp. 1–39).

Games of Soviet Democracy 211

the voting at the representative scholarly meeting and by Stalin's support, the victory of Michurinist biology became final.

One can recognize behind this pattern the model provided by another game of "democratic centralism": the party congress, or s"*ezd*. The first important feature is that, officially, the decision adopted by the representative collective body had more strength than a decision of any individual leader. Even Stalin could later be declared fallible by Khrushchev, but none of the decisions of party congresses could be. Second, everyone knew from the party *Short Course* history that congresses had served several times in the 1920s as the method for final resolution of the most divisive disputes that threatened to split the party. Factions and propaganda on behalf of opposing views were allowed before the congress, but after the ballot further polemics were forbidden. The opposition, or the losing side, had to "disarm itself" through self-criticism and to cancel all

August 1948. The Session of the Lenin Academy of Agricultural Sciences convened by Lysenko to close discussion in the field of biology and issue a verdict against Mendelian genetics as a wrong and ideologically foreign science. The participants did not know yet that Lysenko had secured the backing of Stalin and the Politburo, but many of those invited were already supporting Lysenko and his "Michurinist biology." The slogan above their heads quotes from Stalin's speech at a party congress in 1939: "The humblest Soviet citizen, being free from the fetters of capital, stands head and shoulders above any high-placed foreign bigwig whose neck wears the yoke of capitalist slavery." The meeting took place in the club of the Ministry of Agriculture in Moscow.
[*Courtesy*: RGAKFD.]

organizational activity. For the Central Committee, preparing such a *s"ezd* was a challenge: the election of deputies on the local level had to ensure the necessary majority.

Lysenko proceeded in a similar way. His difficulty was that the Agricultural Academy, where he had many supporters, was not the only natural authority for adjudicating theoretical problems in biology. Early interference from the Academy of Sciences could have spoiled the smooth scenario. Hence preparations were made very quickly, and most of Lysenko's opponents from the outside did not know of them and did not attend the session. Iosif Rapoport learned about the meeting only by chance and at the last moment. With some difficulty he managed to get into the building to become one of the very few who raised a dissident voice (Manevich 1993, 124–125). These few were just enough to create the impression of a militant, but numerically insignificant, opposition to the majority. One cannot say that almost everybody in the hall was a convinced Lysenko follower, but many who in a different setting would have preferred to remain aloof from the polemics or even take the opposite side joined the common chorus at the August Session.

This behavior was to all intents and purposes enforced by the genre of discourse set by Lysenko's main talk and the subsequent initial speeches. Opponents tried unsuccessfully to change the game being played, and therefore the style of polemics. They argued that the dispute had not been organized properly and that the other side had not been informed and given time to prepare and explain its views. Boris Zavadovsky reminded the audience that the party had fought an "ideological struggle on the two fronts ... against right- and left-wing deviations" and "against mechanistic vulgarization of Marxism, on the one hand, and against Menshevistic idealism, formalism and metaphysics, on the other." Zavadovsky argued that the meeting ought to defend the middle line of "correct Darwinism" against both Neo-Lamarckian and Neo-Darwinian deviations (Situation 1949, 338, 345). "We have to hold another free *diskussiia* in a different place," demanded Piotr Zhukovsky (Situation 1949, 391). Many other speakers, however, made it clear that the game was different and the time was up.

> "*Diskussiia* had been finished after the meeting at the editorial office of the journal *Under the Banner of Marxism* [a 1939 discussion, which ended rather unfavorably for genetics—AK]. Since then ... on the part of formal genetics ... there is not a scientific creative *diskussiia*, but factionalism and struggle, which took most unnatural and useless forms,"

proclaimed the Lysenkoist Nikolai Nuzhdin. The message was that geneticists had failed to meet the basic rule of loyal party opposition: to "disarm" after being defeated during the *diskussiia*. Their status therefore changed, from that of tolerable partners for dispute to that of disloyal saboteurs who needed to be suppressed administratively, rather than verbally (Situation 1949, 101, 165, 233, 254, 510).

According to the rules of the game of *s"ezd*, the voting at the session resolved the dispute forever. Further *diskussiia* was off the agenda. The only possible games to play were *obsuzhdenie* and *kritika i samokritika*, which had already started on 7 August, the last day of the Session, and continued on 24–26 August at the Presidium of the Academy of Sciences. The local authority subjected to criticism was the secretary of the Biology Division, Leon Orbeli (1882–1958). The president of the Academy, Sergei Vavilov, presided over the meeting as moderator and opened it with a portion of self-criticism, reproaching the Presidium for "neutrality" and its attempts to preserve parity between two directions in biology. In the discussion that followed, Orbeli failed to convince the audience of the sincerity of his repentance. Vavilov then suggested the election of a new secretary of the Division, but instead of the natural choice, Lysenko, he proposed a more neutral candidate, Aleksandr Oparin (Rasshirennoe 1948b). In later speeches, Vavilov had to talk about the successes of "Michurinist biology." However, his talent for praise failed at these moments, and he avoided mentioning Lysenko by name (Vavilov 1949b).

While the Academy was allowed the privilege of purging itself, a dozen directors of large agricultural institutes and biological departments were replaced after the August Session by direct decision of the Central Committee Secretariat, and over a hundred professors were fired by an order of the Ministry of Higher Education. The minister's proposal to remove a number of biology textbooks from public libraries gathered support from Agitprop but was rejected by the Secretariat. In most biological institutions, non-Michurinists had to "disarm themselves" by performing *samokritika*. Teaching and research plans were changed according to the results of the controversy.[23] Contrary to a common opinion, scientists were not arrested or executed for having followed an "ideologically wrong" theory, but those who refused to admit a "mistake" and persisted

---

[23] See the Protocols of the Central Committee Secretariat from 6, 9, 11, 16 and 20 August (CPA, 17-116-364 through 369) for personnel decisions; (CPA, 17-132-66) for the proposal to remove books; (Krementsov 1997, ch. 7) for other administrative consequences of the August Session.

in defending their views faced losing their jobs. An entire field of research, once one of the most advanced in Soviet science, was thus abruptly terminated and survived only underground, in few laboratories under the cover of other research activities or behind the high walls of secret atomic institutions. A slow revival of genetics would start after Stalin's death in the mid-1950s, leading to a gradual erosion of Lysenko's power, although the 1948 decision remained officially in place and the new Soviet leader, Nikita Khrushchev, continued to support Lysenko personally and protect him from criticism. The party recognized the blunder behind closed doors in 1964, when the Central Committee conspiracy voted Khrushchev out of office and listed among his numerous political mistakes his erroneous support of Lysenko. Without a public admission of the error, the Soviet government then allocated huge resources for the development of molecular biology and genetics, trying to compensate for the lost time.

6. Soviet Ideology and Science

The series of post-war ideological discussions in sciences came as a consequence of an increased political importance and status of science in Soviet society and of politicians' conscious efforts to stimulate scientific progress by available cultural means. In particular, several rituals of party life that were thought to provide mechanisms for change and repair of local defects were applied in academic fields. The choice of these rituals reveals a characteristic distribution of authority between politicians and experts in Stalinist society. The politics prescribed certain operative procedures with open agendas and outcomes, which provoked grassroots initiatives, criticism, and conflicts. Scholars were invited to fill them with more substantive matters and concrete scholarly content. Although politicians rarely had their own agendas in sciences, they reserved the right to intervene if and when some important political, philosophical, or ideological issue was at stake. This possibility had the effect of stimulating appeals to them to serve as referees. In order to make politicians understand and intervene, scholars competed in translating conceptual, institutional, group, and personal agendas and conflicts into the language of the current politics and ideology. Such behavior was not an unknown phenomenon in the Soviet Union at least since the 1920s—nor was it uniquely Soviet—but in the 1940s it reached an unprecedented scale.

Soviet ideology, like any rich ideology, was inconsistent enough to allow the presentation of a great many academically meaningful positions in ideological terms. Still, the ideological language was not sufficient to ensure adequate translation. Scholars and politicians thus participated in Wittgensteinian language games, communicating by means of a language with severely limited resources (Wittgenstein 1958, § 7). Some of the confused results of the campaign of ideological discussions in the sciences can be ascribed to the indeterminacy of translation.

An important feature of the party games was that they closed with a single definite resolution, even though at the initial stages pluralism and freedom had been encouraged. This offers an explanation of why policies announced as, in Mao's later words, "let a hundred flowers bloom" usually ended in the opposite result. Actually, this is characteristic of many political games in general, in contrast to many regular academic ones. Stalinist culture, however, was particularly strong in its belief in the single truth, as well as in the desire to reach a conclusion without delay. No matter how strongly the struggling parties diverged in their specific views, they usually agreed in their denial of an even temporary pluralism of truth and in their intolerance to the opposing opinion.

In communists' own theories, the party and the state had the obligation and power to decide all politically important issues. This idea of omnipresence and total control was, of course, utopian and impossible to realize in practice, and it often resulted in sporadic interventions in arbitrarily selected cases. In the events discussed above, the instances when the leadership did actually interfere were determined by peculiar constellations of circumstances rather than according to any consistent logical criterion. It was impossible to predict which of the thousands of letters addressed to Stalin would manage to reach his desk, attract his attention, and stir his emotions. But once this had happened, the case would immediately be declared supremely important. The Stalinist system thus reacted on a random basis but with excessive power, producing outputs that were quite inadequate to the level of the incoming "signal from below." In modern physics, systems with such behavior are called "chaotic": they can be deterministic on the microscopic local level but produce unpredictable global results.

Each of the important political decisions, however, including those resulting from internal bureaucratic chaos, had to be publicly presented as a logical outcome of high principles. Portraying itself as an ideologically governed and effectively controlled society, Stalinism developed ideological rationalizations for all its major actions. The notion of ideology as determining the master plan and of the totalitarian regime as capable of directing society towards its implementation has been a very powerful explanatory model. Insiders were often deceived by it, therefore miscalculating the consequences of their moves. Outsiders and critics who opposed the ideology and politics of the regime depended upon the very same rationalizations in their constructions of the enemy as a logical and powerful "Manichean" evil, rather than as an controversial and chaotic "Augustinian" one.[24] Such simple pictures of the relationship between Soviet ideology and science, and of the ideological nature of the regime in general, provide only an illusory historical explanation and should be abandoned.

The five main discussions—which saw higher politicians acting as referees and which brought about an effective resolution and official conformity—were, although the most publicized, still exceptional cases. Many scholars tried to gain the support of the political leadership, but only a few succeeded. The chances of organizing a scholarly meeting that would be representative enough to definitely settle a serious academic controversy and, even more, of getting Stalin to intervene and adjudicate, were very small. In the majority of fields, discussions were held but their impact was either indecisive or limited. The following chapter will discuss two other important cases that illustrate the variability of outcomes: the 1950 discussion in linguistics that ended with a result diametrically opposite to that of the Lysenko case and the 1949 discussion in physics that did not produce a definite result at all and was cancelled without the convening of a major meeting.

---

[24] On images of the enemy, Manichean and Augustinian, see (Galison 1994).

Chapter 9

Modernist Science, Ideological Passions

The important discipline of physics is conspicuously absent from the list of major ideological debates in sciences discussed in the previous chapter. It was not forgotten, however. An All-Union conference on philosophical and ideological problems of physics was scheduled for spring 1949 but was called off at the last moment. There was no public announcement of either the preparations or the cancellation, but rumors about them circulated for decades among Soviet physicists. In the course of forty years those rumors crystallized into a folklore story that managed to appear in print for the first time around 1990, during Gorbachev's perestroika, and has since been repeated many times in somewhat varying formulations. The core of the story can be summarized roughly as follows: Ideological authorities were prepared to ban quantum mechanics and Einstein's relativity as contradictory to Marxism. Fortunately, Soviet leaders decided at the last moment that physics, unlike biology, should not be sacrificed to ideology. The credit for this is usually attributed to Kurchatov, scientific director of the atomic bomb project, who had reportedly contacted his political supervisor, Lavrenty Beria, or possibly Stalin himself, and explained that the bomb could not be built without reliance on modern theories.[1]

Around 1990, the opening of Soviet archives helped uncover extensive documentary evidence about the preparation of the 1949 physics conference

---
[1] See for example (Frish 1992).

but relatively little about its cancellation (Sonin 1994; Kojevnikov 1996). Although no documentary indication of Kurchatov's or Beria's role in the latter has been found, recent historiography has for the most part accepted the folklore story in some variation.[2] The idea that "the bomb saved Soviet physics" appeals to today's mentality as logical and believable, but it is actually illusory.

Several essential parts of the folklore story cannot be sustained and need to be rejected or changed. To start with, the theories of relativity and the quanta were *not* in danger of an ideological ban in 1949. Debates about their acceptance had long been settled in the Soviet Union, and their validity was no longer questioned in 1949. The question that remained was about their correct philosophical interpretation, a discussion that could still have some negative consequences for individual scientists but did not put the theories as such in doubt. Second, the argument that biology could be sacrificed but physics was too important would not have been acceptable at the time. From the perspective of communist authorities, neither science could be allowed to be harmed. Biology for them was also extremely important because agriculture depended on it. Linguistics, another field to be considered in this chapter, was likewise politically important because of its relevance to very complex ethnic relations in the multinational Soviet state. In the views of the authorities at the time, the 1948 decision about biology brought no damage or sacrifice—on the contrary, it contributed to the progress of biology. Communists used ideology with the intent of helping good science, and physics was no different from biology in this respect.

Thus, whether or not Kurchatov discussed the issue with high politicians in 1949, he could not tell them that Marxism was hampering his physics or ask them to make a choice between ideology and the bomb. Any such assertion would have been utterly politically incorrect and self-defeating for him. Today we usually take for granted that pragmatic choices are different from ideological ones and that one has to choose between them, but communists believed otherwise, that true ideology and true pragmatism were in perfect agreement (Joravsky 1983). The folklore

---

[2] (Sonin 1994; Holloway 1994, 211; Gorelik 2000, 161–162). Other hypotheses rendered include that the meeting was canceled for security reasons (Akhundov 1991), that at the late sessions of the Organizing Commitee the 'true physicists' turned the discussion in their favor thus making it impossible for the meeting to achieve the presumed goal of subordinating physics to ideology (Tomilin 1993), and that Vavilov appointed the top 'ideologizer' Topchiev as the Academy's scientific secretary in return for cancellation of the meeting (Esakov 1991). For a debate on this issue see (Gorelik and Kozhevnikov 1999).

story essentially relies on today's mentality and thus looks believable today, but it is highly unlikely in historical terms.

In order to find a historically more sensible explanation, this chapter will review the history of the relationship between physics and Marxist philosophy in the Soviet context and then follow more closely the peripeteia of the 1949 discussion in physics. At stake then was not the fate of modern physical theories but institutional power in the discipline and some influential careers. Kurchatov, Beria, and the atomic bomb apparently played no role in the cancellation of the meeting. Whatever slim archival evidence there is extant points towards a different scenario that includes Sergei Vavilov, president of the Academy of Sciences, and Dmitry Shepilov, head of Agitprop. While the debate in physics ended inconclusively, a similar struggle in linguistics produced a definite outcome, though almost directly opposite to the one in biology. By comparing these various cases one can discuss why different sciences, in particular biology and physics, fared so differently under Stalinism.

1. Debates about the New Physics

The early-20th-century revolution in physics—brought about by the theories of relativity and the quanta—implied a radical break with the classical tradition and claimed profound philosophical consequences. It produced divisions and serious polemics both within the discipline and among wider audiences. The specific forms of these debates in different countries reflected a variety of religious and ideological traditions. In the Soviet Union they inevitably came to be colored by the existing official ideology, although not in the way this was usually portrayed by anti-communist propaganda.

Retrospectively, the Soviet debate about modern physics was often described as an opposition between ideology and science, a struggle between Marxist "ideologizers," or "ignorant philosophers," and "true physicists." In reality, however, both sides included physicists and philosophers—with physicists usually playing much more active roles— and both used philosophical arguments quite heavily. In terms of real philosophical commitments, there were some 19th-century materialists among the partisans of classical physics and some positivists among "modernists," while convinced Marxists were present in both camps. Both sides proclaimed themselves "dialectical materialists," of course, the only philosophy that was considered officially true. What is usually not realized, though, is that the overall attempts to oppose relativity and quantum mechanics in the Soviet Union were weak and isolated compared to similar ones in other countries.

Probably the strongest Soviet opponent of the new theories was physicist Arkady Timiriazev (1880–1955), son of the biologist Kliment Timiriazev. He was a student of Lebedev's in Moscow and left the university together with many other academics during the Kasso affair of 1911, in protest against tsarist educational policies. He returned, like many other former protesters, immediately after the revolution of 1917 to help rebuild the Physico-Mathematical Department and in 1921 became a Bolshevik party member, which was then extremely rare among university professors. This political affiliation helped Timiriazev acquire important administrative powers in the department, and, though officially not in charge, he was also pulling strings in the university's newly organized Physical Research Institute. Like his father a stalwart philosophical materialist, Timiriazev in physics adhered to atomistic mechanical models, and for both of these reasons he became deeply suspicious of the theories of relativity and the quanta (Andreev 2000, 250–251).

Timiriazev took an open stand against these theories at the 1926 Congress of Russian physicists where he tried to convince others of the validity of experiments by an American, Dayton Miller (Andreev 2000, 33–35). Miller claimed to have disproved the theory of relativity by repeating the famous Michelson–Morley experiment and recording a nonzero ether drift. Timiriazev's colleagues felt rather embarrassed by what they considered a scandalous display of scientific conservatism, and in 1928 Vavilov, then a younger teacher at the university, published a book that reviewed and explained the solid experimental foundations of Einstein's relativity (Vavilov 1928). Having found little support among physicists, Timiriazev was turning to philosophical audiences and publishing his critiques in Marxist journals. His opposition to recent developments in science, however, was used against him in Marxist circles as well, and in 1929 an authoritative meeting of Soviet philosophers declared his views—and those of the "mechanist" philosophical faction to which he belonged—to be not correctly Marxist (Joravsky 1961, 205–214).

A simultaneous letter of complaint by university physicists against Timiriazev's authoritarian administrative style resulted in a departmental review by a higher party commission and deprived him of his institutional power. A Marxist philosopher from a rival "dialectical" faction, Boris Hessen, was appointed director of the Physical Research Institute at Moscow University and threw his support behind modern physics, favoring in particular Leonid Mandelstam and his group of advanced researchers.[3] Deprived of almost any authority in either physical or philosophical circles, Timiriazev subsequently occupied himself mostly with teaching the history of physics and found consolation in the company of a sole ally, Nikolai Kasterin (1869–1947). Kasterin abstained from Marxist argumentation but tried to develop an alternative to relativity on the basis of classical mechanics (Andreev 2000, 169–185). He was as thoroughly anti-Soviet as Timiriazev was pro, but the two were united in their stubborn opposition to Einstein, as well as in their virtually complete isolation. Soviet physicists firmly embraced the theory of relativity and quantum mechanics; expressing doubts was the surest way towards marginalization in their community (Vizgin and Gorelik 1987).

The revolutionary intellectual and ideological climate of the 1920s contributed to the overenthusiastic early acceptance of the radical new breakthrough in science. Einstein's relativity was equally popular at the time among avant-garde artists and the lay public as among professional physicists. A similar revolutionary association also helped the reception of another modernist new science, Mendelian genetics in biology, which underwent an explosive development in Soviet Russia of the 1920s, second only to that in the United States. By the late 1930s, however, the general intellectual climate became considerably more conservative, or rather mixed. In the arts, the radical avant-garde gave way to more traditional naturalistic painting in recognition and fashionableness, and generally it became possible, while still praising the revolution, to celebrate the pre-revolutionary Russian tradition, which in the sciences also meant the tradition that preceded the breakthroughs of the earlier part of the century. This gave the previously silenced critics and skeptics of the newer theories some delayed opportunities to counterattack.

In biology, a combination of this partial conservative revival with an available powerful alternative—Lysenko and his theories—allowed a temporary reversal: criticism of genetics grew stronger in the late 1930s and led to its institutional suppression in 1948. In physics, the key powerful positions remained firmly with those who supported modernist theories, while the critics were fighting an uphill battle in occasional publications. Having by that time at their disposal fewer scientific arguments, they tried to compensate for this weakness by stronger philosophical and ideological critique. In 1937 the Academy of Sciences planned a discussion of the philosophical issues of modern physics with Timiriazev,

---

[3] On Mandelstam and his work see (Livanova and Livanov 1988; Mandel'shtam 1979). On Hessen as director of the university physics institute see (Andreev 2000, 67–83).

Kasterin, and Vladimir Mitkevich—another proponent of classical physics—as speakers, but other physicists blocked the initiative, downgrading the discussion to several polemical articles in a philosophical journal (Gorelik 1993).

Philosophers took some part in that exchange, though their main problems lay elsewhere. They rarely questioned the physical content of the new theories but were very much concerned about the philosophical implications that could be derived from them. Both relativity and quantum mechanics were, of course, laden with philosophy, but by no means were those philosophies Marxist dialectical materialism or even simpler 19th-century-style materialism. A Soviet philosopher's typical repertoire thus included: praising scientists such as Einstein or Bohr for their great revolutionary contribution to science, criticizing them for their "philosophically wrong" statements, and trying develop better, dialectico-materialist interpretations of modern physics. The proportions in which a philosopher used these available options essentially defined his

Marxist philosopher A. A. Maksimov, fighter against idealist interpretations of modern physical theories. Maksimov took inspiration from Lenin's critique of the philosophical followers of Ernst Mach and other positivist reactions to the crisis in the foundations of physics in the early 20th century. Around 1950, he was actively accusing Soviet physicists of consciously or unconsciously deriving similarly non-materialist philosophical conclusions from relativity theory and quantum mechanics. Physicists retaliated in 1952 with accusations that Maksimov did not understand physics, and after the mid-1950s he could no longer publish his views in the Soviet press.
[*Courtesy*: IPPI.]

position in the controversy. Probably the harshest critic among the philosophers was Aleksandr Maksimov (1891–1976), who had some background in physics and devoted his life to struggle against "physical idealism," or philosophically wrong conclusions derived from science. He actively searched for philosophical "mistakes" in the writings of physicists, both foreign and Soviet, and demanded that the latter join him in this critique rather than mindlessly follow the misguiding "idealist" philosophy of Einstein, Bohr, and the like (Vucinich 2001).

The other line in Soviet Marxist debate was best represented by the philosopher Boris Hessen (1893–1936) and, after Hessen's death in Stalinist purges, by Sergei Vavilov. During the 1920s Vavilov's arguments in favor of new physics were mainly empirical (Vavilov 1925; 1926; 1928); by the 1930s, as the debate shifted towards philosophy, he started using Marxist dialectics. Simple logical consistency was on the side of philosophical critics like Maksimov, but since dialectics can be used to circumvent conclusions required by strict logic, it helped Vavilov and other authors dress up modern physics in dialectico-materialist clothing. A typical argument proclaimed that Einstein and Bohr, their personal philosophical wavering aside, had through their great contributions to science helped reveal the objective dialectics of nature. Vavilov then proceeded to elaborate on these dialectics—for example between waves and particles—and concluded that recent developments in physics had absolutely confirmed the ingenious predictions of Lenin's *Materialism and Empiriocriticism* and contributed to the general progress of Marxist dialectico-materialist worldview (Vavilov 1934b; 1934c; 1939; 1944; 1950a). While this protective rhetoric continued in philosophical and political publications, Soviet physicists made quantum mechanics and relativity a core part of the university curriculum and the basis for their research.

Generally, Soviet reasoning about science included as an essential part the argument of social constructivism that has recently become very popular in the West. In the views of Soviet Marxists, science and its concepts, like any other kind of human activity, were related to economic, political, and class interests. At the same time, they also strictly adhered to the principle of scientific realism, or the view that science provides true knowledge about nature. Today these two views are often considered to be contradictory, yet Soviet authors managed to marry them, at least in theory. The solution is brilliant in its triviality: interests can be either right or wrong; having the right (progressive) interests helps achieve the truth (or limited truth) in science. In his 1949 presentation on the occasion of Stalin's seventieth birthday, Vavilov refined this general epistemological thesis for the particular case of science: "*Partiinost'* [or party-mindedness—AK] of science is the expression of its correctness" (Vavilov 1949b, 12).[4]

This formula can be read in both directions, either that the correctness of science suffices to guarantee its *partiinost'* or the other way around. Although the statement was considered to be true on the general theoretical level, it was not straightforwardly applied to specific practical cases. Instead, Soviet Marxists typically tried to follow Lenin's example from *Materialism and Empiriocriticism*, distinguishing between the special problems of science, reserved for experts, and matters of interpretation, where philosophers and politicians had their say (Lenin 1909, 252). Vavilov often quoted Lenin's statement and strongly insisted on this sort of demarcation. In his papers he used physical arguments to demonstrate that modern physics represents truth and philosophical arguments to show that it proves dialectical materialism, but he did not infer one thesis from the other.

As part of his argument in favor of modern theories, Vavilov showed that philosophical criticism could be directed at classical physics as well. Although Newtonian theory was based on philosophically incorrect notions of space and time, he observed, it remained useful. More physical, Einsteinian notions of space and time were methodologically more sound, but not entirely free of difficulties either. Both classical and modern physics reflect objective knowledge about nature (limited truth), but one should not infer wrong philosophical conclusions from them (Vavilov 1950a, 89–90).[5] When Stalin came to publish his own paper on linguistics, he opened it with a similar demarcation: as a Marxist philosopher he was entitled to speak only on the question of Marxism in linguistics, not on special problems (Stalin 1950, 407).

This demarcation helped but did not make easy the task of philosophical evaluation of a particular science. A failure to develop a philosophical interpretation compatible with dialectical materialism could lead to the criticism and possible rejection of a scientific theory—for example, relativistic cosmology that declared the possibility of a finite universe—but such cases were rather rare. For most sciences, interpretations compatible with Marxism could be proposed, but often several ideological interpretations competed, each reflecting the particular interests of the actors

---

[4] See (Joravsky 1961, 25) for the explanation of the word *partiinost'* as referring to the Marxist sociology of knowledge.
[5] According to the reconstruction by Loren Graham, Boris Hessen's social constructivist criticism of Newton served a similar rhetorical purpose of defending modern physics (Graham 1985).

involved. This sort of theoretical interaction and competition can be viewed as playing with the boundary between the domains of "special problems" and "philosophical interpretations"; different authors could draw the line differently, emphasizing as philosophically relevant those aspects which supported their side in the controversy.

In 1947, when the party put forward the slogan of "creative disputations," Vavilov attempted to shift the philosophical balance even further in favor of modern theories. He encouraged Moisei Markov, a young theoretical physicist from FIAN who had an interest in both quantum and Marxist philosophies, to write down his views on the philosophy of physics. Markov's paper, "On the nature of physical knowledge," attempted a "consistent materialistic interpretation ... of the theory of complementarity" (Markov 1947, 142). This was a risky move because Bohr's complementarity was often criticized as an idealistic interpretation, and it was much more common for the Soviet authors to exclude it from the body of accepted theory of quantum mechanics (for example, the volume on quantum mechanics from the famous Landau–Lifshitz course of theoretical physics, now a standard textbook in many countries, avoids mentioning complementarity altogether). Markov remembers that he had doubts concerning the wisdom of publishing the piece, but Vavilov discussed the matter with Agitprop officials in the party's Central Committee and received approving signs (Markov 1990).

Bonifaty Kedrov welcomed Markov's paper and published it in the new philosophical journal *Voprosy Filosofii*, which he edited. Vavilov wrote a short introduction, expressing his hope that the paper would "become the starting point for a large serious dispute (*diskussiia*)" which would not "be reduced to branding with infamous labels—a detailed and practical analysis of the essence of the question is needed. Hopefully ... similar studies in other fields of natural science, especially in astronomy and biology, will appear in *Voprosy Filosofii*" (Vavilov 1947). Markov's paper indeed immediately became a subject of a fierce polemics. Maksimov accused it of idealism and attacked Kedrov for publishing it. In response, Kedrov printed in the June 1948 issue of his journal a collection of polemical letters whose authors mainly sided with Markov and denounced Maksimov's critique as dogmatic (Diskussiia 1948a).

The course of further discussion was influenced by the August 1948 Session that brought Lysenko to the pinnacle of power in biology. As a political event, the August Session set a new model and a new stimulus for imitation and provided fresh inspiration. "Creative disputations" were giving way to more militant styles of polemics. The criticism of idealism—always a welcome rhetorical resource—became especially fashionable in the following season. This had implications for physics, too.

2. Professors vs. Academicians

The physics community was split along different lines than the biological one: the dominant controversy was institutional rather than conceptual. Scientific disagreements, as well as various philosophical and political issues, played a subsidiary role in the main conflict, at the core of which was the opposition between Moscow University and the Academy of Sciences. Members of Joffe's, Kapitza's, and Vavilov's institutes dominated the Physics Division of the Academy. Leonid Mandelstam's group once represented Moscow University there, but Mandelstam had died in 1944 and his students Landsberg, Papaleksi, and Tamm gradually moved from the university to Vavilov's FIAN, mainly because of tensions among the faculty. None of the remaining physics professors achieved a full membership in the Academy, the most prestigious position in Soviet science. Since the Academy was well above the universities in the scientific hierarchy, professors often complained of academicians' monopoly on journals, privileges, and resources and were accused in return of providing institutional protection to low-level science or even "antiscience."[6]

One of the most open struggles broke out over the university chair of theoretical physics. Tamm had organized the department's *kafedra* of theoretical physics and headed it before the war, but the department refused to reelect him to this position in 1944. The roots of that conflict probably go back to internal departmental tensions during the 1930s, and it was in this time that Tamm published a devastating critique of Aleksandr Predvoditelev's ventures into theoretical physics. Predvoditelev (1891–1973), whose primary field was experimental molecular physics, later became the department's dean and in this position enjoyed the support of the majority of the academic

---

[6] There were similar conflicts in other fields and in other universities, obviously reflecting a general problem. In trying to solve them, the Stalinist bureaucracy in the late 1940s discussed proposals for serious reform aimed at raising the status of the universities, both socially and academically. Only restricted measures were taken, however. See (CPA, 17-125-342, pp. 42–84; 17-125-361, pp. 66–140). Vavilov typically supported plans to improve the state of university teaching and research, but not the critique of the Academy. See for instance Vavilov, "Vstupitel'noe slovo," *Vestnik AN SSSR* 1947, 1:23-25, on p. 25; (CPA, 17-125-543, pp. 57–75, 154–158).

council. During the war Tamm accompanied FIAN in its evacuation and thus separated from the university. When both institutions returned to Moscow from their wartime locations he applied to be reappointed to his university chair, but the academic council voted to promote instead a younger associate at the *kafedra* of theoretical physics, Anatoly Vlasov (1908–1975) (Andreev 2000, 102–106).

Vlasov defended his dissertation not long before the war and was not tarnished by the earlier history of departmental skirmishes. In the dissertation, and in an accompanying article (Vlasov 1938), he made a landmark contribution to the theory of plasma, currently known as the Vlasov kinetic equation. After returning from evacuation with the University in 1943, he embarked on a much more ambitious program, using the fact that his non-linear kinetic equation allowed a variety of non-trivial solutions. Encouraged by his earlier success, he aspired to derive from plasma "the spontaneous origin of a crystal structure" and to reconsider the very conceptions of the 'gaseous,' 'liquid,' and 'solid' states of matter (Vlasov 1945a, 25; 1945b).

He was punished for his inflated ambitions, some mathematical mistakes, and an unjustified assumption by a severe collective critique, "On the fallacy of the works by A. A. Vlasov on the generalized theory of plasma and the solid state," which came from a group of four high-ranking Soviet theoretical physicists, including two full members of the Academy (Ginzburg, Landau, Leontovich, and Fock 1946). The dogmatic and uncompromising tone of the attack was only partly attributable to the vicious style of polemics common in Stalin's times. It also reflected the continuing state of administrative warfare over the university chair of theoretical physics. Vlasov was not even given an opportunity to publish his reply in the same journal, controlled by the Academy, and had to send it to a less influential periodical of the university.

A group of concerned members of the Academy complained to the minister of higher education about the university's decision to replace Tamm as the chair of theoretical physics with a physicist of much lesser reputation. The ministry created a commission to review the department and asked Vavilov to chair it. Vavilov probably sympathized more with the Academy's physicists, but he also felt allegiance to his alma mater and tried to be fair to both sides. The Academy achieved partial success in 1946 when the ministry overruled the departmental election, replacing both Predvoditelev as the dean and Vlasov as the chair of theoretical physics. Yet the new appointees managed to hold their posts for less than

The first major post-war elections to the Academy of Sciences in 1946. The enormous political importance acquired by science in the age of atomic weapons resulted in the dramatic expansion of the Academy's institutes and financing during the immediate post-war years. The membership of the Academy increased, while the privileges and prestige associated with the status of academician grew even more strongly. In the photo, A. F. Joffe proposes candidates for membership selected by the Division of Physical and Mathematical Sciences for the final vote by the Academy's General Assembly in November 1946.
[*Courtesy*: RGAKFD.]

a year, as the majority of the department's academic council were strongly against them.[7]

The polarization between the Academy and university physics continued to grow stronger, with a number of other related and unrelated conflicts adding fuel to the fire. Predvoditelev complained to the authorities,

---

[7] For more on the conflict see (Andreev 2000, 106–115; Gorelik 1991; Andreev, Kozhevnikov, and Yavelov 1994; CPA, 17-125-361, pp. 19–63).

accusing the Academy members of prejudicially blocking his promotion from the corresponding to full member. Other professors had their own, mostly personal agendas and grievances against the Academy or some of its members. After the 1948 August Session, the university physicists saw a chance to counterattack.

3. Anatomy of a Discussion

The earliest document pertaining to the discussion in physics is a November 1948 draft of a letter from the Minister of Higher Education, Sergei Kaftanov, and the President of the Academy of Sciences, Sergei Vavilov, to the Central Committee Secretary, Georgy Malenkov.[8] It expressed concern about defects in physics teaching and research due to the lack of a criticism of idealism and asked for permission to organize a conference on this topic. Since the draft talked about the "All-Union Council of the directors of physics *kafedras* at the institutions of the Ministry of Higher Education," it must have originated in the ministry. Vavilov edited the document and added "with the participation of the Physico-Mathematical Division of the Academy of Sciences." One can understand his move as an effort to retain a voice for the Academy in the proceedings. After all, Lysenko's success had owed a great deal to his ability to convene the Agricultural Academy and fix the official result before the Academy of Sciences could react.

With preliminary approval from the Central Committee Secretariat, the Ministry and the Academy put together the joint Organizing Committee of the "All-Union Council of Physicists" with Aleksandr Topchiev, deputy minister, as chairman.[9] The council (*soveshchanie*) was a genre of meeting that usually had a practical rather than a political function: administrators listened to the instructions of a higher official and talked about how they might do their jobs better. With Vavilov's participation, some work had been done on the earlier proposal in order to turn the discussion in a practical direction, away from polemics. The topic was changed to a politically neutral one: "the present state of physical science in the USSR and

---

[8] (ARAN, 596-2-173, pp. 7–11). Malenkov regained Stalin's favor as Zhdanov lost it in July 1948.
[9] "Postanovlenie Kollegii Ministerstva Vysshego Obrazovaniia i Prezidiuma Akademii Nauk SSSR, 17 dekabria 1948 g" (ARAN, 596-2-173, pp. 1–6, 12–14, and *Tsentral'nyi Gosudarstvennyi Arkhiv Rossiiskoi Federatsii,* hereafter *GARF,* 9396-1-244).

improvements in the teaching of specialists in physics"; the philosopher Maksimov, with his talk "The Struggle against Idealism," was dropped from the program. The only keynote speech on questions of general political importance was to be Vavilov's talk "On the State of Modern Physics and the Tasks of Soviet Physicists," scheduled for the plenary session on the first day. This was to be followed by general discussion; later sessions would take up more specialized and practical problems. The Central Committee Secretariat officially approved the proposal on 31 January 1949 and set the date for 21–26 March. Vavilov, who was busy putting together three other Academy meetings on unrelated topics, was not a member of the Organizing Committee.[10]

Meanwhile, the committee worked hard. The university representatives, with some support from philosophers, ministry officials, and provincial physicists, wanted the conference to be conducted "on the level of the discussion at the August Session and ... on the high *ideinyi* (high-principled) level." Talks on neutral topics were approved without long dispute, but many speakers wanted to deal with questions related to politics and ideology and to turn the discussion towards *kritika i samokritika*. Even before the first draft of Vavilov's main talk became available, they had started presenting the texts of militant discussion comments on his topic. The transformation of the genre and of the prevailing mood was not achieved overnight. The Organizing Committee worked for three months, holding forty-one long sessions that were attended by concerned activists as well as committee members. The university party proved to be more active and better prepared for the political discussion.[11] They presented episodes of the previous struggles between academicians and professors as politically charged and zealously searched for "philosophically wrong" quotations in books and talks by their academic opponents.

The second dominant ideological topic emerged after 28 January, when *Pravda* published the editorial "On One Anti-Patriotic Group of Theater Critics." This brought to a climax the campaign of exposing and criticizing

---

[10] A meeting on the history of Russian science, another one on the occasion of the 25th anniversary of Lenin's death, and the regular annual meeting of the Academy. Vavilov's talks at these meetings are published in *Vestnik AN SSSR* 1949, 1: 3–4; 2: 6–10, 38–53, 124–125; 3: 5–6.

[11] The statistical summary of the Committee activities shows that university representatives attended at least twice as much and spoke at least three times as frequently as their opponents. See "Otchet o rabote orgkomiteta" (ARAN, 596-2-173, pp. 31–64). In total, 106 people participated in 35 open sessions. Typical argumentation is quoted at length in (Sonin 1994, 114–160).

so-called cosmopolitans, or insufficient patriots, many of whom were found among ethnic Jews. On 2 February Nikolai Akulov was already speaking to the Organizing Committee on this topic. At an unrelated meeting in the Institute of Philosophy, Kedrov, Vavilov's main ally among philosophers, was obliged to repent his "cosmopolitan," or too internationalist, pronouncements, subsequently losing control over *Voprosy Filosofii*.[12]

Although the division into two academic parties was quite noticeable, nobody mentioned it explicitly, and nobody accused the Academy itself, any particular institute, or a theory like quantum mechanics or relativity of idealism. Vavilov's authority was not openly challenged, and criticism did not touch those who worked on the atomic bomb project. All accusations were raised by individuals and directed against individuals. The main targets were several influential academicians, most of them Jewish. The defense was personal, too: Aleksandr Andronov managed to defend his deceased teacher Mandelstam against Akulov's accusations that he had worked for Germany. But nobody could help Yakov Frenkel in the face of quotations drawn directly from his open polemics with dialectical materialism—even if these did date from 1931. Consensus on his philosophical mistakes was reached quickly.[13] Markov's paper on complementarity also received much criticism.

Vavilov knew what was going on in the committee when he passed the first draft of his talk to Topchiev on 10 February. He had already changed the title to "Philosophical Problems of Modern Physics and the Tasks of Soviet Physicists." The manuscript criticized idealistic statements, chosen mainly from popular writings of Western authors, Soviet physicists' (in particular Landau's and Frenkel's) uncritical attitude and reluctance to write on philosophical questions, and the use of the philosophically "incorrect" terms "spontaneous" and "annihilation." Vavilov gave a balanced critique of Markov's "dogmatism" in presenting quantum mechanics as a completed theory but defended him against accusations of idealism.[14]

---

[12] (GARF, 9396-1-256, pp. 167–175). Simultaneously, Akulov brought his charges against Mandelstam, Joffe, Kapitza, and others to the attention of Malenkov (CPA, 17-132-211, pp. 105–115). Vavilov was mentioned in his denunciation letter as the one who was under the influence of antipatriots. On the events in philosophy and accusations against Kedrov see (Za bol'shevistskuiu 1948) and (CPA, 17-132-160, pp. 46–52).

[13] Organizing Committee sessions on 27 and 28 January, presentation by P. E. Zrebny (GARF, 9396-1-256, pp. 1–77; and ARAN, 596-2-174).

[14] Vavilov, "Filosofskie problemy sovremennoi fiziki i zadachi sovetskikh fizikov," (ARAN, 596-1-80, pp. 27–72).

Vavilov opening the founding meeting of the All-Union Society for Dissemination of Political and Scientific Knowledge, July 1947. Sitting left in the front row are A. V. Topchiev and G. F. Aleksandrov.
[*Courtesy*: RGAKFD.]

The Organizing Committee discussed Vavilov's manuscript for two days. Though Topchiev succeeded in getting a formally polite passed resolution granting approval "of the paper in general" but with the request that Vavilov improve it and take into account comments by other speakers, the resolution did not mask serious disappointment with the talk's "narrative rather than militant character." Suggestions included changing the title to "Ideological Problems of Modern Physics," strengthening the critique of Soviet physicists, removing the defense of Markov, and making explicit mention of the August Session and cosmopolitanism.

In further drafts of his talk Vavilov gradually acceded to many of these demands.[15]

By the end of February the Organizing Committee was approaching consensus. Frenkel and Markov were instructed to rewrite their speeches and to criticize their own philosophical mistakes (*samokritika*). Accusations against Fock and Tamm failed to gather enough support, but the two were advised to sharpen their critique of others (*kritika*).[16] A large conference room was reserved, quotas for representation of groups among the 600 participants were agreed upon, and invitations were printed. By mid-March the committee had approved texts of almost all the major talks and short discussion comments, compiled the manuscripts into a volume intended for eventual publication, drafted the general resolution, and sent all the material to the Party's Central Committee for final approval. In its critical part, the resolution blamed Landau and Joffe for *rabolepie* (obsequiousness) before the West, Kapitza and Kedrov for cosmopolitanism (although neither of the two was Jewish), and Frenkel and Markov for uncritical use and propaganda in favor of idealistic aspects of physical theories. In its constructive part, the resolution suggested establishing an All-Union Physical Society, revising the editorial boards of journals, and increasing the number of graduate students and funds for research at the universities.[17]

Everything seemed to be ready, except the final version of Vavilov's main presentation. The scheduled opening day, 21 March, passed and nothing happened, as there was apparently no reply from the Central Committee. On 2 April Kaftanov submitted another letter to the Central Committee proposing to reschedule the meeting for 10 May. This time, however, Vavilov did not add his signature to second Kaftanov's proposal, as he had done earlier with all similar documents. The matter was reported to the Secretariat on 9 April by the director of Agitprop, Dmitry Shepilov. Shepilov added a brief note that also suggested postponing

---

[15] Organizing Committee sessions on 16 and 18 February (GARF, 9396-1-249, pp. 245–275; 9396-1-250, pp. 1–33; ARAN, 596-2-174). Vavilov, "Filosofskie..." (note 14), handwritten insertions, and "Ideologicheskie problemy sovremennoi fiziki i zadachi sovetskikh fizikov" (ARAN, 596-1-80, pp. 73–129), typed text and handwritten insertions. An abridged version of Vavilov's paper was published after his death in (Maksimov *et al.* 1952). It is unlikely that Vavilov had given permission for the publication.

[16] Organizing Committee Sessions on 21, 22, 25, 28 February and 4 March.

[17] "Proekt Postanovleniia Vsesoiuznogo Soveshchaniia Fizikov" (GARF, 9356-1-244; ARAN, 596-2-173, pp. 65–73).

the meeting, which he said had not been sufficiently prepared for. Shepilov's note differed from Kaftanov's proposal in one important detail: it did not specify any particular date. Vavilov was not officially involved in these machinations, but he managed to be present when the Secretariat discussed the issue because he was scheduled to report on another matter. He was probably prepared to blame his own overwork for the unfinished state of his manuscript, if need be. The Secretariat had to validate many decisions every day, and it approved Shepilov's formulation apparently without long consideration. Thus the seemingly unstoppable movement of a huge bureaucratic machine was derailed by means of a subtle bureaucratic intrigue involving Shepilov and Vavilov.[18]

The decision had the effect of postponing the dispute forever. Kaftanov's detailed proposal of 5 April "On Serious Shortcomings in the Preparations of New Cadres of Physicists and on Measures to Overcome Them," which drew heavily on the Organizing Committee's drafted resolution, remained unanswered by Agitprop until October, when it was passed to the archive with a note that the meeting had been "canceled" by the Secretariat.[19] Had the carefully rehearsed performance been played publicly, it would have resulted not in the ban of a certain theory but in personal promotions and demotions, shifting the balance in the existing hierarchy of the physics community. Instead, the new editors of *Voprosy Filosofii* published another collection of replies to Markov's article, this time declaring it wrong, but a journal publication did not have the same authority as a resolution of the representative meeting would have had and did not have administrative consequences (Diskussiia 1948b). The institutional conflict in Soviet physics between the Academy and the University effectively remained unresolved until Stalin's death and until nuclear physicists emerged from their secret work with the new authority bestowed upon them by the success of the atomic bomb project.

---

[18] "Protokol no. 426 zasedaniia sekretariata T_sK VKP(b), 9 aprelia 1949" (CPA, 17-115-806, 426/334, p. 202); Kaftanov to Malenkov, 2 April 1949, and Shepilov to Malenkov, not dated, in "Materialy k protokolu..." (CPA, 17-118-360, pp. 168–170). Besides Malenkov, the Secretariat consisted of P. K. Ponomarenko, G. M. Popov, and M. A. Suslov. Vavilov presented to them the proposal to elect Topchiev as a full member of the Academy, which could have been part of the entire intrigue (CPA, 17-115-806, 426/333, p. 200; 17-118-360, pp. 159–167).

[19] Kaftanov to Malenkov, 5 April 1949, "O krupnykh nedostatkakh v podgotovke kadrov fizikov i merakh po ikh ustraneniiu" (CPA, 17-132-211, pp. 77–94); P. Zherebtsov to the Central Committee Technical Secretariat, 6 October 1949 (CPA, 17-132-211, p. 95).

4. Paradigm Shift, Soviet Style

The year 1949 thus passed without a major *diskussiia*, and the Lysenko's August Session was not eclipsed by similarly important political event in sciences until 1950, when two discussions occurred almost simultaneously. The first of them, the "Free Discussion on linguistics in *Pravda*" broke out in May 1950 completely unexpectedly, even for Agitprop. The controversy shattered the emerging order and reversed the consensus that had nearly been achieved in the field, which already had passed through several consecutive rounds of *kritika i samokritika*.

A figure in Soviet linguistics who was in some respects similar to Lysenko, Nikolai Marr (1865–1934), was a mixture of genius and insanity, with a tendency to develop from the former towards the latter. He spoke an enormous number of languages, in particular those of the Caucasus and other linguistically complicated parts of the world. The Caucasus remains a problem for standard systems of linguistic classification even now. Marr's groundbreaking studies of this area challenged the accepted Indo-European theory. In 1923 he announced a complete break with that theory and started developing what would become known as the "new doctrine on language." In place of the existing picture of multiple languages developing from a few common ancestors, Marr substituted a reverse evolution from an initial larger variety, through mixture, towards the future unification of languages. In Marr's scheme, independent languages passed through common stages which corresponded to the level of the development of the society. This later offered him an opportunity to connect his theory with Marxism, declare it materialistic, and oppose it to bourgeois Western linguistics (Alpatov 1991, ch. 1–2).

In the battles of the cultural revolution around 1930, Marr and his school defeated their non-Marxist and Marxist opponents and achieved a monopoly in the field. Upon his death in 1934, Marr was beatified as one of the "founding fathers" of Soviet science along with Michurin, Pavlov, and the soil scientist Williams. The "new doctrine on language" became the official Soviet linguistics. Its keeper and the heir of Marr's position in the Academy, Ivan Meshchaninov (1883–1967), adopted a conciliatory approach: heresies and pluralism in actual research were tolerated, so long as ritualistic loyalty was expressed and the political status of Marrism as the Marxism in linguistics was not challenged (Alpatov 1991, ch. 3–4). Alas, this compromise did not survive the test of the discussion campaign.

The genres of discussions in linguistics from 1947 to 1950 were dictated by the need to respond to and hold *obsuzhdeniia* of the model events: Zhdanov's 1946 critique of the literary journals *Zvezda* and *Leningrad*, the 1947 Philosophical Dispute, and the 1948 August Session. Correspondingly, linguists reviewed the work of their own journals, discussed the quality of their own textbooks, and criticized idealism. But, driven largely by the aspirations of two deputy directors—of the Moscow Institute of Language and Thought, Georgy Serdiuchenko, and of the Institute of Russian Language, Fedot Filin (1908–1982) these ritualistic performances were suffused with exposé and criticism of those who deviated from Marr (Obsuzhdenie 1948c; 1948d; Ocherednye 1948).

The titles of the two main talks at a joint meeting of the Leningrad branches of these institutes in October 1948, "On the Situation in Linguistic Science" and "On the Two Trends in the Study of Language," were borrowed from Lysenko's address at the August Session. In fact, there were not just two but three main trends in linguistics, for Marrists attacked modern structuralism as well as classical Indo-European linguistics, but the ritual of imitation proved to be stronger than logical considerations. Meshchaninov, who spoke first, took a softer theoretical approach, which showed his reluctance to fight. After all, he could not avoid some self-criticism for having tolerated idealists so long. Trying to draw parallels between linguistics and biology, he equated Wilhelm Humboldt's "spirit of the nation" with the "hereditary substance" and both Indo-European theory and genetics with racism. The second speaker, Filin, provided more militant and practical criticism, calling for the "total scientific and political exposure" of open and hidden non-Marrists and arguing that peace in Soviet linguistics was only illusory and that the struggle between materialism and idealism had to break out (Meshchaninov 1948; Filin 1948).

Besides conceptual considerations, institutional ones were obviously in play, since the main target of criticism was Viktor Vinogradov (1895–1969), who was not the most open non-Marrist but the most highly placed one. He directed the Philological Department of Moscow University and had recently become a full member of the Academy. At the 1947 discussion, Vinogradov's textbook *Russian Language* had been criticized (Obsuzhdenie 1948c). Now Filin accused him of sticking to his views even *after* that dispute. "Undisarmed Indo-Europeanists among us have to think carefully! They must abandon incorrect methodological principles not only in words, but also in deeds," he concluded (Filin 1948, 496).

Several similar local battles took place during 1949, in which Marrists gradually suppressed heretics one by one, institute by institute (Alpatov 1991, 149–167). Vinogradov was driven to engage in *samokritika* a couple of times, repented in words, and resigned as the department's dean, but survived as chair of the university's *kafedra* of Russian language. Only a few were fired, but many more were forced to denounce former views and at least formally subscribe to the prevailing orthodoxy. As the Marrists' major administrative success, the Ministry of Higher Education ordered changes in the curriculum, and the Academy changed the research plans of the institutes. Only on the periphery, in particular in Georgia and Armenia, had a few open dissidents not yet been disciplined. (Topchiev 1950; Samarin 1950; Alpatov 1991, 150–151, 157, 167).

Thus, even without the participation of politicians and without a major meeting, the linguistics community was straightening itself out. A real political event, however, would have certainly helped to fix the consensus. Starting in July 1949 the Academy of Sciences sent reports to the Central Committee about its decisions against anti-Marrists and about the continuing struggle. Agitprop supported its position and was quite willing to host a meeting with linguists (all Marrists) "in order to finish the work of discussing the situation in Soviet linguistics and to submit to the Central Committee a proposal on the improvement of work." The Secretariat, however, responded in January 1950 that the discussion should be organized by the Academy itself, without party organs.[20]

Meanwhile, disagreements were developing among Marrists. Meshchaninov was still trying to keep to the middle ground, accepting that there were mistakes in Marr's doctrine too and that it needed *creative* development. But his position as institutional leader was becoming shaky as radicals criticized him ever more often and more openly. On the other hand, on 13 April 1950, Suslov received a report that referred to information received from the Academy of Pedagogical Sciences and accused Serdiuchenko of intolerance, lack of professionalism, denying any mistakes in Marr's works, and opposing *samokritika*. Suslov showed a willingness to distinguish between what was ideologically right and wrong in Marrism: his draft stated that "scientific problems cannot be solved administratively" and mentioned the need to organize a *diskussiia*.[21] But the crucial moment had already occurred three days before, when leaders of the Republic of Georgia presented Stalin with a new *Encyclopedic*

---

[20] I. P. Bardin (Academy of Sciences) to Malenkov, 30 July 1949 and November 1949; Kruzhkov and Zhdanov (Agitprop) to Suslov, 19 August 1949, November 1949, and 10 January 1950 (CPA, 17-132-164, pp. 16–114).

[21] P. Klimov and P. Tret'iakov to Suslov, 13 April 1950, and Suslov's remarks (CPA, 17-132-336, pp. 4–9). See also I. I. Meshchaninov, "Dokladnaia zapiska," 12 April 1950, and Academy of Sciences to Malenkov, 17 April 1950 (CPA, 17-132-336, pp. 10–76).

Dictionary of the Georgian Language. They also introduced him to the dictionary's editor, Arnold Chikobava (1898–1985). Probably the most open fighter against Marr's "new doctrine on language," Chikobava had called it anti-Marxist and racist because it placed Indo-European languages higher than Georgian on the developmental scale. Supported by republican party leaders and enjoying a stronghold in the Georgian Academy of Sciences and the University of Tbilisi, Chikobava remained one of the few who had not yet been subdued (Chikobava 1985).

As a result of his meeting with Stalin, Chikobava got the commission to write down his views as a discussion note: "You will write, we will consider," said Stalin. They met twice more to discuss the text, and on 9 May 1950 the linguistic order was broken again:

> "In connection with the unsatisfactory state of Soviet linguistics, the editors consider it essential to organize an open *diskussiia* in *Pravda* in order to overcome, through *kritika i samokritika*, stagnation in the development of Soviet linguistics and to give the right direction to further scientific work in this field... Arn. Chikobava's article 'On Certain Problems of Soviet Linguistics,' is printed as a matter of dispute" (Chikobava 1950).[22]

In his essay Chikobava accepted Marr's early works on the theory of Caucasian languages, but not the general linguistic theory; he praised Marr's desire to become a Marxist but denied the thesis on the class nature of language, thus accusing Marr of being "unable to master the method of dialectical materialism and to apply it to linguistics" (Soviet 1951, 10).

Reportedly, *Pravda* received over two hundred letters in response to the article (Alpatov 1991, 169). In numbers Marrists should have prevailed, but the letters selected for publication constituted a very symmetrical set. In articles as long as Chikobava's, Meshchaninov praised Marr, while Vinogradov was inconclusive. The same structure of one positive, one negative, and one opportunistic letter was preserved in three other issues. Every Tuesday readers—workers and peasants, intellectuals and policemen—received a sophisticated scholarly-ideological discourse on linguistics, knowing neither why the subject had suddenly become a matter of general political importance nor what the truth was. Then, on the seventh week came the following message: "We continue to print articles sent to *Pravda* in connection with the dispute in Soviet linguistics. Today, we

---

[22] "As a matter of dispute" (*v diksussionnom poriadke*)—a necessary remark to let readers know that, despite publication in *Pravda*, the text should not be considered authoritative.

publish articles by I. Stalin "Concerning Marxism in Linguistics," and Prof. Chernykh "Toward a Critique of the 'New Doctrine on Language'" (*Pravda*, 20 June 1950).

It may be that Stalin originally planned to participate and gave himself some time to develop an opinion, or that his contribution was triggered by one of the articles of the previous week, which was devoted almost entirely to the question of class and language. Having admitted in the beginning that he was "not a linguist and cannot of course, satisfy comrades fully," Stalin continued: "As to Marxism in linguistics, as well in other social sciences, this is a subject with which I have a direct connection." (Stalin 1950, 407) From the linguistic point of view, Stalin's paper consisted of trivial but surprisingly competent statements; from the point of view of orthodox Marxism, it certainly would have been considered heretical had the author been anybody else. Stalin denied not only that language was a class phenomenon but also that it had a place in the superstructure, which none of Marr's harshest critics dared to do. The stress on the class issue, once a very powerful ideological resource, proved to be a misfortune for Marrism. By the 1940s internationalist class rhetoric had lost its central role in Soviet ideology to nationalist themes, although it continued to receive lip service. In the end, Stalin approved *Pravda's* (in fact his own) decision to open the dispute and accused Marr's school of suppressing critics and free discussion, which could have revealed the mistakes and the non-Marxist nature of the theory. "Elimination of the Arakcheyev [=police—AK] regime in linguistics, rejection of N. Ya. Marr's errors, and the introduction of Marxism into linguistics are, in my opinion, the way in which Soviet linguistics may be put on a sound basis" (Stalin 1950, 430).

The "Free Discussion in *Pravda*" lasted another few weeks, but the discourse changed from *diskussiia* to *obsuzhdenie* (commentary, praise, and further applications) and *kritika* and *samokritika*. Then the time came for more practical meetings in ministries and institutes and for administrative changes. Meshchaninov, Filin, and Serdiuchenko lost their administrative jobs and became ordinary scholars. Their institutes were merged into the Institute for the Study of Language, with Vinogradov as its director and the new leader of the field (Postanovlenie 1950). "Stalin's doctrine on language" was the hottest ideological topic until 1952, when the "corypheus of science" wrote another theoretical piece on political economy. Dozens of volumes and hundreds of articles commented on Stalin's paper and were "introducing Marxism into linguistics." The result of this party involvement in science and of the suppression of a scientific theory by Stalin's heavy hand was, in the case of linguistics, the opposite of what happened in biology—the rehabilitation of the classical and international, comparative Indo-European approach and the defeat of a maverick homegrown rival. One older academic even spoke of Stalin's piece as of a "sobering voice of reason" (Tolstoi 1950). The rehabilitation of structuralism would have to wait a few more years until Khrushchev's liberalization.

5. Changing Guards

Vavilov had little role in the linguistic controversy, but a month later he presided over the second major dispute of the year, the Pavlov Session in physiology. Unlike the Free Discussion, that meeting had been in preparation for about a year. The main moving force behind it, Yury Zhdanov, claimed later that he had wanted to counterbalance the August Session with something more reasonable (Zhdanov 1993, 88). It is clear from the archival documents, however, that his goal was to end the monopoly of Leon Orbeli (1882–1958), who had inherited from Ivan Pavlov the main physiological institutions. Other pupils of Pavlov's agreed to criticize Orbeli and receive their share of the institutes. Zhdanov had learned the lessons of the Lysenko affair and rehabilitated himself. He prepared the Pavlov Session without haste, in a professional bureaucratic way, and secured Stalin's approval for it (Nauchnaia 1950). Every politically important event in those days needed an ideological rationalization; the high principle applied in this case was strict faithfulness to Pavlovian doctrine, despite the fact that it did not belong to the body of Marxism–Leninism. This also brought under fire several other unorthodox physiologists and psychologists and resulted in another monopoly in the field.[23]

Vavilov's heart disease intensified that year, and he spent several months in a sanatorium. He probably understood that he did not have much time left, and he started writing an autobiography. In October he had a meeting with Joffe, after which Joffe submitted a letter of resignation from the directorship of Leningrad Physico-Technical Institute. A party commission had been criticizing Joffe for allowing nepotism, or the tendency for members of the same family to hold academic appointments within the same laboratory or institute, and for hiring predominantly Jewish research staff. The public campaign against 'cosmopolitans' was

---

[23] (CPA, 17-132-161, pp. 43–73, 180–186; 17-132-177, pp. 144–162; 17-132-347, pp. 1–10). On Orbeli see (Grigor'ian 1997).

complemented by the revision of another consequence of the cultural revolution—the high level of Jewish representation in responsible administrative positions. The Academy, too, quietly reduced the number of Jewish directors of research institutes.[24]

One of the last known of Vavilov's actions concerned another colleague, Piotr Kapitza, who fell into disgrace. In his official role, Vavilov passed through the Presidium a decision confirming the government 1946 decree and firing Kapitza from the directorship of his institute. Kapitza formally retained only a couple of less significant positions, but by 1950 he lost even them.[25] Vavilov then privately asked the director of the Institute of Crystallography, Aleksei Shubnikov, to formally appoint Kapitza senior research fellow (Shubnikov 1991). There was not much sympathy between them, and Kapitza was surprised to receive an invitation to Vavilov's home and even more so to find the host dangerously open and astonishingly bitter in that private conversation (Kapitza, S.P. 1990). The practical reason for that meeting was probably the struggle over Kapitza's former institute. Vavilov tried to resist changing its profile from low-temperature physics to nuclear research, and on 24 January he met with the directors of the Soviet atomic project, Igor Kurchatov and Avraamy Zaveniagin (Frank 1991a). The next morning, 25 January 1951, he died.

Vavilov's death may have emboldened Maksimov to bring his personal crusade against 'physical idealism' to its highest. In 1952 he published with some revisions the draft of Vavilov's undelivered report of 1949—in which Vavilov was pressed to include ideological criticism of Markov and other Soviet physicists. The text appeared in a volume co-edited by Maksimov and three other university authors, in which a younger contributor actually dared to suggest that Einstein's wrong philosophy corrupted the general theory of relativity (Maksimov et al. 1952; Vucinich 2001, 107–115). Another article by Maksimov, "Against the reactionary Einsteinianism in physics," apparently met serious difficulties in publication, and he was able to have it printed only in an inappropriate place, a newspaper of the Red Navy. It still made physicists worry, and this time the leaders of the countercharge were Vladimir Fock as the country's top specialist on general relativity (and also an expert on Marxism) and Kurchatov as the new authority in dealing with politicians.

Unlike Vavilov, Fock became a committed Marxist in addition to being a convinced believer in Bohr's complementarity and Enstein's relativity.

---

[24] (Sominsky 1964; Kosarev 1993, 155–156); (CPA 17-132-169, pp. 180–184).
[25] Kapitza's personal dossier (ARAN, 441-3-445); (Kapitza 1989a, 291).

He took the matter of developing the correct Marxist interpretation of physical theories to heart, and his views on the topic are worthy of special investigation (Graham 1987, 337–343). They apparently formed the basis of Fock's original approach to the general relativity, from which came the theory's major mathematical advances of the time, including the solution of the equation of motion for finite bodies. Fock accused Maksimov's critique of being based on simple ignorance of physics and wrote a letter to the Central Committee, receiving in response assurances that the 1952 university volume and views expressed there reflected personal opinions of its editors, not the view of the Party. Fock sent his devastating reply to Maksimov, "Against the ignorant criticism of modern physical theories," to the main philosophical journal *Voprosy Filosofii*, where Maskimov served as a member of the editorial board (Ilizarov and Pushkareva 1994).

Kurchatov forwarded Fock's manuscript to Beria and enclosed a letter signed by several top physicists of the atomic bomb project asking Beria to help its publication. Changes in Kurchatov's position by 1952 allowed him to talk to politicians with authority. Three years earlier, in 1949, he had not yet proven his ability to build the bomb and was under constant suspicion from Beria working on the plutonium-producing nuclear reactor in the Urals. In those very months when the Organizing Committee in Moscow was preparing the ideological meeting of physicists—January through March 1949—Kurchatov was in serious trouble himself, struggling with the worst accident of his project, leaks in the reactor's cooling pipes that threatened the entire operation (Kruglov 1994, 72–75; 1998; Larin 2001). The gravity of that emergency, the threat of severe punishment, and the enormous, round-the-clock efforts and risks required from him to have the reactor repaired by the end of March, all made involvement in—or even knowledge of—the Organizing Committee sessions in Moscow almost impossible for him. He lacked then the authority to interfere in a matter—the criticism of ideological mistakes committed by several physicists from the Academy—that was clearly outside his administrative responsibilities and the scope of his project.

By 1952, Kurchatov and his collaborators emerged triumphant from the atomic bomb work, having justified the Party's trust in them and strengthened the country's defense with the "nuclear shield." The success gave them much more prestige and authority, and, according to various recollections, Kurchatov decided that year to start carefully using that authority outside the sphere for which he was directly responsible (Gorelik and Kozhevnikov 1999). Beria's responsibilities did not include ideology either, so he took no risks but simply forwarded physicists' letters to Malenkov at the Central Committee's Secretariat. The Secretariat's ideologists Suslov and N. A. Mikhailov, together with Yury Zhdanov from the Science Department, agreed with the physicists, scolded the Navy newspaper for printing controversial material that was beyond its competence, and instructed *Voprosy Filosofii* to publish the article by Fock (Ilizarov and Pushkareva 1994). This 1952 episode with Kurchatov's and Beria's participation apparently became one of the main sources of the legend, that credits them with the cancellation of the 1949 discussion in physics. For later memories, three years' time difference does not mean as much as it did in those years.

Even if nuclear physicists did not postpone the discussion in 1949, they certainly helped the subsequent resolution of the conflict in physics community. In 1954 Kurchatov took part in the special commission of the Central Committee that investigated complaints against the Moscow University Physics Department (Andreev 2000, 286–290). After that review by party officials, the department received new leadership and a number of physicists from the Academy started teaching there part-time. The next major meetings on physics and philosophy took place in the second half of the 1950s in the already very different political situation of Khrushchev's liberalization and its new intellectual atmosphere. Nobody questioned the authority and hierarchy of the Academy physicists any more. Both physicists and philosophers united in proclaiming perfect agreement between modern physics and dialectical materialism (Filosofskie 1959).

After the bomb and after Stalin's death, physicists enjoyed their new power and status in Soviet society and no longer needed to develop philosophical justifications for their discipline. It subsequently became the job of Soviet philosophers to write philosophical works with the goal of demonstrating the compatibility between Marxist principles and the ever-changing developments in science. Not that critics like Maksimov suddenly disappeared—theories with profound philosophical consequences continued and still continue to generate some controversy. But the new cult of science, and of physics above else, no longer allowed criticism of Einstein and Bohr in print, not even for their philosophy (Vucinich 2001, ch. 5). Maksimov was effectively ostracized and banished from the press, though he continued to write manuscripts against idealism in science.

As much as they aspired to, opponents of relativity and quantum mechanics never came close to representing the official Soviet position. Timiriazev was the only one among them who had considerable power and influence, but after his defeat in 1929 the dominant views in Soviet

Marxism with regard to the new physics were defined by such highly placed and authoritative authors as Vavilov, Fock, and their allies among philosophers, who argued that modern physics and dialectical materialism agree and support each other. Could a reversal have happened? It is not entirely impossible, given that a similarly unlikely and sudden reversal occurred in linguistics. By the same token and even likelier, Lysenko could have lost his fight against genetics after World War II had events developed somewhat differently. It was not anything in the nature of biology or physics or their perceived importance that determined their different fates under Stalinism. If one has to choose just one most important factor, it was probably the Soviet physicists' skills—and some luck—in playing the rhetorical, ideological, and political games of that culture.

Chapter 10

Collective Excitations

"Fermions are individualists, while bosons are collectivists."
M. I. Kaganov and I. M. Lifshitz (1989, 14)

Many Soviet physicists who went to work on the atomic bomb project felt that they were sacrificing opportunities to accomplish more original research in fundamental science. Landau was among those few who agreed to do the bomb work out of fear—in a precarious position after his release from prison, he felt that classified research provided additional protection. He still kept his involvement rather limited and withdrew at the first available opportunity after Stalin's death. Others, including Khariton, Kurchatov, Sakharov, Tamm, and Zeldovich, participated out of a sense of moral and patriotic duty. Several years of complete engagement for some, bomb work became a lifelong commitment for others, but no matter how strongly they felt about the necessity of the assignment they could not help regretting their diversion from fundamental physics. Not only did they view the atomic project as more technological than scientific, the ultimate goal of the work consisted in large part of repeating and reproducing what had already been accomplished overseas. The strategy advocated by Kapitza in 1945 of having it both ways—follow an original path and build the bomb—was rejected in favor of one that minimized time and risks and could be summarized precisely by the slogan "To Catch Up and To Surpass!" Implementing this slogan in practice meant more often than not following the American path, adopting many of the same strategic choices and goals, spending huge efforts and funds on catching up, usually trailing, while occasionally using chances to actually surpass, as happened with the hydrogen bomb of the Sloika design.

The ubiquitous Stalinist slogan was valid not just for the nuclear field but also for the majority of other branches of postwar Soviet science, military and civilian alike. In other European countries after the end of World War II scientists faced similar dilemmas as they contemplated their strategies of coming to terms with the predominance of American science, methods, and money, and usually made similar choices of following overseas trends. The alternative of competing along different paths, as had often been the case in scientific rivalries between leading European nations during the 19th and early 20th centuries, was occasionally discussed but more often than not deemed less promising of success or too risky. For international science, the overall effect of the Cold War was thus a significant loss of diversity.

The strategy of favoring self-reliance in scientific research over following foreign models obviously entailed serious risks, which were particularly high in the Stalinist Soviet Union and were demonstrated there so disastrously by the Lysenko case. Though Lysenko succeeded for a time in convincing a large following that his theories were actually superior to foreign ones, for the majority of Soviet physicists he remained even during the period of his official dominance a charlatan and scientific imposter. To them then as to us now, Lysenko's political rapport with Stalin and the rejection of international genetics meant great damage inflicted upon the whole of Soviet science. The tug-of-war between two possible post-war strategies in science found reflection in the choice of political slogans. Stalin's call to "catch up" was obviously not very suitable for Lysenko's purposes, and when the latter convened his Session in August 1948 he decorated the room with a different quotation from Stalin, one not originally meant for science:"[T]he humblest Soviet citizen, being free from the fetters of capital, stands head and shoulders above any high-placed foreign bigwig whose neck wears the yoke of capitalist slavery!"[1] Though the Organizing Committee for the planned 1949 conference in physics was ostensibly inspired by the Lysenko example, it chose the quotation "To Catch Up and To Surpass!" as its main slogan and printed it on the official invitations. Despite vicious conflicts among

---

[1] The quote is from Stalin's report to the 18th party congress in 1939 (Stalin 1972, 368). Parts of this slogan can be seen in the photograph of the August Session in Chapter 8, 211.

themselves, physicists mostly abstained from using "slavishly following foreign models" as an ideological accusation. The few who indicated some inclination towards such rhetoric did not find much support either in the Organizing Committee or in the discipline at large. Then and for decades after, the negative example of Lysenko made most Soviet scientists particularly wary of arguments in favor of seeking original socialist paths in science.

But such paths could nevertheless develop even regardless of whether scientists consciously desired to seek them, assisted in part by the international isolation of Soviet science and scarcity of foreign contacts. Some of those who resisted the rhetoric of Soviet specificity in science most actively were in fact following research strategies that differed fundamentally from those developed further west. Some Soviet scientists were aware of these differences, others less so, as they pursued their original directions in research with varying degrees of success. This chapter presents one such case, and a particularly important one, that of a distinctively Soviet fundamental contribution to theoretical physics—collectivist methods and the notion of quasiparticles. The beginnings of the collectivist approach in the earlier work of Frenkel and Tamm have already been discussed in Chapter 3; here the story continues into the post-war period and includes accomplishments by Landau and his school and an analogous approach developed by David Bohm in America.

The four chief protagonists of collectivist methods in physics just mentioned had certain things in common. All were socialists of various kinds, mostly unorthodox, and cared about politics almost as much as about science. They also lived through the existential experience of persecution and deprivation, to various degrees, of personal freedom, which left an impact on their thoughts about freedom and how much of it could actually be achieved by people in real life and by particles in real bodies. Their socialist worldview and collectivist understanding of freedom led them into disagreement with the existing mainstream tradition in solid state physics—the theory of free electrons in bands, which was developed mainly in Britain and the United States—and its analogs in the theories of the liquid and plasma states of matter. Instead of following the dominant trend of the day, they followed their basic political philosophy and intuitions and, as a result, introduced collectivist terminology and models into quantum physics. Initially, collectivist methods were a minority approach pursued by a few socialist, mostly Soviet, theorists. In the 1950s, however, they overcame ideological hostility and gained widespread international recognition even among physicists who were neither socialists nor leftists. They have since come to underwrite much of the language and many of the basic notions of contemporary physics.

1. Electrons Free and Trapped

In 1931 Alan H. Wilson (1906–1976) arrived from Cambridge to study with Heisenberg in Leipzig and tried to make sense of Bloch's far from transparent papers on the theory of metals. Having interpreted them in his way, Wilson extended Bloch's methods from metals to insulators and semiconductors and explained why these substances differ. For Bloch (and also for Frenkel) the difference had been merely quantitative, electrons in insulators being more tightly bound to their respective atoms than electrons in metals. Wilson, however, invented a way of preventing electrons from conducting current without binding them to atoms. If the number of electrons in the crystal just suffices to fill completely some of the allowed bands in the energy spectrum, these filled bands form "closed shells" of zero net current, with equal numbers of electrons moving in opposite directions. The energy gap between the highest filled band and the next available empty band makes the crystal an insulator: only a sufficiently strong excitation or force can make an electron jump across the gap to the empty band of higher energy and thus become a conduction electron. In metals, by contrast, one of the allowed energy bands is only partially filled, and its electrons have room to accelerate easily even in a weak electric field (Hoddeson et al. 1992, 119–123).

Heisenberg quickly combined Wilson's concept of filled bands with Dirac's idea of electron holes into a new interpretation of the anomalous Hall effect. Peierls had explained the effect by an unusual relationship between energy and momentum for electrons in the tight-binding approximation, a relationship equivalent to a negative "effective mass." Heisenberg pointed out that the same mathematical formulas could be easily reinterpreted as referring to vacancies in nearly filled Wilson bands, which behave like normal particles but with opposite electric charge, just like Dirac's holes. Bloch suggested further that this combination can also explain the photoelectric current in insulators: the absorption of a light quantum by an electron allows the latter to jump across the gap of forbidden energies. A pair is generated in the process, a conduction electron in the formerly empty band and a hole in the formerly filled one, both of which contribute to the resulting current (Heisenberg 1931; Bloch 1931).

Wilson drew even more profound lessons from his accomplishment. The generalization of the Bloch method allowed him to generalize the notion of quantum freedom, too. According to Wilson, the classical Drude–Lorentz theory confused electrons' freedom with their ability to carry electric current. But in the quantum theory, where all electrons are described as traveling waves in a perfectly periodic lattice, "we cannot assume, as we do in the classical theory, that only valence [conduction] electrons are free." "[A]ll the electrons in a metal are free," as they are also in semiconductors and insulators, regardless of whether or not they transport electric current (Wilson 1931–1932, 458–459). This interpretation constituted an important departure from the 1928–1932 attempts by Bloch and other Weimar physicists of the Heisenberg school to comprehend the unclear situation of a state that was neither bound nor free. Wilson's view that the periodicity of the lattice makes all electrons free in all crystals, whether metals or insulators, was further extended and developed after 1933 into the band theory of solids.

Band theory assumed that abrupt, strong, and very complex forces acting upon an electron in a crystal from all the atoms and other electrons could be accounted for, summarily, by a smooth periodic potential, the effective or self-consistent field, through which the electron could glide as a nearly free particle. That this not very realistic model, sometimes called "electrons in a jellium" (Ziman 1963, 20), could deliver results close to the experimental characteristics of real substances initially came as a surprise, but Eugene P. Wigner and Frederick Seitz demonstrated this in their prototypical 1933 calculation of sodium metal, which became the model for many subsequent studies (Wigner and Seitz 1933–1934; Hoddeson *et al.* 1992, 182–202). Mathematical sophistication and the liberty to choose whichever form of the effective potential worked best allowed band theorists to reproduce the properties of many real materials. The main results of Bloch and Peierls were also reinterpreted and incorporated into the band theory: the Bloch electrons in the loose-binding approximation were accepted as they were, while electrons in the tight-binding approximation were replaced by nearly free holes.

These events coincided chronologically with the geographical transition of the center of the solid state community from Germany. The band theory was developed primarily in the universities of Britain and the United States, with a particularly visible role played by scientists who had fled to free countries from Central European dictatorships. The fact that the refugees were also eager to see electrons in solids as free is a suggestive coincidence that might be worthy of a separate investigation. Within the English-speaking community of physicists, the band theory quickly developed into a dominant approach in the quantum theory of the solid state and remained practically unchallenged throughout the rest of the 1930s and '40s. In the Soviet Union, the reaction to it was ambivalent.

An early report of Wilson's achievement was brought to Leningrad by Ralph H. Fowler from Cambridge in September 1932. He presented it to a conference on the physics of metals together with a new lesson about freedom:

> If we discuss quantum mechanically the motion of an electron in a perfect lattice we are forced to conclude that all the electrons in all the atoms are free to move through the lattice—are "free electrons" in the classical sense—except in so far as they are prevented from moving by Pauli's exclusion principle (Fowler 1932, 508).

The audience included most of the young Soviet quantum theorists who worked on the solid state: Frenkel, Tamm, Fock, Landau, and also Dmitry Blokhintsev, Matvei Bronstein, Semion Shubin, Sergei Vonsovsky, and a few others. They listened politely to the foreign luminary and no objections were recorded in the conference proceedings, but deep in their bones they disagreed with Fowler's thesis.

The small Soviet community was not coherent but was split on a variety of scientific, political, and personal issues. Only a very few, Frenkel and some of his students, were already using collectivist language and developing corresponding models of solids. Most Soviet physicists initially accepted at least parts of the band theory, albeit with some significant reservations. Their varying social experiences had already taught them that not all public declarations of freedom should be taken at face value. As far as electrons in solids were concerned, Soviet theorists knew that the adjective "free" could not be applied in its true meaning. This attitude was the common denominator of their various approaches and reactions to band theory. Even those who used the models of free or nearly free electrons rarely failed to add a reservation or remark about the limited validity of their basic assumption. When referring to the Bloch electron, Frenkel used the adjectives "free" and "collectivized" interchangeably, the former as a conventional term, the latter as a description of the actual state of affairs. Other Soviet physicists tended either to put "free" in quotation marks or to replace it with "conduction" or other terms, depending on the context.

Their shared intuition about electrons' freedom can be seen in a confused polemic between Tamm and Frenkel in 1931. Frenkel, then in Minneapolis, attacked a new theory of the photoelectric effect in metals put forth by Tamm and his student Shubin for allegedly "introducing...the notion of free electrons of two types, completely free 'Sommerfeld electrons' and partially bound 'Bloch electrons,' " which he found incompatible with "the fact that Sommerfeld's 'completely free' electrons simply do not exist and are but an approximation to the more real 'relatively bound' electrons of Bloch" (Frenkel 1931c, 315–317). Tamm replied that Frenkel had misunderstood their work. For him, too, Sommerfeld's model was only a rough approximation of restricted applicability, and conductivity electrons were not absolutely free. Indeed, Tamm and Shubin opened their paper with a reference to the "well known fact that the photoelectric effect cannot take place with free electrons in free space" and on this basis criticized as inconsistent an earlier theory of the photoeffect by Sommerfeld's student Gregor Wentzel. Frenkel's admission of error ended the confusion (Tamm and Schubin 1931; Tamm 1932a).

The sensitivity of Soviet theorists to the limited applicability of band theory's basic assumption motivated them to explore specifically the limits of the formalism. Rather than following the mainstream approach of the theory of free electrons, they studied boundary situations and cases in which electrons become bound or trapped, and from this came arguably their most important contributions to the electron theory of metals during the 1930s. The theory by Tamm and Shubin of 1931 took into consideration two types of binding acting upon the electron in metal—the potential barrier at the surface of the body and the periodic potential inside it— which allowed them to distinguish and describe two different mechanisms for the photoelectric effect. In another paper of the same year, Tamm showed how these two potentials combined can produce new bound states of electrons in a metal. He studied the behavior of Bloch's electrons in a periodic lattice limited on one side by the body's surface. Tamm's calculation showed that adding the surface potential allows the electrons to occupy some of the formerly forbidden energy levels, but these new levels correspond to bound states of electrons trapped near the surface and capable of moving only along the surface. His prediction remained unconfirmed for several decades until Tamm's levels, as they are currently called in the physics of surface phenomena, were finally observed (Tamm 1932b; Keldysh 1985; Volkov 1995).

2. Arrested Electrons and the Polaron

A different possibility for the trapping of an electron in a crystal lattice was suggested by Landau in 1933. Landau accepted Wilson's notion of electron bands and considered it to be a mathematically proven theorem that all electrons in a strictly periodic lattice should be able to move without resistance. This, however, did not justify to him declaring electrons in solids free, and he consistently refused to use this term in his publications. In the 1930s, Landau's favorite approach to electrons in solids was statistical. He described an ensemble of electrons by means of a kinetic, or transport, equation, always abstained from assuming that an electron's energy and momentum are related by the free-particle formula $E = \mathbf{p}^2/2m$, and also forbade his students to use this common assumption.[2] At the same time, Landau was also very critical of, and occasionally openly hostile to, many ideas of Frenkel, his former teacher. He disagreed with, among other things, Frenkel's hypothesis of the excitation quantum and proposed a different excitation scenario instead. That part of band theory that Landau accepted sufficed for him to conclude that a small excitation or deformation of the lattice at some point leads to the scattering, but "does not mean yet the electron is trapped at this point." Exploring further the limits of the theory, he suggested that trapping may occur if the deformation is large and distorts the periodicity of the lattice. Landau proposed some materials in which such a trapping of an electron near the distorted area might possibly be observed and estimated that the formation of these bound states would require an activation energy (Landau 1933a).

Frenkel replied to Landau's implicit criticism by simply adding this new scenario to his picture of solids and developing it further. In a paper presented at a conference in Kiev in 1935, he pointed out that trapping can occur even without an activation energy, spontaneously, in the deformation of the lattice caused by the electron itself. He insisted, however, that this combination of an electron with a "trapping" of the lattice disturbance would not remain fixed but could move through the crystal, though much more slowly than a "free" electron. The possible states of an electron in a solid thus included a wide spectrum of complicated situations, from collectivized, almost "free" particles to slowly moving, "trapped" ones, but Frenkel consistently avoided the two poles: that of electrons bound to individual atoms and localized in fixed places, and that of electrons free like atoms of an ideal gas (Frenkel 1936).

Several years later in his book *Kinetic Theory of Liquids*, Frenkel metaphorically described a trapped electron or electronic hole as "visiting [a lattice cell] and becoming 'self-arrested' during its movement through

---

[2] Oral history interviews with A. I. Akhiezer and M. I. Kaganov (AIP).

the atom." The shocking word "arrested" (*arestovannye*) (which in Russian has only the meaning of forceful deprivation of freedom) was replaced in the British edition with "trapped," but for Frenkel's Soviet audiences, who had behind them the nightmare of the worst years of Stalinist terror, it was a familiar and frequent concept of everyday language, if not of personal experience.[3] Traumatic memories of the 1937–38 purges also paled in their minds in comparison with the even more inhumane conditions of the Great Patriotic War. When Frenkel wrote his book during the harsh winter of 1942/43, evacuated to Kazan with the larger part of his institute, he had already become an indirect witness not only to the pre-war arrests and disappearances of millions, including many of his friends and colleagues, but also to the wartime deaths and starvation of millions, in particular in his home city Leningrad, then in the second year of the tragic siege that virtually exterminated its inhabitants.

To those who had lived through that cruel existential experience, freedom did not appear as an unproblematic gift and a natural state of life, but neither could they accept its absolute impossibility. Freedom was for them a difficult challenge and a serious problem to be solved by everyone: some portion of it, some complicated state of it, had to be achieved *de facto* even under the most terrible conditions. Even when "arrested," Frenkel noted, electrons or holes

> can liberate themselves again; the liberation, however, requires an increase of potential energy ... and can occur after a certain period of time. ... The freedom obtained on such release will, [however], be of extremely short duration; liberation being followed by a new "self-arrest" [*samoarest*—AK] near one of the adjacent atoms (Frenkel 1945, 62–63).[4]

Neither Landau nor Frenkel developed mathematical models for their proposals of "trapped" or "self-arrested" electrons. The solution was found by the war's end in 1945 by Solomon Pekar (1917–1985), a professor

---

[3] (Frenkel 1945, 62–63; Frenkel 1946, 52). In the same book, holes and dislocations in the atomic lattice are called "offenders of order" ("*narushiteli poriadka*," the Soviet militia's term for hooligans) (Frenkel 1945, 25).

[4] Incredible as it seems, the phenomenon of "self-arrest" could occur in Stalinist Russia during the insane chaos of the Great Terror. A case of a woman who, unsure of her own political loyalty, denounced herself to the state prosecutor is described in (Hellbeck 2000). The cycle of consecutive arrests and releases was a more common occurrence, in particular among scientists and engineers who worked in so-called *sharashki*, privileged prisons for arrested specialists put to work on secret military projects. Frenkel might well have been aware of this social phenomenon, known in the folklore of radio engineers as the "Mints cycle" after Aleksandr Mints (1895–1974), who reportedly went through this experience thrice, each time he was in captivity building another more powerful radio transmitter.

in Kiev and former student of Tamm's. Pekar realized that the deformation of the crystal lattice can be described mathematically as the electrical polarization of the lattice by electron's Coulomb field. He managed to solve the corresponding wave equation and find energy levels for an electron in a polarizable dielectric medium, which had bound-state solutions. Tamm brought Pekar to report his results at Landau's seminar in Moscow, where the new particle was baptized during the discussion as the "polaron," the name that eventually became standard. Pekar's publications of 1946 contained the basic theory as well as the name of the polaron, and an account of its movement through the lattice, its dissociation into heat or phonons, and the electron's transitions between "free" and "polaron" trapped states (Pekar 1946a; 1946b).

3. The Exciton and the Collectivist Alternative to Band Theory

Band theory accepted and absorbed some of the products of the earlier collectivist approach, such as phonons and holes, by reinterpreting these concepts as free particles. On the other hand, it did not agree with Frenkel's 1931 concept of the excitation quantum. The exciton was the solid-state collectivist analog of what in a gas would have been a bound excited state of an electron, but the band theory explicitly denied the possibility of bound electron states in a periodic lattice. Even its restricted versions accepted by Landau and many other Soviet physicists seemed to leave no room for the excitation quantum state. The concept of filled bands and the energy-gap explanation of the difference between metals and insulators carried with them a particular vision of an excitation of a crystal by a light quantum. An electron, when excited, was supposed to jump from a filled band into an empty one, thus creating a pair consisting of a conduction electron and a hole. The absorption of light by an insulator produced carriers of electric current, which explained the phenomenon of photoelectricity. Frenkel's excitation quantum, on the contrary, had zero electric charge and could not transport any electric current. When explained in these terms, Frenkel's hypothetical mechanism of light absorption seemed to contradict that of band theory already at the level of phenomena available for experimentation and thus "appeared...completely paradoxical" even to most of his Soviet colleagues, both theorists and experimenters, including those at his home institute in Leningrad.[5]

---

[5] S. I. Pekar's recollection in (Frenkel 1986, 198). On excitons being "practically incompatible" with band theory see also (Rashba 1985, 792).

Facing widespread skepticism, Frenkel defended his proposal of the excitation quantum in a paper of 1936, where he also introduced the name "exciton." Not only the name but also his understanding of the particle had changed somewhat in the interim. Frenkel accepted part of Peierls' and Landau's critique and agreed that the excitation would be localized in a small area rather than shared uniformly by all atoms of the crystal, but he criticized them in turn, insisting that it would not remain fixed to this location but could move through the lattice. "[T]he collectivization of the exciton" in the 1936 version of Frenkel's theory was closer to his "collectivized electrons" of conductivity of 1924 and "empty spaces" of 1926 than to Tamm's phonons of 1930: excitons moved in an already familiar fashion of jumping from one atom to a neighboring one (Frenkel 1936).

On this occasion, Frenkel also publicly presented his criticism of the band theory for the first time. The agreed-upon shared ground was the mathematical model of the Bloch electron in metals, although band theorists called it "free" while Frenkel preferred "collectivized." Their disagreement about metals remained mainly terminological, but when extended to insulators it evolved into contrasting, physically different approaches. In insulators, Frenkel insisted, almost all electrons remained "uncollectivized," or more tightly bound to individual atoms than in metals, and therefore could not easily move and transport current. Rather than move, it was easier for electrons of neighboring atoms to exchange their energy of excitation, or exciton. The picture of a moving collectivized excitation quantum, according to Frenkel, represented processes in non-metallic solids better than did the picture of a nearly free moving electron. Applying the latter model to insulators was, for him,

> an inexcusable abuse of Bloch's method. It has to be regretted that such an abuse has been practiced by nearly all writers on the electron theory of the solid state, leading them occasionally to wholly erroneous results. One such mistake... consists in the exclusion of nonconducting excited states of a crystal, i.e. such states which... are characterized by moving excitons (Frenkel 1936, 161).

Frenkel's defense of the exciton thus led him so far as to oppose the applicability of band theory to non-metallic crystals and the band gap explanation of the difference between insulators and conductors, which were then already almost universally accepted.

Frenkel went further than just defending the exciton. He summarized the entire collectivist approach and the sorts of new particles that had been proposed and presented it as a full-blown alternative to band theory.

His synthetic picture shunned both extremes—practically free and absolutely fixed electrons—and described instead various processes in solid bodies as movements of particles with varying intermediate states of freedom. The exposition started with an electron liberated from its atom by an incident light quantum to become a "free or collectivized electron" of conductivity that moved by jumping from one position in the lattice to another. The liberated electron left behind a "collectivized positive hole, or positron" (actually a positive ion), which could also move through the lattice by "capturing the missing electron from one of its neighbors, and thus converting the latter into a 'positive hole.'" If the energy of the light quantum was not sufficient to liberate the electron fully into a collectivized state, the electron remained bound to the hole, but the entire complex—an atom in an excited state, or a "collectivized electron and positron pair"—could still travel through the crystal in a similar way and could thus be called a "collectivized exciton" (Frenkel 1936, 158–159). Any "free or collectivized" particle—electron, hole, or exciton—could cause a local deformation of the crystal and become trapped in it. These "trapped" ("*prilipshie*") particles would also be able to move through the lattice, though much more slowly, and carry the deformation with them. Frenkel suggested that these slowly moving complexes might be observable, pointing in particular to the reports of the apparent trapping of photoelectrons in the experiments of Robert Wichard Pohl and his collaborators in Göttingen (Frenkel 1936, 171).

His presentation furnished the qualitative physical picture with only rudimentary basic calculations. In less than a year, John Clarke Slater and William Shockley at MIT and finally Gregory Wannier at Princeton developed a strict and consistent quantum mechanical formalism for the exciton. Without mentioning its collectivist philosophy, they attempted to eliminate, at least partly, the contradiction between the exciton and band theory.

> "For several years, there have been two competing pictures in use to describe the behavior of electrons in crystals," explained Wannier. "The one adopted in most theoretical calculations and especially successful for metals describes each electron by a running wave, but Frenkel has shown that in many cases the more elementary atomic picture may be the better approximation. This apparent contradiction has been removed by Slater and Shockley, who showed with a simplified model that the two types of states actually coexist in crystal" (Wannier 1937, 191; Slater and Shockley 1936).

Wannier succeeded in constructing a complete set of orthogonal wave functions for the electron in a crystal that was not limited to the periodic solutions representing Bloch's electrons, as had been usually assumed before. Wannier's functions included Frenkel's excitation waves in the lower part of the energy spectrum and another series of solutions for intermediate energies, which corresponded to hydrogen-like states of an electron and a hole bound together by the Coulomb field. The following year Neville Mott in Bristol favorably referred to the "collective electron" treatment as a more convenient approach in some problems of the solid state theory.[6] Mott's review covered Tamm's surface levels, Landau's electron trapped in a lattice distortion, and the bound states of an electron in the Coulomb field of a distant hole. These investigations helped clarify the difference between two distinctive types of excitons, as they are now called: the original "Frenkel exciton," a tightly localized excited state of the atom, and the "Wannier-Mott exciton," or "mega-exciton," a bound pair of an electron and a hole that can be far apart inside the crystal.

Even furnished with appropriate mathematical apparatus, the exciton remained a hypothetical construct and contradicted the most widespread interpretation of the band theory. Only a very few physicists accepted it until 1951, when the hydrogen-like spectrum it predicted was observed, apparently accidentally, in cuprous oxide by Leningrad spectroscopist Evgeny Gross (1897–1972). Gross also had to overcome serious opposition from colleagues, both theorists, especially Landau, and experimenters, which caused a year's delay in the submission of his paper. It took several more years before the exciton was finally recognized as a natural fact.[7]

The collectivist approach formed the basis of Frenkel's lecture course at Leningrad Polytechnical Institute in 1946 and of the resulting textbook *Introduction to the Theory of Metals* (1948). By that time, the collectivization of electrons was already to Frenkel an "experimentally proven" fact, demonstrated by X-ray diffraction in metals. The simplest example of collectivization, the sharing of electrons by two atoms in a diatomic molecule, leads to the so-called exchange forces. Frenkel considered exchange forces more appropriate than the model of self-consistent field as a mathematical way of describing the collective behavior of particles. He reproduced on his model the main results obtained by Sommerfeld, Bloch, and Peierls in

---

[6] "We find it most satisfactory to treat electrons in the closed shells as belonging to the ions, and reserve the collective electron treatment for the electrons in the...conduction band" (Mott [1938] 1995).

[7] (Gross and Karyev 1952). The same year the exciton spectrum was also reported in (Hayashi and Katsuki 1952). On the initial rejection of the experimental results on excitons and difficulties with publication see (Zakharchenya and Frenkel 1994, 473) and the recollections by Gross in (Ioffe 1973, 147–148).

the late 1920s through early 1930s, but he largely ignored the later band theory, whose "level of mathematical complexity is not matched by the level of the importance of results it delivers." Instead, his picture of the solids featured the entire family of "collectivized particles"—electrons, holes, excitons, and polarons—plus the phonon, which belonged to a somewhat separate category (Frenkel [1948] 1972, 8, 124).

Frenkel included more about band theory in the textbook's second edition in 1950, but also added a new preface with a stronger critique (simultaneously satisfying stricter ideological demands imposed on textbook authors by Cold War hawks) of the

> "formalistic tendencies of some Western-European and American ... physicists, who often diligently develop the theory in its formal mathematical aspect while paying little attention to the question whether or not its basic assumptions correspond to reality" (Frenkel [1948] 1972, 7, 153).

While accepting and interpreting from his perspective some of the results of band theory, he continued to reject as "radical" the idea of "wholesale collectivization," or the thesis that all electrons in all solids, including insulators, are free or collectivized, which did not leave room for the existence of either excitons or polarons. Until the end of his life, Frenkel maintained his quixotic opposition to the band-gap explanation of the difference between metals and insulators.[8]

Frenkel died in early 1952 of a heart attack and did not have an opportunity to see the first photographs of exciton bands and to learn of the success of his prediction. His last years were difficult for him: politically, rising anti-Semitism and ideological accusations of philosophical idealism had hit him rather badly, while in the professional community he suffered from the intolerant attitudes of Landau's school. The band theory of free electrons still dominated in the physics of the solid state. Outside its sphere remained most magnetic phenomena, which were usually accounted for by models with bound electrons. The collectivist approach, with its more complex understanding of freedom, existed only at the margins. Had Frenkel lived a few years more, until the mid-1950s, he would have seen an intensification of criticism of band theory and a strong increase in interest in collectivist

---

[8] "Wholesale collectivization" was a term in Soviet political language denoting admittedly wrong and violent "excesses" of the collectivization campaign in agriculture of 1930. Another careless phrase of Frenkel's, "forced collectivization of electrons," was criticized at the Polytechnical Institute's academic council as mocking the Soviet collective farm system (Suris and Frenkel 1994, 371).

models. Some of the most influential new models came from a somewhat different approach developed by Landau.

4. The Roton and Collective Excitations

A year under arrest changed many of Landau's views. He had nearly died of malnutrition by the time Kapitza managed to bail him out of prison in April 1939 with the written promise to Beria to prevent any further "counterrevolutionary" acts by the theorist. After he escaped from the "negative energy state," Landau withdrew from politics and public activity outside physics and reduced his political talk to private conversations with students. Even before his arrest he had hated Stalin, but after his direct encounter with Stalinist penitentiary system his vision of the Soviet regime changed altogether: he no longer regarded it as a socialist society of liberated labor but saw it as a fascist dictatorship (Kaganov 1998, 48). Landau no longer considered himself free either, even after he had been released from jail. Despite all the privileges, public recognition, and cult worship he received in the later years after he became a full member in the Academy of Sciences in 1946, he continued to refer to himself as a "scientific slave" of the repressive regime. In fact, until the end of his life he must have been aware that the charges against him had not officially been withdrawn and that somewhere in the KGB files he was still classified as a political criminal.[9]

Some of his other proclaimed principles also changed, and after his release from prison Landau married Konkordia (Kora) Drobantseva, his first and late love from the Kharkov years, who taught him when he was twenty-seven what sexual love means in practice. He did not, however, give up his earlier revolutionary theory that marriage must not limit either partner's sexual freedom and insisted on his wife's agreeing to this principle as a "marital non-aggression pact," which in the beginning was merely theoretical. Later on, with the growth of his fame and status in the post-war years, Landau would compensate himself abundantly for the lack of sexual experience in his younger days (Landau–Drobantseva 1999). In a personal sense, he was not particularly close to Kapitza and even claimed sometimes that he had never visited the personal laboratory of his institute's boss. But he remained forever grateful to Kapitza for

---
[9] KGB memorandum "Spravka po materialam na akademika Landau L'va Davydovicha," (1957) (CCA, 89-18-42).

The prison photo of L. D. Landau, 1938. Landau was jailed at the height of the Stalinist purges and released one year later thanks to the efforts of P. L. Kapitza. His political and scientific views changed after that experience. Landau no longer believed that anyone, including himself, could remain free while living in Stalinist society, and he no longer accepted the view that atoms and electrons could be treated even approximately as free particles when placed inside dense solid or liquid bodies.
[*Credit*: AIP, Emilio Segrè Visual Archives.]

saving his life, always behaving correctly and politely—not among his usual virtues—toward him and remaining respectful of Kapitza's authority (Kaganov 1998, 49). He was also aware that in some sense he "owed" Kapitza a theory of superfluidity in helium, which Kapitza had used as an argument to save Landau from prison.

While Landau was still in jail, his former student Laszlo Tisza (1907–), working with Fritz London in Paris, developed a "two-fluid" theory of liquid helium in which the atoms of the liquid were divided into a normal fraction and a superfluid one (Tisza 1938). Landau objected to Tisza's assumption that the superfluid fraction can be identified as the Bose-Einstein condensate in helium gas. It seemed to him totally unjustified to model the behavior of tightly packed atoms in helium liquid as analogous to that of the freely moving atoms of a rarefied gas. Some of Landau's scientific convictions regarding the freedom of quantum particles had also changed in the interim in a way somewhat akin to the changes in his political views. Already in the 1930s he expressed reservations about the model of free electrons in crystals, judging it unrealistic and approximate at best:

> The existing theory of metals does not take into account the interaction of the electrons of the metal with one another, although the latter must be very large and cannot be ignored to any extent whatever (Landau and Pomerantschuk 1936, 171).

But understanding that no better alternative was available at the time, he still allowed for some use of the model of nearly free electrons on the grounds that "[s]uch an examination at least provides the possibility of elucidating the limits of applicability of the existing [band] theory" (*ibid.*). However unnatural, this model offered what seemed to be the only way of simplifying a hopelessly complicated physical system to a state in which mathematics could be applied and at least some calculations performed.

After his arrest, Landau began to reject free-particle models in condensed matter theories on principle, both the model of an electron gas in solids and the model of an ideal Bose gas for liquid helium. Having experienced imprisonment firsthand, he no longer permitted himself and his students to assume, even as a an artificial and rough approximation, that atoms packed into the states of high concentration and strong interactions were somehow capable of retaining their freedom. He set out to develop a theory of superfluid helium not as a gas but as a "quantum liquid" and needed to find some different simplifying assumption to reduce the complexity of the system and allow the use of mathematical methods. He found inspiration in the model of phonons developed by Tamm ten years earlier. Phonons, quanta of sound or elastic vibrations in solid bodies, could be described mathematically as an ideal gas, thus becoming amenable to all the powerful mathematical tools of the kinetic gas theory. At the same time, they were just a phenomenological (some said fictional, or formal) model for a system of strongly interacting atoms fixed in the crystal lattice—and one did not have to pretend that those "real" atoms were free. A model with the similar two features would have satisfied Landau's criteria for a theory of liquid helium, and in order to explain the phenomenon of superfluidity he found it necessary to generalize Tamm's approach and postulate, in addition to phonons, one more kind of collective quantum.

The new quantum that did not exist in the case of a solid was described by Landau as the quantum of vortex movements in the liquid. Tamm participated in some of the early discussions of the idea and suggested naming these quanta "rotons" (Landau 1941, 75; Ginzburg 2001, 285–286). Landau's theory of superfluidity published in 1941 proceeded from the main assumption that at temperatures near absolute zero, or not far from the ground state of lowest energy, liquid helium could be effectively described as a mixture of "elementary excitations" of two sorts, phonons and rotons. Elementary excitations behaved in many ways like particles of an ideal gas, with their own effective mass and momentum. If the relationship between the energy and the momentum satisfied certain conditions the system could have superfluid properties. The gas of phonons and rotons constituted the "normal" component of the liquid. At sufficiently low temperatures the total mass of elementary excitations was less than the total mass of helium, the rest of the mass left to the "superfluid" component.[10] Increasing helium's temperature, or adding energy to it, increased the number of elementary excitations. At the point when their total mass equaled the total mass of available helium, the entire liquid became "normal" and the phenomenon of superfluidity disappeared (Landau 1941).

Landau's and Tisza's theories agreed in some of their mathematical predictions as well as in dividing helium into a normal and a superfluid component. They differed, however, in the basic physical picture: in Tisza's theory, helium atoms belonged to one or the other fraction, while according to Landau the normal component consisted of quantum excitations, which were not to be confused with particular atoms or groups of atoms, being instead the units of collective motion of the entire system, or all atoms in the liquid together. London and Tisza saw this as a strange and unacceptable hypothesis:

> There is unfortunately no indication that there exists anything like a "roton"; at least one searches in vain for a definition of this word ... The same measurement of Kapitza could ... have served as a good, though empirical, basis for the macroscopic hydrodynamics, had Landau not preferred to present his theory as based on the more than shaky grounds of imaginary rotons (London 1947, 13).

---

[10] Colloquially, the normal and superfluid components were also referred to as "live" and "dead" water from Russian folktales.

London admitted that an ideal gas was not a very realistic approximation for a dense liquid but still judged his assumption more sensible than Landau's proposal of a completely new kind of hypothetical quantum. Landau believed exactly the opposite, that in dense bodies collective phenomena were more real than individual atomic movements:

> It follows unambiguously from quantum mechanics that for every slightly excited macroscopic system a conception can be introduced of "elementary excitations," which describe the "collective" motion of the particles ... It is this assumption, indisputable in my opinion, which is the basis of the microscopical part of my theory. On the contrary, every consideration of the motion of individual atoms in the system of strongly interacting particles is in contradiction with the first principles of quantum mechanics (Landau 1949).

The two rival theories of superfluidity disagreed in some of their concrete predictions, which after the war allowed experimentalists to choose in favor of the Landau version. The importance of Landau's solution, however, was not limited to the specific case of helium and superfluidity. It suggested a new general way of treating dense systems with strong forces of any kind. The basic assumption that the lower excited states of any such system can be described by means of "elementary excitations" representing collective movements of atoms or electrons has since become a commonplace among physicists. The hypothesis has been applied successfully to numerous problems in the physics of the solid state, plasma, and liquids. New kinds of collective excitations have been proposed in the process and confirmed in many experiments, becoming so familiar and well recognized that physicists today may find it hard to understand why initially they were seen as counterintuitive by Landau's critics.

Guided by this general—but in his time still very controversial—hypothesis, Landau waged his battle against the ideal-gas models not only as applied to liquid helium but on several other fronts in condensed matter physics, sometimes against the odds. He was assisted in this by his pupils Pomeranchuk, Lifshitz, and Akhiezer among others. They introduced some new fictive particles, or "collective excitations," which resembled phonons on the one hand but also possessed new and strange properties and did not necessarily have well-defined classical analogs. Already in 1941 Pomeranchuk, referring to Landau, quantized Bloch's spin waves in crystals and baptized the resulting quantum "magnon" (Pomerantschuk 1941; Walker and Slack 1970, 1386). In Landau's original theory of liquid helium, excitations obeyed Bose-Einstein statistics. He later developed another theory of the quantum liquid in which excitations resembled electrons by obeying the statistics of Fermi-Dirac (Landau 1956). Landau's Fermi-liquid theory has since replaced the model of electron gas as the main paradigm in the electron theory of metals. It also allowed his student Lev Pitaevsky (1933–) predict superfluidity in helium-3.

Landau's influence, probably more than anybody else's, elevated quasiparticles to fundamental objects in physics. This happened not only because of his personal contributions or his multi-volume *Course of Theoretical Physics*, but also because he commanded a strong following of younger physicists who, calling themselves the Landau school, consistently and diligently applied the spirit of his approaches to a great many of the important problems in the field. By agreeing to perform some calculations for the atomic bomb project Landau created positions for several new theoreticians at the Institute of Physical Problems, and Kapitza tolerated this enlarged theory group after he returned to the directorship in 1954. A new generation of Landau's students included Isaak Khalatnikov (1919–), Aleksei Abrikosov (1928–), Igor Dzyaloshinsky (1931–), and a few others. Together, they applied the method of collective excitations to

Landau's post-war school grew around him in the Institute of Physical Problems. Back row (left to right): S. S. Gershtein, L. P. Pitaevskii, L. A. Vainshtein, R. G. Arkhipov, I. E. Dzyaloshinsky. Front row (left to right): L. A. Prozorova, A. A. Abrikosov, I. M. Khalatnikov, L. D. Landau, E. M. Lifshitz. All but two—Vainshtein and Prozorova—in this photo of 1956 were members of the Landau school.
[*Credit*: AIP, Emilio Segrè Visual Archives, *Physics Today* collection.]

such a wide spectrum of problems in condensed matter physics, and with such impressive success that it became a standard method in many fields.

Clan structure, or the clustering of the academic community into "scientific schools" around top-ranked scientists became a common social phenomenon in post-war Soviet science. Schools typically did not mix; their members tended to work in the same or friendly institutions. The existence of such schools was often perceived by Soviet scientists as a natural and necessary feature of science itself, universal and beneficial for the very progress of knowledge. Institutional leaders of Soviet science were expected to create such schools, while new members were usually recruited into one of the (often rival) schools in late student years and remained associated with the same school for the remainer of their careers. The collectivism of the Landau school displayed a characteristic mixture of camaraderie, cohesion, and hierarchy, and also allowed a strong "personality cult" to develop around Landau. Group solidarity was strong, and new members were accepted after a special process of initiation, a series of exams called the "Landau minimum." Their collective work and strong interaction within the group helped advance its coherent research program, in particular in condensed many physics, but the same intellectual cohesion could also be blamed for some missed opportunities in research (Khalatnikov 1989; Kaganov 1998).

Tragedy struck in 1962 when Landau was at the height of his influence and fame. He barely survived a car accident and was never able to work again. Later that year his scientific contributions were recognized with the award of the Nobel Prize for his pioneering theories of condensed matter, particularly liquid helium. Both the accident and the prize boosted his aura even further. Landau died in 1968 at the age of 60, but his school continued its productive existence after his death under the "collective leadership" of several of his students, who established a new institutional base, the Landau Institute of Theoretical Physics near Moscow (Khalatnikov and Mineev 1996). Meanwhile, Soviet collective models and approaches to condensed matter physics influenced many scientists in the United States and Europe, starting with the American pioneer of such methods, David Bohm.

5. The Plasmon and the Collective Movement

David Joseph Bohm (1917–1992) was born in Wilkes-Barre, Pennsylvania, to a dysfunctional family of Jewish immigrants. He started graduate school at Caltech but later transferred to Berkeley, where he received his

On a pedestal—a strong "cult of personality" grew up around Landau in his later years. Landau is shown here in 1961 at Palanga, a Baltic Sea resort in Lithuania, shortly before his near-fatal accident.
[*Credit*: AIP, Emilio Segrè Visual Archives, *Physics Today* collection.]

physics Ph.D. in 1943. By that time his official adviser, J. Robert Oppenheimer, was away most of the time, absorbed by his duties in the Manhattan Project. Oppenheimer asked to have Bohm transferred to Los Alamos, but the security officers denied him clearance, primarily because of his left-wing political activity. With several other physicists from the Berkeley Radiation Lab, Bohm participated in FAECT (Federation of Architects, Engineers, Chemists, and Technicians), a Marxist labor union, and in 1942 he joined the Communist party. Bohm dropped out of the party after about nine months but remained a convinced Marxist with a particular interest in dialectical philosophy. Together with a visiting British experimental team, he started working on plasma during the war as part of an effort to solve the problem of isotope separation. He continued to be interested in the peculiar properties of this strange medium even after the war's end, as he moved from Berkeley to Princeton as an assistant professor in 1947.[11]

In May 1949 Bohm along with several former FAECT members was subpoenaed to appear before the House Committee on Un-American Activities, which was suspicious of possible wartime espionage at the Berkeley lab by a certain "Scientist X." He invoked the Fifth Amendment and refused to testify about his or others' political beliefs. As the anticommunist witch-hunt intensified in the wake of the first Soviet atomic test in 1949, Bohm was arrested on charges of contempt of Congress. Bailed out of jail, he continued to work in Princeton, although the anticommunist and possibly anti-Semitic university administration suspended him from teaching. Even after he was acquitted in court, Princeton refused to renew his contract once it expired in 1951. The politically tainted physicist could not find another academic job in the United States and accepted a professorship in São Paulo, Brazil (Olwell 1990). His emigration subsequently also cost him his American citizenship. As these political developments unfolded, Bohm was developing his famous unorthodox interpretation of quantum mechanics and, together with several graduate students, laying the foundation of plasma theory.[12]

Bohm's initial approach to the problem of electrons in plasma, as well as in metals, was based on the metaphor "organized movement." Like all socialists and those who had some experience in unionizing, he was familiar with the philosophy of collective action and of the organized workers' movement and, like Frenkel before him, relied on this philosophy when expressing a hope that the idea of organized movement could explain the yet unsolved riddle of superconductivity. Frenkel's language is, as usual, more metaphorical:

> Let us imagine a crowd of electrons moving in the same way ... through the crystal lattice of a metal. Because of the electromagnetic mutual-action of the electrons ... the motion of each electron will be affected by an external perturbation ... to a much lesser degree than in case it moved alone ... Indeed, if an electron is knocked out of the crowd ... the resulting change of the total momentum of the whole crowd will not be equal to the

---

[11] On Bohm's biography see (Peat 1997; Hiley 1997).
[12] Here I only deal with Bohm's work on plasma. His "hidden variables" interpretation of quantum mechanics is analyzed in a vast body of literature, physical as well as historical and philosophical.

David Bohm in 1949. The original caption on the photograph: "David Bohm, Princeton University physics professor who worked on the wartime development of the atomic bomb, shown outside the House Un-American Activities Committee today, where he refused, under oath, to state that he was—or was not—a member of the communist party." The original photograph is damaged in the lower part.
[*Credit*: Library of Congress, New York World-Telegram and Sun Collection. *Courtesy*: AIP, Emilio Segrè Visual Archives.]

individual [momentum] $m\mathbf{v}$ of this electron, there will be, in addition, [the] collective term ... So long, therefore, as the electrons in a metal move collectively as an organized crowd of sufficiently large size, their motion can remain unaffected by the heat motion of the crystal lattice, the energy and momentum quanta of the heat waves being insufficient to knock out even a single electron (Frenkel 1933).

Bohm's formulations are less explicitly political, but the basic intuitive idea is exactly the same, that an organized workers' demonstration is harder to scatter than an unorganized crowd and that it can therefore move more easily through the police lines:

> [I]f superconductivity is caused by interactions between electrons, it is probably due to a somewhat localized tendency for electrons of the same velocity to move together as a unit, which is held together in some way by the inter-electronic forces. In order to stop such a group of electrons, it would be necessary to scatter all of them at once. Such a process would be enormously less probable than one in which electrons are scattered individually by lattice... the current carrying state would then be very long-lived because of the small probability of scattering (Bohm 1949, 504).

Despite the complete agreement, Bohm was apparently unaware of Frenkel's sixteen-years-earlier paper in the same journal, *Physical Review*, for otherwise he would have known also that Frenkel's proposal had then been refuted by critics. A common political worldview explains why the two came to the same hypothesis independently. Bohm's proposal of 1949 also did not work for superconductivity—at least not immediately—but the same approach proved extremely fruitful for his works on plasma, where he also had Soviet predecessors.

Frenkel's models of collectivized particles concerned primarily the state of electrons' freedom vis-à-vis atoms of the crystal. Landau's main critique of the model of free electrons in solids focused on another obstacle, the strength of electrons' interaction among themselves. According to his estimate, the energy of that interaction had the same order of magnitude as the electrons' kinetic energy, which made the main assumption of the band theory unjustified (Landau and Kompaneets 1935, 804). For Landau as for Bohm later, the main challenge was to describe the collective interactions and movements in an ensemble of particles, whereby the main difficulty was the lack of mathematical methods capable of handling the multi-particle interaction. Admitting in 1936 that a strict solution of the problem seemed impossible at the time, Landau tried several palliative remedies and turned from metals to the similar but somewhat simpler case of electrons in plasma. He tried to extend the basic methods of classical kinetic theory from the gas of neutral molecules to the gas of electrically charged ions and electrons and proposed a generalization of Boltzmann's kinetic (transport) equation with a new expression for the collision integral. Unlike atoms in a regular gas, which interact only if they collide at very short distances, particles in plasma interact through long-range Coulomb forces. Landau's new term in the collision integral corresponded to the case when particles scatter each other at long range along small angles and with small changes in their velocities (Landau 1936).

Two years later Anatoly Vlasov from Moscow University achieved a real breakthrough. In his paper "On the vibrational properties of electron gas," Vlasov (1938) pointed out an inconsistency: Boltzmann-type kinetic equations work when particles are rare and collide one-on-one, but in plasma a sphere with the radius of the electrons' effective interaction included many particles at the same time. Long-range Coulomb-type forces made every electron in plasma feel the impact from a large number of other remote particles simultaneously. Individually these interactions were much weaker than a direct collision, but in the aggregate they produced considerable overall effects:

> In the classical conception of 'paired' collisions the weak though actually existing forces of interaction at 'long' distances (exceeding the mean distance between particles) are disregarded. [This] also disregards the collectivizing effect and together with it a great deal of phenomena... [Taking them into] account reveals totally new dynamic properties of polyatomic systems (Vlasov 1945a, 25).

Vlasov replaced the collision integral in Boltzmann's kinetic equation with a different term, describing the movement of particles in the field jointly produced by them all (the approximation known as the method of self-consistent field) (Vlasov 1938). The collective interaction of electrons was responsible for new effects; for example, a fluctuation in the density of particles did not dissipate quickly, as in the normal gas, but oscillated with a characteristic "plasma" frequency $\omega_p = \sqrt{4\pi N_0 e^2/m}$, where N_0 is the mean density, e the charge and m the mass of the electron. The existence of such oscillations in plasma was already well known from the classic experimental studies by Lewi Tonks and Irving Langmuir (1929) but was not properly understood.

In his post-war papers (1945a; 1945b) Vlasov tried to generalize his set of equations even further to include not only Coulomb but also other kinds of central forces, including short-range ones, which was not really justified in the approximation of the self-consistent field. He thus came under severe criticism from Landau and coauthors as part of the 1946 administrative struggle for the Moscow University chair of theoretical physics.[13] In a separate paper written at the same time, Landau used refined mathematical techniques to derive a new solution of the Vlasov equation, which corresponded to the damping of electromagnetic oscillations in plasma owing to collective interactions. As a consequence of this

---

[13] (Ginzburg et al. 1946). See discussion in Chapter 9, 227.

Collective Excitations

damping, external electromagnetic radiation cannot penetrate deep into plasma (Landau 1946). Meanwhile, at yet another level of mathematical sophistication, bordering on mathematics proper, Nikolai Bogolubov provided strict justification for both the Vlasov equation and the Landau collision integral as two different approximations in the statistical description of a gas with Coulomb forces.[14] These Soviet accomplishments constituted the first theoretical treatments of collective phenomena and effects in classical plasma, which were further developed in Bohm's more general approach.

According to one of his students, Eugene Gross, Bohm had started developing a collectivist treatment of plasma independently, even before he came across the papers by Vlasov and Landau (Gross 1987, 48). The acquaintance with Soviet work shifted his language subtly but significantly. If earlier he had conceptualized the behavior of electrons using the unionist notion "organized movement," starting in 1950 he added the word "collective" to his physics vocabulary and adopted for his approach the name "collective description [which] is based on the organized behavior produced by the interactions in an electron gas of high density" (Bohm and Pines 1951, 625).

In a later oral history interview, Bohm gave the following explanation for his peculiar interest in plasma:

> First of all, it was a sort of autonomous medium; it determined its own conditions, it had its own movements which were self-determined, and it had the effect that you had collective movement, but all the individuals would contribute to the collective and at the same time have their own autonomy.[15]

Indeed, Bohm's early publication on plasma reveal his desire to understand how free particles can act together and participate in a collective movement without losing their individuality. One senses a personal agenda here, as he was trying to solve mathematically with regard to electrons a serious life problem, the conflict between his uncompromising intellectual independence and his longing to become part of a genuine collective. A resolution of this difficulty evaded him in his social life, but electrons in plasma apparently managed to combine both: they were practically free particles, independent of one another, but as a result of subtle

---

[14] (Bogolubov 1946a; 1946b; 1946c). For a contemporary comparative analysis of the classical papers by Vlasov and Landau and the controversy about them see (Aleksandrov and Rukhadze 1997; Klimontovich 1997).
[15] Oral history interview with David Bohm by Lillian Hoddeson (1981), 3–4 (AIP).

interactions within a large group they developed patterns of organized coherent movement.

Bohm's further thoughts developed in a general philosophical direction, concerned with the problems of society and freedom, most notably the major sociological problem of coherent action, or how stable and conformist patterns of behavior emerge in a society with individuals who consider themselves free. In plasma, he saw "a model of society where I wanted to begin to understand the relation of the individual and the collective. Where one did not greatly interfere with the individual freedom and yet could understand collective action."[16] Motivated by such concerns and assisted by Eugene Gross and David Pines, Bohm from 1948 to 1952 produced a series of highly sophisticated mathematical papers that laid the foundations of the modern theory of plasma—both classical and quantum—based on the notion of collective interaction. Among other results, this approach eventually delivered one more quasiparticle to the list, the plasmon or the quantum of plasma oscillations.[17]

6. The Prose of Collectivism

Versions of collectivist approaches, besides those of Frenkel, Tamm, Landau, and Bohm, continued to accumulate in the field over the years. In the mid-1930s Shubin and Vonsovsky in Sverdlovsk tried to develop a collectivist alternative to Heisenberg's theory of ferromagnetism, which viewed electrons as bound to particular atoms. A few years later Edmund Stoner in Leeds proposed a "collectivist electron" theory of ferromagnetism. In the mid-1940s, the method of collective excitations received a thorough mathematical justification in the work of Nikolai Bogolubov. Starting in 1950, Ilia Lifshitz with Moisei Kaganov and other collaborators in Kharkov created a more sophisticated formalism for the electron theory of metals, treating electrons as quasiparticles. Also in the early 1950s, Ben Mottelson and Aage Bohr in Copenhagen developed collective description in the theory of nuclei.

The new collectivist models were usually less explicit on the political side but better developed mathematically. Although not entirely independent of

---

[16] Oral history interviews with David Bohm, conducted by Maurice Wilkins (1986), 16 tapes, unedited transcript (AIP and Birkbeck College Library, London), 253–254.
[17] For detailed analysis of Bohm's program in the theory of plasma, its philosophical background and results, see (Kojevnikov 2002b).

one another, they were pursued by different authors driven by somewhat different motivations. Practically all of the early authors were socialists—of all possible stripes—in their political views, with the majority coming from the Soviet Union. The family of approaches developed by them used, to various degrees, collectivist terminology and shared an understanding of real freedom as a complicated issue and a distaste for models with either completely bound or totally free particles. They added new members to the list of quasiparticles and collective excitations, of which now there are over a dozen major species and a great many more varieties. Together these efforts produced in the mid-1950s a qualitative change of mind among condensed matter physicists and brought the collectivist approach and the method of quasiparticles to general recognition.[18]

When Frenkel first applied collectivist notions in physics during the 1920s, he was acutely aware of the political meanings of the terms and treated them as conscious metaphors. By the 1940s and 1950s, however, Soviet physicists had gotten so accustomed to the use of collectivism-related words in everyday Soviet language that the implied meaning of these words for them was not so much a controversial political ideology, but a familiar and ubiquitous reference to their ordinary way of life. They correspondingly used these terms ever more often—in physics as in life— as words from ordinary language. To the best of my knowledge, Soviet physicists did not try to gain any political advantage by explicitly connecting their collectivist approach in science to the official ideology or by presenting it as a specifically Soviet or socialist way in physics. Their awareness of the ideological roots of physical terminology and sensitivity to the possible political meanings of the terms greatly decreased with time.

On the other side of the Atlantic, however, collectivist words still sounded suspicious to many and bespoke a political—and foreign— ideology. After Bohm's emigration, David Pines remained the main representative of his approach in the United States. Pines found a new patron in John Bardeen and worked for several years in Bardeen's group at the University of Illinois, which required an adaptation of the collectivist

---

[18] Among the earliest review articles that reflect the growing awareness of the generality of the quasiparticle approach and its universal recognition are (Vonsovsky 1952; Bonch-Bruevich 1955). For an updated list of major quasiparticles and excitations one can check, besides standard textbooks and encyclopedias, the appropriate heading in the *Chemical Abstracts Service* database. Most individual physicists mentioned in this chapter were given separate credit and recognition for their particular discoveries and proposals of quasiparticles, but the cumulative result of their collective effort and the fundamental change it produced in the concepts of modern physics, have not as yet received proper acknowledgement.

approach to a politically much more conservative environment. Mathematical methods were developed further and some collectivist terminology remained, but all allusions to subversive political meanings disappeared. Some additional image control and rhetoric were applied to ensure that the approach did not appear dangerously "radical." American physicists in the field much more commonly worked for business and private corporations than for leftist political causes, and at least some of them felt a little uncomfortable hearing and using collectivist terminology. As a reflection of these subtle concerns, the more neutral term "quasiparticles" gradually became more popular and is currently used as a synonym of, interchangeably with, and probably overall somewhat more often than "collective excitations."[19]

The basic characteristics of quasiparticles are in many ways similar to those of ordinary particles: they carry energy and momentum, possess effective mass and charge, and can scatter, emit, or absorb other particles and quasiparticles. Quantitatively, their behavior can be unusual: some values of energy E can be forbidden, momentum $\mathbf{p}$ is not conserved exactly but can change by a quantum, and the relationship between energy and momentum is often described by a mathematical function more complicated than $E = \mathbf{p}^2/2m$ of the usual mechanics, either classical or quantum. These features, however, are not considered at all problematic by today's physicists, for whom quasiparticles have become familiar and uncontroversial natural objects, fundamental elements of our physical picture of the world. They offer to physicists an indispensable method of describing processes that involve movements of inconceivably large numbers of particles, providing much more convenient models with relatively few participating constituents. Some physicists have even considered quasiparticles, rather than atoms or molecules, the real elementary constituent parts of practically all material objects in the world surrounding us (solids, fluids, and plasmas), except rarefied gases and the high vacuum inside particle accelerators and out in interstellar space.

For physicists, the existence of quasiparticles in nature is no less real than the existence of the electron. Indeed, the philosophical criteria of the reality of electrons, for example those formulated in Ian Hacking (1983), equally apply to quasiparticles. They have become so real that some physicists even wonder whether there is a fundamental distinction

---

[19] The earliest appearance of the word "quasiparticle" to my knowledge is in (Bogolubov 1947).

between quasiparticles and ordinary particles. Those who had thought the matter through concluded that:

> In all dynamic properties, quasiparticles are just like ordinary particles (although the laws of their movement may be significantly more complicated). However, in contrast to ordinary particles, quasiparticles cannot appear in vacuum; they need a certain medium as the background, because, being elementary units of movement, they are different from the elementary building blocks of the medium. This is the main difference between particles and quasiparticles; in all other major characteristics they are the same (Lifshitz 1958, 402).

In other words, one needs to postulate other particles first in order to construct out of their movements new combinations, or quasiparticles, which are thus explicitly entities of the second ontological order. This solution, however, does not exclude possible doubt about the ontological status of ordinary particles, which may, in the end, also turn out to be artifacts of a medium.

The language of contemporary science includes many phrases and terms such as "collective excitations," "collective phenomena," "collective coordinates," "collective modes," and "collective oscillations." Their scientific meanings and usage have become sufficiently separated from related political language to receive separate treatment in some encyclopedic dictionaries.[20] Most of today's physicists know, and probably more than half of them actively use, quasiparticles and collective excitations as a fundamental concept of physics. They speak unwittingly in the prose of collectivism, without realizing that the idea and its roots have something to do with a political philosophy, that many of them regard as controversial and discredited. Retrospectively, the family of collectivist approaches can be regarded as the most important contribution of Soviet and, more generally, socialist thought to the conceptual development of physics in the 20th century. The triumph of these methods required, however, that their connections to their parental philosophy had to be made virtually invisible.

---

[20] See, for example, "Kollektiv" in: *Brockhaus Enzyklopädie*, 19th ed. (Mannheim) 12, 170.

Chapter 11

Dialogues about Knowledge and Power in Totalitarian Political Culture

Dictatorships are not known to provide friendly environments for open discussion of the distribution, limitations, and boundaries of power. Even when such discussions do occur, the chances are slim that they will surface in public or survive in historical documents. One can, nevertheless, find in the history of the Soviet Union records of several episodes in which politicians and scientists argued about their respective spheres of competence and authority. Some such dialogues that will be analyzed below are rather well known; others have been hidden in forgotten and obscure sources. The conversations were very restricted, because they took place in the context of supposedly total dictatorship, with its strong limitations on even internal and unofficial political talk. Very often participants had to resort to metaphorical language, and each quotation separately allows different interpretations. But taken together and juxtaposed in a sequence, they reveal a specific pattern of relationship between knowledge and power.

This relationship was not stable, but rather subject to negotiation and compromise, with terms that shifted over time. Politicians and scientists were two privileged and mutually dependent elite groups in Soviet society. The partners in this relationship, though of course not equal, exerted influence upon each other. Politicians had a share in deciding on matters related to science. At the same time, scientists had *de facto* access to political decision making, although the nature of this access was not easy to formulate in acceptable Soviet political language. Even more problematic was the process of drawing a boundary between those topics which, in Soviet society, were to be labeled 'scientific' and those which were considered 'political.'

Societies differ in their exact methods of solving this problem of labeling, or demarcation between the scientific and the political, and in their culturally specific ways of establishing a relationship between knowledge and power. Quite often, such a demarcation is difficult to make, with no obvious or generally agreed upon solution available at hand. This happens characteristically in important cases involving serious interests: What belongs to the sphere of scientists' professional expertise and what to the competence of politicians? What are their relative shares of authority? This chapter attempts to specify more precisely Soviet patterns of dealing with these issues, and to do so in a form that is convenient for comparisons.

1. The Trinity of Higher Knowledge

Loren R. Graham has recently noted that one can hardly imagine a case better suited for testing our contemporary ideas and theories regarding science, technology, and society than the historical experience of Russia and the Soviet Union (Graham 1998; 1999). Indeed, the Soviet case was characterized by, on the one hand, an exceptionally high development of science and, on the other hand, a distinctive social and cultural milieu artificially isolated from most international contacts by political barriers. This exceptional combination offers historians a perfect opportunity for genuinely comparative studies of science and society.

In the past, such comparisons have relied on an easily available ideological attitude that opposed East and West. The usual answer stated that the Soviet regime refused to separate science from politics and ideology and thus distorted the practice of science and violated the norms of objective research. Indeed, as a corollary to the Marxist thesis about theory and praxis, Soviet communists understood science as rooted in human beings' material and social life. They correspondingly declined to view scientific knowledge as independent of either industry and technology or politics and values.[1] For this they could be and usually have been condemned on

---

[1] Examples of this general attitude abound. See for instance (Bukharin 1931; Hessen 1931), in particular the following quotation: "The economic position is the foundation. But the development of theories and the individual work of a scientist are affected by various superstructures, such as political forms of class war and the results, the reflections of these wars on the minds of the participants—political, juridical, philosophic theories, religious beliefs and their subsequent development into dogmatic systems." (Hessen 1931, 177).

the basis of an explicit or implicit assumption that science as a free intellectual activity should be protected from such external influences.

This assumption—the ideology of pure and apolitical science—is no longer as popular as it was in the heyday of the Cold War; it does not adequately represent the social realities and practice of democracy and has disappeared from most current discussions of science and society. Strangely enough, however, it has survived in the discourse about science under totalitarian regimes. Soviet experiences with science are still often evaluated against an outdated and utopian conception of scientific knowledge. Although once partly justified by the needs of continuing Cold War propaganda, such discourse has now become an anachronism. Even if the ideological solutions offered before no longer seem satisfactory, the question remains of whether, and how, different political systems—dictatorships and democracies, among others—differ in their attitudes to the problem of knowledge and power.

A good starting point for changing the discourse can be found in Carlo Ginzburg's account of the metaphysics of higher knowledge in early modern Europe. Ginzburg starts from the distinction, in the Christian tradition, between profane knowledge, intended for everybody, and the higher knowledge that constituted the key to power and was considered inappropriate for lay persons. Aspirations to the latter were typically condemned as sinful pride, and only partial exemptions were allowed for specially designated mediators and interpreters. According to Ginzburg, early modern Europe recognized three higher secrets of this sort: the secret of God, the secret of Power, and the secret of Nature. The corresponding spheres of religion, politics, and philosophy-science were neither considered totally separate nor fully conflated: they constituted different reflections in the mundane world of what in the metaphysical hierarchy was the trinity of higher knowledge (Ginzburg 1989).

The metaphysical programs of modern societies have tended to reject the assumption of privileged knowledge, assuming the accessibility and separability of different kinds of knowledge and of the religious, political, and scientific authorities associated with them. This belief led in practice to shifting frontiers and to the establishment of far-reaching divisions. Many formerly restricted domains of knowledge came to be regarded as open to inquiry. Furthermore, their assignment to one of the three spheres has generally been agreed upon. In a number of other important cases, however, such consensus or conventions are fuzzier or lacking. The program of modernist division has been very successful as a guiding ideal, but not as something that can be ultimately completed.

The precise paths chosen by modern societies and, correspondingly, their results varied: some societies invested more effort in isolating science from religion, while others concentrated more on separating science from politics.

Soviet society, representing a particular version of the program of modernity, developed its own approaches to the problem of the metaphysical trinity of higher knowledge and its mundane divisions. Having discussed the role of ideology in its relation to scientific knowledge in earlier chapters,[2] here I will concentrate primarily on the interplay between science and politics. The logical sequence of the episodes is as follows. First, as background, I will describe the Soviet metaphysics of Power and Knowledge, or the general theoretical formulation of the relationship between them. The practical solutions that were tried were noticeably different from, though influenced by, the existing theoretical views. In the first approximation, one can speak of a Bolshevik pact of the 1920s between communists and specialists, which was broken off by the Cultural Revolution in 1928. The latter was an intermission, a radical and violent attempt to do away with the compromise in favor of the envisioned ideal. After an indirect acknowledgement that such utopian goals could not be fully achieved, a new order started to emerge in 1932, establishing a new, Stalinist pact between politicians and the intelligentsia. Despite attempts to renegotiate its basic terms after World War II, it remained in force at least until Stalin's death and the success of the Soviet nuclear weapons project. By 1960, however, the earlier compromise was finally replaced by a modified post-Stalinist or late Soviet pact that reflected the increased importance and power of scientists.

2. Elements of Soviet Metaphysics of Knowledge and Power

The Russian Academy of Sciences celebrated its 200th anniversary in 1925. At the jubilee meeting, the party was represented by Politburo member, Chairman of the Communist International, and Leningrad party boss Grigory Zinoviev, who delivered the following message: "We [communists] know that quite a lot of divisions exist between us and scientists." The party, however, "likes to hope that its program is totally grounded in the conclusions of science ... It is indeed very strange that there exist some scientists who are against the revolution." And even if many scientists

---

[2] See Chapters 8 and 9 and (Kojevnikov 2000).

look at Marxism with skepticism, "the era will come... when our two camps converge" (Zinov'ev 1925).

That the relationship between scientists and communists was far from rosy was quite obvious, so much so that mentioning the differences could not be avoided even on a festive occasion. Zinoviev even had to recognize two distinct parts of his audience by greeting them with the phrase "Citizens and Comrades!" Although a significant portion of the scientists liked the idea of socialism, they with very few exceptions sympathized with political parties that opposed the Bolshevik regime. And despite their professed respect towards science, Bolsheviks with very few exceptions did not possess even basic scientific literacy and could be highly suspicious of scientists in real life. What is important, however, is that Zinoviev could not allow these unpleasant empirical realities to destroy the ideal picture of the relationship, for that was part of the Bolshevik metaphysical world. The ideal picture described how things should be, even though in reality they were not.

Zinoviev's way of saving the ideal image of the world from confrontation with the hard realities of everyday experience is a very common one, indeed used by practically everybody. When confronted directly, one can acknowledge the uncomfortable reality but add that things are different "at some higher level," "in essence," "in most cases," or "in the future." When not pressed, one can simply proceed arguing without modalities, replacing "should be" with "is." Thus, Zinoviev acknowledged discrepancies in public while addressing an audience of non-believers in 1925. In an audience of fellow communists, or in a later period with stricter limitations on public talk, a politician of his rank, instead of "lik[ing] to hope," would plainly state that "Party politics and ideology *are* scientific," and also advance the symmetrical statement "Science is *partiina*," which Zinoviev refrained from making to the Academy in 1925 in order not to upset it.

These two complementary equations represent the core of the Bolshevik solution to the problem of Knowledge and Power. In the center of their ideal world was an imagined agency, called Party. This imagined Party should not be confused with the real one: the latter, for example, was ridden by conflicts and intrigues, while the former had a single will and indivisible power. The Bolsheviks, however, were rarely allowed to acknowledge this difference in public talk, where the word "party" usually had to be used in its ideal meaning. The ideal Party was supposed to command single-handed the knowledge of true politics and therefore justifiably monopolized political power. In many Bolshevik phrases, Party substituted for Power as a philosophical category (the unofficial Soviet language reversed the substitution, thus confirming the synonymy).

The same Party was also the bearer of another kind of higher knowledge, the knowledge of true values and higher meanings of the world, society, and history. This kind of knowledge required believers, guardians, and proselytizers, but it was not called Religion, because Soviet society officially had no religion. Instead, it went by the name Ideology and was thought to be contained in the specified body of classical texts by Marx, Engels, and Lenin. Although in practice the dominant meanings extracted from those texts fluctuated drastically and often unpredictably over the years, this inconsistency, like the difference between the ideal Party and the real party, could not be explicitly acknowledged in public discourse.[3]

Even in the world of communist dreams, however, the ideal Party was not expected to command the third kind of higher knowledge, knowledge of the natural world. For this, Bolsheviks imagined a separate ideal agency, which they called Science. In the Russian language the word *nauka* (science) has a higher meaning than the word *znanie* (knowledge), and it has been used—since long before the Bolsheviks—as an elevated philosophical category in contexts where Anglophone philosophers would normally speak of Knowledge. Science as an ideal agency enjoyed a very high status in the Bolshevik world, compared to that of Party, People, or Proletariat. Bolsheviks knew that Party could not rule without relying on Science—in other words, that Power depended on Knowledge. On the other hand, they insisted that Science could not be separated from politics and ideology—in other words, that Knowledge could not be independent of Power. These two symmetrical epistemological theses were represented in the Bolsheviks' language as the two formulaic phrases about Science and Party quoted above.[4] In the Bolsheviks' metaphysical world, these were two different but related agencies. The relationship between them could be nothing short of the preexisting harmony, as pictured in the socialist realist painting shown in the next page.

---

[3] The ideal world of communists had its inverse mirror image in the notion of totalitarian society professed by anti-communists. Despite being motivated by diametrically opposed judgments, both were based on structurally identical metaphysical pictures, in which all power effectively belonged to the single agency, Party, and was executed according to the preexisting master plan, Ideology. Both views had some meaningful relationship with an immensely more complicated reality, but they equally failed to keep in mind the difference between imagined and real worlds.

[4] On *"partiinost'* of science" as a thesis about sociology of knowledge see (Joravsky 1961, 25). The clearest exposition of social constructivism in Marxist terms is the frequently criticized but rarely read classical text "Science and the Working Class" by Aleksandr Bogdanov (1918). On Bogdanov see (Sochor 1988); on his constructivist views on science see (Kozhevnikov 1994) and the conclusion to this book.

The genre of socialist realism prescribed the use of naturalistic images to represent the world as it should be rather than as it is. True to its canons, the painting depicts real historical characters and a meeting that actually took place, in January 1921 when Lenin received representatives of the United Council of Learned Institutions (sitting left to right): the vice-president of the Russian Academy of Sciences, the mathematician V. A. Steklov, the president of the Military Medical Academy, professor V. N. Tonkov,

А. М. Горький с учеными у В. И. Ленина (В. А. Стеклов, В. Н. Тонков и С. Ф. Ольденбург). Репродукция с картины художника В. А. Серова.
Музей А. М. Горького

Socialist realist—i.e. naturalistic in form and ideal in content—representation of the relationship between Party and Science. "A. M. Gorky and scientists meeting with V. I. Lenin," painting by V. A. Serov.
[*Source* (Mendelevich 1964, 116).]

and the permanent secretary of the Russian Academy of Sciences, the orientalist S. F. Ol'denburg. Standing on the right is Maxim Gorky, the proletarian writer who during the early phase of the Bolshevik regime played the role of the mediator and patron of the Russian intelligentsia. Gorky pleaded with the authorities on behalf of starving and freezing artists and scholars unfit for the struggle for existence in the big cities devastated by the Russian Civil War. He convinced Lenin to establish the system of special "academic rations," thus literally saving many lives. In the ideal world of socialist realist painting, his figure is representing classical Russian literature, of which he was the last living exemplar.

The real-world purpose of the visit was a protest by academics against their miserable economic conditions and the communist-enforced radical reform of higher education, which led to a series of strikes by Moscow and Petrograd professors. The painter satisfied the rules of the genre, however, by placing the meeting in the ideal world and making it symbolize the perfect understanding between Science and Party. The larger-than-life stature of Russian literature, so typical of the 19th century, dwindled in the course of the 20th. After Gorky's death in 1936, no writer would even remotely approach his status as the Soviet Union's main cultural icon and authority. Science would replace literature as the center of Soviet culture and its dominant symbol.

3. The Bolshevik Pact with the Specialists, and Its Failure

The practical relationship between Soviet politicians and scientists (in contrast to the ideal one, between Party and Science) was a far cry from preexisting harmony. It was rather a kind of compromise, the terms of which changed over time. The first Soviet pact of the 1920s was characterized by unusually high participation of scientists and engineers in government agencies. Old Czarist ministries were dissolved: of the former Mining Department only its Geological Committee survived, of the Agricultural Ministry only the Agricultural Scientific Committee. The real continuity between former bureaucracies and new Soviet commissariats was that of scientific expertise. During the first post-revolutionary years, as inexperienced communist officials were trying to learn the skills of everyday management, scientists and other professionals enjoyed much stronger political influence than they had had before the revolution. They were not just providing expertise but directly participating in decision making on politically important issues, such as metric reform, calendar and spelling reforms, the grand project of electrification, and matters of everyday politics. The central state commission for planning the nation's economy, Gosplan, had a communist engineer at the top but consisted otherwise of mostly non-communist professors and engineers. Non-party professionals also figured prominently in Soviet military, industrial, medical, agricultural, and educational bureaucracies.[5]

These professionals within Soviet government offices were called "bourgeois [sic!] specialists." The word "bourgeois" referred to the undeniable: not only were they non-communists, they did not speak, behave, or dress like communists, or proletarians, and they were not even expected to be sympathizers. "Specialists are unavoidably bourgeois *en masse*, due to the very circumstances of the social life that made them specialists," Lenin plainly admitted in early 1918 while formulating the basics of party politics toward experts.[6] Yet the second word "specialists" transformed the label from negative to positive and implied a need for toleration and compromise, which in fact could reach very far. Vladimir Ipatieff, monarchist, general of the Tsarist army, and chemistry professor, was appointed *de facto* chief of the Soviet chemical industry and sent to Germany in 1923 to conduct top-secret negotiations about producing chemical weapons.[7]

The condition for such an unusual collaboration was that specialists keep their private political views separate from their public professional service. The Bolsheviks' belief in preexisting harmony between true science and true politics worked to the extent that it helped to justify something rather improbable: the assumption that if specialists would leave their political values at home and enter public office as professionals only, they would act as the communists' "natural allies." The most important boundaries between science and politics to be maintained and guarded during the period were, within government offices, that between their two constituents—political commissars and professional experts—and, within the minds of individual scientists, that between their political views and professional service. Communists were aware that these boundaries were not impermeable, that some specialists could fail to maintain the internal separation between politics and science and could

---

[5] On the Geological committee see (Kleopov 1964); on the Agricultural Scientific Committee and its experimental stations see (Bastrakova 1973, 187–194); on electrification see (Coopersmith 1992); on metric reform see (Kozlov 1981); on Gosplan see (Ipatieff 1946).

[6] In the official Soviet translation, "[T]he specialists, because of the whole social environment which made them specialists, are, in the main, inevitably bourgeois" (Lenin 1918, 248). For more on the notion of "bourgeois specialists" see (Graham 1993, ch. 4).

[7] On Ipatieff's political views and his collaboration with the Bolsheviks see Chapter 2 and (Ipatieff 1946, 246–393).

also mislead technically illiterate commissars. They monitored government bodies and individual specialists for specific failures of that kind but did not abandon the general terms of the compromise.

The more radical among the communists, however, worried that the boundary was even more permeable and that the "wrong" politics could be sold to them under the guise of scientific expertise. In plain words, they suspected that bourgeois experts were guiding communist officials, not the other way around.[8] These views, which were not entirely unjustified, received official support in the spring of 1928, when the existing compromise with "bourgeois specialists" was declared a failure because of an alleged anti-Soviet conspiracy of engineers in the small mining town of Shachty. This event signaled the outbreak of the so-called Cultural Revolution, which attempted to force the real world into correspondence with its ideal picture. Speaking in 1925, Zinoviev had cautiously expected the "convergence of our two camps," Party and Science, to occur in some unspecified future. The Cultural Revolution of 1929 strove to achieve this in the course of just a few years. On the one hand, communist activists and radical students imposed political differentiations upon the existing body of specialists, separating so-called "wreckers," to be unmasked, fired, and often purged, from loyalists, to be quickly "forged" into sincere communist sympathizers. A further, even more ambitious goal was to create an army of new "Red" specialists. While older professionals were forced to learn and internalize "correct" politics, hundreds of thousands younger communists and proletarians were sent to colleges to learn and internalize scientific and technical knowledge.[9]

---

[8] For an open expression of the concern that Narkompros can become an agency for "specialists," and that they could use government offices rather than the reverse see *Piat' Let Sovetskoi Vlasti* (Moscow 1922), p. 508, quoted in (Alekseev 1987). On internal opposition to the party's official policy of cooperation with specialists see (Fitzpatrick 1992, ch. 5).

[9] On Red experts see (David-Fox 1997).The application of the term "Cultural Revolution" to the social upheaval in the Soviet Union between 1928 and 1931 is due to (Fitzpatrick 1978). In a enlightening recent discussion, Michael David-Fox pointed out that the term's origins in Bolshevik language preceded 1928 and that its actual historical meanings differed from the later, more famous Chinese usage, as well as from Fitzpatrick's (David-Fox 1999; Fitzpatrick 1999). It is also true that among the slogans of the Stalinist revolution around 1930, "industrialization", "collectivization," and "socialist reconstruction" figured more prominently than "Cultural Revolution." The importance of Fitzpatrick's notion, however, is that it helped to conceptualize the events as a specific historical phase different from both the preceding society of the New Economic Policy (NEP) and the later Stalinist society. The alternative and also somewhat conventional notion, the Great Break, which in the language of historical actors referred to one specific year, 1929, is in the professional historical discourse usually associated with models of a direct transition from NEP to Stalinism that did not acknowledge the existence of a socially distinctive intermediate stage.

The purge in the Academy of Sciences at the height of the cultural revolution in 1929. The commission chaired by the Workers' and Peasants' Inspection representative Yu. I. Figatner, and composed of both academics and political activists, reviewed the credentials of the Academy's staff. The commission found that far too many former nobles and Tsarist officers without special academic credentials found employment and institutional protection in the Academy after the revolution. It fired 128, or about 11 percent, of the Academy's total staff, and a significantly larger number of adjunct personnel (Perchenok 1995, 211). 1929 was also the year when the first few Marxist scholars and communist party members were elected to membership in the Academy, and many workers and peasants without proper academic credentials were encouraged to join the staff of academic and scientific institutions. Sitting on the left at the commission's table are Professor P. M. Nikiforov, academician A. E. Fersman and permanent secretary of the Academy S. F. Ol'denburg.
[*Courtesy*: RGAKFD.]

This new class of students was encouraged to graduate as quickly as possible, in two or three years, ignoring the traditional curricular requirements and discipline. The revolutionary chaos, during which the boundary between the scientific and the political was plainly rejected, lasted about three years as well. Once the first cohorts of hastily trained engineers entered the workplace, their lack of qualifications was tacitly acknowledged. In June 1931 Stalin signaled the closing of the radical campaign by criticizing "excesses" of specialist-baiting. Although utopian goals were not realized, Stalin proclaimed the main objective to have been achieved. In his view, the old intelligentsia had learned its lesson and was "turning toward Soviet power."[10] From that point onward it became obligatory to state in public that Soviet scientists were devoted to the party's goals. This change in public speech was accompanied by changes in the practical relationship between politicians and scientists.

4. The Stalinist Pact with the Intelligentsia

Some things returned to where they had been before the Cultural Revolution, or even before the Revolution of 1917. Old specialists were restored to high positions in professional—though not political—offices. Class preferences for students were abolished and replaced by insistence on high educational standards and discipline. Professional boundaries between scientists and politicians were restored: no longer would a person without academic qualifications be sent to direct an academic institute, no longer would a non-communist specialist be appointed to direct a responsible government office. Yet, along with the stricter division of labor, one can observe a rapprochement between professional politicians and professional scientists: the two groups were starting to resemble each other in some important aspects. Consider first the politicians.

One could no longer find in government offices two collaborating but distinctly different species, bourgeois specialists and political commissars. Both were wiped out "as a class"—the former by the Cultural Revolution, the latter finally by the purges of 1937–38—and were replaced by a new generation of political managers. A typical Stalinist *apparatchik* was raised as a communist and educated as a specialist; the Stalinist bureaucracy typically recruited its members from the graduating classes of engineering schools, not those of law or politics. Many who received their technical education during the Cultural Revolution went into politics and management (Fitzpatrick 1992, ch. 7). Even if their scientific training was less than adequate for professional work in science or engineering, it was certainly far superior to that of a conventional politician.

That is why their political and managerial skills were declared superior to those of "old Bolsheviks," the brilliantly eloquent but technically illiterate political agitators and revolutionary conspirators. Here again, the practical solution reflected the ideal picture: the combination of communist upbringing and technical training was supposed to guarantee the

---

[10] On the 1928–1931 campaign of purges against bourgeois specialists and its termination see (Bailes 1978, 69–156).

"scientific" quality of political management. The purges of the old communists opened breathtaking career opportunities for the younger generation of Stalinist politicians. Some of them, by the late 1930s—while themselves in their thirties—rose to become heads of government commissariats. By the time of Stalin's death in 1953 they constituted the bulk of the second echelon of politicians from which new Politburo members were recruited.

From their academic mentors the new generation of politicians learned not only basic science but apparently some values as well. They wanted to appear "cultured," valued the aesthetics of a well-off middle-class lifestyle (previously branded "bourgeois"), abandoned egalitarian values and proletarian appearances, discovered the simple pleasures of hierarchy and privilege, and generally referred to themselves as belonging to the "Soviet intelligentsia." Sheila Fitzpatrick has characterized "Soviet intelligentsia" as the self-description of the new Soviet elite that came to power in the 1930s (Fitzpatrick 1992, ch. 9). Professionals—engineers and scientists among them—were also considered part of the Soviet intelligentsia, and even though their position in the hierarchy of elite groups

General meeting of the Academy of Sciences, November 1946. Upon Vavilov's proposal, academicians unanimously elected V. M. Molotov to *honorary* membership. Molotov became the second and the last Soviet leader to receive this symbolic honor—Stalin had been elected to such membership in 1939. In the late Soviet years, a number of mid-level politicians and government managers were elected to the Academy as *regular* members for their contributions in the field of "organization of science." At the same time, a number of top scientists who had demonstrated managerial skills were elected as regular members to the Communist Party Central Committee.
[Source: *Vestnik AN SSSR* no. 11–12 (1946).]

was certainly lower, they too enjoyed certain material privileges and prestige. But they also had to learn and internalize new values.

A new generation was rising within their ranks, too, many of whom were sincere communists. Since purges hit the academic circles less harshly than the political elite, the generational change proceeded more slowly among academics. The unwritten rule that required party membership for promotion to a high managerial position in the scientific establishment, beginning at the level of an institute director or chief editor of a journal, did not become effective until the post-Stalin era. Throughout the entire Stalin period, non-party scientists continued to figure prominently at the top of the academic institutional hierarchy. But even older scientists who had lived through the period of harassment and purges had to declare themselves conscientious supporters of and sympathizers with the party and to live up to this image.

It became easier for them to do this because some of the new political values reminded them of their own older ones. They, too, liked the hierarchy and the restoration of academic titles and degrees, with corresponding prestige, which the early egalitarian Soviet regime had abolished as medieval relics. They welcomed the improvements in educational standards, discipline, and academic criteria. Even in official Soviet ideology they could recognize some themes borrowed from the ideology of older academics: the shift from radical internationalism toward more nationalist priorities was accompanied by propagandistic claims for Russian (pre-revolutionary) priority in the sciences. Such claims initially took root among Russian scientists in the years preceding World War I, when all European nations were preoccupied with this kind of cultural nationalism. Suppressed but not entirely forgotten during the revolutionary and internationalist period of Soviet history, they resurfaced in official discourse by mid-1930s and inflated to caricature dimensions during the 1940s.[11]

Although one should not take at face value senior scientists' public declarations of love for the party, still some significant changes in this direction did occur. One such case has been widely discussed: Ivan Pavlov included in his address to the 1935 international Physiological

---

[11] The prototypical example was the claim by the Russian Physico-Chemical Society on behalf of A. S. Popov regarding priority in the invention of the radio (Berg 1945; Kaptsov 1970). Although many claims had at least some factual basis, gross overstatements and entanglement with nationalistic and political propaganda subsequently undermined their credibility.

Congress in Moscow public praise for the Soviet regime's generous support of science, proposing a toast to "the great social experimenters":

> As you know, I am an experimenter from head to foot. My whole life has been given to experiment. Our government is also an experimenter, only in an incomparably higher category. I passionately desire to live, to see the completion of this historic social experiment.

Pavlov had enjoyed such a high status that not even the terror of the Cultural Revolution had prevented him earlier from publicly ridiculing communist policies. In 1935 he agreed, for the first time, to change his tune.[12] An even more striking example is an entry in the private diary of the geochemist Vladimir Vernadsky. In 1937, during the frightening year of great purges, Vernadsky dared to write down life-endangering comments about "police communism," about the horrors of the terror and dictatorship, and—not without some gloating—about the demise of old Bolsheviks, whom he genuinely disliked. To these, he added,

> we see here that positive creative work is done by "non-party" intelligentsia and by such men as Stalin, Molotov ... —but not by that vast mass of communists who are morally and intellectually below average. A conviction clearly understood and spreading among the intelligentsia is that the politics of Stalin-Molotov is Russian and is necessary for the state. Their party enemies are also enemies of the Russian people.[13]

The restoration of order, hierarchy, and boundaries between science and politics meant a return to a compromise between scientists and politicians—a compromise that can be called the Stalinist pact. This restored relationship has often been characterized in implicitly gendered terms. Soviet publications and statements pictured it as a romantic (but traditional) partnership between Party and Science. The former provided support and leadership, while the latter responded with devotion and assistance, with the two inseparably tied together by true and mutual love. Understandably, jealous ideological rivals denounced the relationship as one of tight control, abuse, and domination, occasionally even implying the metaphor of rape. Far from being either romantic love or forcible submission, the partnership—to carry the gender metaphor

---

[12] Translated in (Babkin 1949, 162). Historians have good reasons to doubt the ultimate sincerity of Pavlov's statement (Todes 1995; Tolz 1997, 123–140). My point, however, is not about sincerity but about the Soviet scientist's public image and Pavlov's late decision to satisfy its criteria.
[13] Vernadsky's private diary entry for 7 July 1937 in (Vernadsky 1993, 234).

Dialogues about Knowledge and Power

further—more closely resembled a traditional marriage of convenience. Based on some shared values and interests and a process of give-and-take, it was not free of occasional domestic violence.

5. Post-War Negotiations about Power

During the war, the Soviet Union could not afford the luxury of grand-scale advanced scientific projects of uncertain promise for the future. Such projects imitating German and American wartime undertakings were launched only at the very end of the war. By that time, the specter of the U.S. atomic bomb dominated the renewed Soviet discourse about science and politics and required changes in the terms of the compromise. According to David Holloway, Stalin feared the bomb as a diplomatic weapon more than as a military one. The bomb became the very symbol of a superpower, a status Stalin was determined to claim for his country (Holloway 1994, 150–171). It also was the symbol of the most advanced science. Thus the idea that Power and Knowledge are at some higher level inseparable lost its abstraction: it had acquired a very powerful representation in a concrete and much desired object, the bomb.

On 20 August 1945, the Politburo ordered the creation of a Special Committee for replicating the Manhattan Project. This institution brought together scientists and top politicians, providing a forum for regular contact and creating opportunities for renegotiating their respective spheres of authority. The first scientist who tried to shift the terms of the existing pact, as early as 1945, was a member of the Special Committee, Piotr Kapitza. Among Soviet Politicians Kapitza was unofficially known for being "non-Soviet," yet the Politburo thought it was possible to work with him. During the war Kapitza had even been appointed to a public office, head of the Oxygen Trust, where he was organizing a branch of industry for the production of liquid oxygen. Normally the Stalinist pact would not have allowed this, but the war was an exceptional time, and an exception was made for Kapitza (see chapter 5).

The famous letters that Kapitza wrote to Stalin and other top politicians usually dealt with concrete problems of his work, but in such a way as to allow him to address more general topics. In the fall of 1945 he sent Stalin two long letters about the Special Committee and his conflict with its head, the terrifying former chief of the state police Lavrenty Beria. Kapitza also addressed general principles of the organization of the Soviet atomic project and the roles of scientists and politicians in the

Soviet Union. He used his own career as an illustration for the present unsatisfactory relationship:

> My turbine oxygen producing installation... only got going when I, quite abnormally for a scientist, became the head of the Oxygen Trust... This was quite an abnormal and indeed absurd situation, and the power it brought weighed heavily on me, but I put up with it because there was a war on... Experience shows that I was able to make people listen to me only as Kapitza, Director of the Trust under the Sovnarkom, but not as Kapitza, the scientist, above Kapitza, the head of the Trust... This is exactly the situation today in solving the problems of the atomic bomb.[14]

Kapitza dared to propose to Stalin a redistribution of power between scientists and politicians, but since the words "party" and "power" were practically synonymous it was difficult to formulate the idea of the division of power in politically acceptable language. He thus resorted to the use of a metaphor:

> There was a time when alongside the Emperor stood the Patriarch; the Church was then the bearer of culture. The Church is withering away, and the Patriarchs have had their day, but the country cannot manage without leaders in the sphere of ideas. Even in the realm of social science, however important were the ideas of Marx, they still have to grow and develop further. Only science and scientists can move our technology, economy, and state forward. You personally, like Lenin, move the country forward as a scholar and a thinker. The country has been exceptionally fortunate to have such leaders; but this may not always be the case, and not in all disciplines. Sooner or later, we will have to raise scientists to "patriarchal" rank. This is necessary because without it, scientists will not always serve the country with enthusiasm. We cannot buy such people. Capitalist America can, but not us. Without that patriarchal position for the scientist, the country cannot grow culturally on its own, as was already pointed out by Bacon in his *New Atlantis*. Therefore, it is high time for comrades like Comrade Beria to start learning more respect for scientists.[15]

Both Kapitza and Stalin must have been aware that the metaphor was wrong as a historical reference. Peter the Great, who was the first to proclaim himself Emperor of Russia, also instituted a church reform that abolished the post of the patriarch. The institution of patriarchy in the Russian Orthodox Church was restored only after the revolution. But

---

[14] Kapitza to Stalin, 25 November 1945, translated in (Kapitza 1990, 372–378).
[15] Kapitza to Stalin, 3 October 1945, in (Kapitza 1989a, 232–235); censored English translation in (Kapitza 1990, 368–370).

Kapitza needed the image of the patriarch alongside that of the emperor to symbolize the level of his claim. His proposal that the Party share power with scientists was not acceptable to Stalin. Although no verbal answer exists—perhaps Stalin, too, lacked the appropriate language—he accepted Kapitza's rhetorical request to resign from the Special Committee.

In contrast to Kapitza, the scientist who was chosen to lead the Soviet atomic effort, Igor Kurchatov, was "very Soviet" in the opinion of politicians, and that was why he was preferred to another candidate for the job, Abram Alikhanov (Kaftanov 1985). Kurchatov's attitude toward his role with respect to the party and the government was that of an exemplary military general: he minded his special business and area of competence and obeyed political orders. Very characteristically, the future scientific head of the Soviet atomic project had already been nicknamed "General" by his academic peers before the war (Aleksandrov 1988, 137). In his perception—and that of other participants—the atomic project was a direct continuation of the wartime effort, necessary to protect the country from possible military aggression, and required equal discipline and self-sacrifice. Kurchatov himself is known for occasionally signing his letters as "soldier Kurchatov." It was very unlikely that he would start bargaining for better terms with Stalin. Perhaps precisely because of Kurchatov's modesty and devotion, Stalin made the move himself in January 1946, at what was apparently his first meeting with the scientist. Stalin reportedly said

> that our scientists were very modest and they sometimes did not notice that they live poorly ... our state suffered very much, yet it is surely possible to ensure that several thousand people can live very well, and several [hundred] people better that very well, with their own dachas, so that they can relax, and with their own cars (Holloway 1994, 148).

Stalin did not offer scientists more political power—the party was not allowed to trade away power—but he extended their privileges to a level comparable to those of the party elite. The word "privileges," too, could not be used in official discourse, but Stalin found a way to make his offer public when, on 9 February 1946, he gave one of his rare public speeches. This was a campaign address in advance of the first post-war elections to the Supreme Soviet, and it contained the necessary promises to the electorate. To Soviet citizens Stalin promised two things: the end of rationing and a "wide scale construction of all kinds of scientific research institutes," assuring them that

> no doubt ... if we render the necessary assistance to our scientists they will be able not only to overtake but also in the very near future to

surpass the achievements of science outside the boundaries of our country (Stalin 1946, 303).

The latter phrase was also a message to foreign observers. Western analysts misinterpreted Stalin's speech—perhaps not so unwillingly—as a declaration of hostility, and this became an important pretext for the beginning of the Cold War. In fact, Stalin was not interested in hostility but offered the former allies competition. Ostensibly, this competition was to be in science. But since the phrase "the achievements of science outside the boundaries of our country" obviously meant the bomb, "science" here was also a parable for "superpower." Stalin was not going to accept American hegemony based on a monopoly of nuclear weapons. He would not give up claims for superpower status and was ready to compete in science and therefore in power.

The phrase "To Catch Up and To Surpass!" became a popular Soviet slogan of the early Cold War period. It typically referred to science—not just nuclear, but any kind—and its motivation was sincerely shared by a great many Soviet researchers, who were also happy to enjoy their newly privileged status. This status did not include much in absolute material terms: living "very well" for several thousand senior scientists meant sufficient quantity and somewhat better-quality food from special rations and grocery stores and, with some luck, a separate apartment for the family. Although contradictory to the egalitarian Soviet mentality, these privileges meant a lot in relative terms and in practice were highly valued in a poor country ruined by the war, with burnt-out villages and leveled cities and factories, with peasants starving and big-city dwellers typically living in shared, one-room-per-family "communal apartments". Perhaps even more important were social prestige and public image in society, in which scientists now ranked higher than engineers, the pre-war elite among the professionals.

Like many other moral and material rewards, such privileges were distributed in late Stalinist society strictly according to hierarchical criteria. The physicist Sergei Frish recalled with irony that as early as 1943 the Soviet Academy of Sciences used to have among its system of special stores and buffets some specially designated "for full members only" (Frish 1992, 332). (Frish was only elected a corresponding member of the Academy). Many post-war apparatchiks and some scientists developed the kind of obesity common among those who had suffered from hunger in earlier years. It also reflected the stressful and sleepless life of the managerial elite, since higher privileges in Stalinist society came together with grave responsibilities, increased dangers, and sacrifices at work.

Valentin Berezhkov noted with some puzzlement that he had never had a cold during all the years he worked as Stalin's personal interpreter and for the first time felt sick and took a day off only after Stalin's death (Berezhkov 1994). Kurchatov, among others, received high doses of radiation while repairing a malfunctioning nuclear reactor in 1949 (Kruglov 1994, 74). Like Sergei Korolev, the head of the Soviet space program, and many Stalinist ministers, Kurchatov died at an early age of fifty-seven. For him and for many others, those sacrifices were not unusual—in comparison to the much greater sacrifices by soldiers and civilians during the war—and were worthwhile for the higher goal, "to catch up and to surpass."

6. The Post-Stalin Settlement

With the test of the fully developed H-bomb in 1955 the Soviet atomic quest could be considered successful. This did not mean any actual military parity, which would not be achieved for at least another decade, after the development of ICBMs as delivery systems. But it meant a lot symbolically, bringing a feeling of improved security and confidence in Soviet superpower status. Scientists had fulfilled their promise, and the government did not stint on honors, privileges, and new investments in scientific infrastructure. However, the public stature and prestige of scientists had risen so high that privileges alone were not sufficient, and a new round of negotiations was in sight. This time it was not a politically sophisticated senior scientist but a young and inexperienced, unsociable genius, Andrei Sakharov, who challenged the boundaries of the existing compromise. This did not come out of his political opposition to the regime—which started later—but arose because Sakharov himself was a believer in the ideal image of the socialist system, its morality and professed goals. Most scientists lose their political naiveté somewhere on the way upward to positions of political importance. Thanks to his concentration on weapons design and his social isolation in the secret laboratory, Sakharov rose so quickly that he still possessed youthful idealism at the time of his first direct dialogue with political power.

Immediately after the successful test of 1955, the deputy minister of defense, Mitrofan Nedelin, hosted a banquet where Sakharov, as the chief contributor to the device, was invited to propose the first toast:

> Glass in hand, I rose and said something like: "May all our devices explode as successfully as today's, but always over the test sites and

never over the cities." The table fell silent, as if I had said something indecent. Nedelin grinned a bit crookedly. Then he rose, glass in hand, and said: "Let me tell a parable. An old man wearing only a shirt was praying before an icon. 'Guide me, harden me. Guide me, harden me.' His wife, who was lying on the stove, said: 'Just pray to be hard, old man, I can guide in myself.' Let's drink to getting hard" (Sakharov 1990, 194).

Marshal Nedelin was a soldier: expressing his professional philosophy on the relationship with political leadership, he advised Sakharov to adopt a similar attitude. The topic of power sharing was as commonsensical as it was, like matters of sex in puritan Soviet culture, unmentionable in serious public discussion. Nedelin thus alluded to it by means of a sexual allegory.

A. D. Sakharov and I. V. Kurchatov, the two top designers of the Soviet "atomic shield," or nuclear deterrent, in 1957. The meeting was probably the occasion on which Kurchatov made a suggestion to Sakharov to write a popular article criticizing the American proposal for a "clean" bomb, one that was supposedly environmentally safe. While working on this article, Sakharov calculated the environmental and health dangers of radioactive fallout from nuclear tests in the atmosphere and turned into a critic of continued atomic testing by both the United States and the Soviet Union. The resulting 1963 Moscow Treaty banned nuclear tests in the atmosphere, oceans, and space. The episode put Sakharov on the path of political activism and eventually led him into open disagreement with and opposition to the Soviet regime.
[*Credit*: AIP, Emilio Segrè Visual Archives, *Physics Today* collection.]

Sakharov understood that the message was about power and about the relative roles of scientists and politicians. He was shocked by what he regarded as obscenity and offended by the rude demonstration of the limits he was not allowed to trespass. The episode provoked in him critical thoughts that his social world was not exactly as it was ideally described. But rather than accepting the terms of the unspoken compromise, he continued to take Soviet metaphysics literally, not separating science completely from politics and values. This meant for him raising an uncomfortable question, "what moral and political conclusions should be drawn from [scientific] figures" (Gorelik 1998), when he was estimating dangers of radioactive fallout from atmospheric nuclear tests. Soviet officials welcomed such a moral approach as long as it helped them to criticize American policies, but no longer when it started to contradict their own decisions about when to resume nuclear testing. On one such occasion, Nikita Khrushchev reprimanded Sakharov in a public, but still friendly fashion:

> Sakharov goes further. He's moved beyond science into politics. Here he's poking his nose where it doesn't belong. You can be a good scientist without understanding a thing about politics.... Leave politics to us—we're the specialists. You make your bombs and test them, and we won't interfere with you; we'll help you.... Sakharov, don't try to tell us what to do or how to behave (Sakharov 1990, 217).

Neither this advice nor Sakharov's growing political maturity prevented him from further interference in political affairs. In 1964, the Academy of Sciences was in the process of approving the results of the election of new members. The biology division presented several elected candidates, among them Nikolai Nuzhdin, a particularly obnoxious associate of Lysenko's. Although Lysenko's reputation and influence among politicians was on the verge of collapsing, he was still officially in favor. Several physicists and mathematicians agreed to join efforts to vote down this Lysenkoist candidate, and so they did. The mathematician Pavel Aleksandrov and the biochemist Vladimir Engelgardt questioned Nuzhdin's academic credentials, and the physicist Igor Tamm criticized him for opposing recent discoveries of DNA structure and therefore hindering practical applications of science. Had Sakharov been informed of the conspiracy, he would have left the matter to more experienced elder colleagues. His entry into discussion was unprepared and emotional, once again breaking the politically acceptable code of public speech: "I urge all present academicians to vote in such a way that all votes in favor belong only to persons who, together with Nuzhdin, together with Lysenko, share responsibility for the shameful, difficult period in the development of Soviet science which is currently coming to an end" (Sakharov 1996, 861).

The phrase provoked immediate scandal and impassioned protest by Lysenko. The Academy's president, Mstislav Keldysh, tried to smooth things over, calling Sakharov's remark "tactless" and urging the meeting to proceed to other topics. The exact wording of Sakharov's comments, however, did not refer to an individual Academy candidate alone but implied a critique of the official decision by the party to back Lysenko in 1948. Khrushchev was outraged by Sakharov's remarks and burst out angrily at the Central Committee Plenum in July 1964: "Comrades, for political leadership, I think, we have our Party and the Central Committee, and if the Academy of Sciences intervenes, we dissolve the Academy to hell" (Afiani and Ilizarov 1999, 168). He was protecting the politicians' sphere of expertise from public criticism, just as some scientists might have protected their field of professional expertise from external interventions. The threat to the Academy was not meant very seriously: Khrushchev was expressing his emotions rather than a potential action. But by letting emotions lead him, Khrushchev, like Sakharov, broke the code of accepted political talk. Three months later, when the Central Committee voted him out of the office, the list of Khrushchev's "serious political mistakes" included his "tactless" tone in relation to the Academy and, further down on the list, his mistaken support of Lysenko (Khrushchev 1964, 8). In post-Stalin and post-bomb times, the party needed to learn polite and non-confrontational ways of addressing science if it wanted to receive the same in return.

In the late Soviet period, polite intercourse became the dominant characteristic of the relationship between political and academic elites. The public prestige and privileges of scientists in Soviet society remained in relative, though not absolute, terms greater than in the West. The scientific establishment was granted considerable degree of autonomy within the walls of the research institutes and inside a number of "scientific towns," such as Akademgorodok in Siberia, which were inhabited and managed largely by scientists themselves.[16] The party learned the lesson of the Lysenko mistake and paid respect to conventional boundaries of scientists' sphere of authority, no longer venturing into the risky business of taking open sides in scientific controversies.

---

[16] On Akademgorodok see (Josephson 1997).

Reciprocally, scientists were not supposed to openly contradict official ideology or the party's political management of society. They had to behave appropriately as polite political animals and pay due respect to social conventions. Those who satisfied these requirements could receive some privileges unthinkable for other Soviet citizens, such as occasional and carefully measured permission to travel abroad. For the most part, the academic community learned how to live up to the terms of the new compromise in a non-confrontational way. Only Sakharov and a few other dissidents continued to insist that scientific expertise made them qualified to speak up on political matters and to complain that "the scientific method of directing policy, the economy, arts, education and military affairs still has not become a reality."[17] The unauthorized circulation and eventual publication in 1968 of his political essay destroyed Sakharov's last ties to the political establishment and turned him into a dissident.

The new polite relationship between science and politics carefully respected their mutual spheres of authority. How, then, could the higher-level harmony between Science and Party be worked out in the mundane world? The solution found in the late Soviet period consisted in reciprocal cooptation. The party still lacked the means to divide political power in any official and explicit way, to grant a portion of it to scientists as a guild. But it found a way to give selected representatives of the scientific establishment the status of politicians, electing some top academicians to the party's Central Committee and thus admitting them to the political elite proper. Reciprocally, the Academy of Sciences was granting some representative managers of science and high-tech industry the status of scholars by electing them regular members of the Academy. This ritual of exchanging members between two tribes helped achieve—for the first time in Soviet history—a long-sought harmonious relationship. From 1964 until the end of the Soviet era, no record exists of any major disagreement between the Academy and the party.

The problem moved elsewhere: since only the few top representatives of the academic hierarchy were able to move into political circles, the majority of scientists felt even more deprived of political power. A rift developed inside the academic hierarchy between those who were admitted to the political elite and the rest. The new location of the boundary between science and politics manifested itself openly during the era of

---

[17] (Sakharov 1968). For excerpts and historical background see *Andrei Sakharov: Soviet Physics, Nuclear Weapons, and Human Rights*, internet exhibit at http://www.aip.org/history/sakharov/.

Gorbachev's *perestroika*, when the Academy's official leadership was among the staunchest defenders of the Soviet status quo while the rank-and-file intelligentsia demanded the most radical reforms (Kneen 1989; Fortescue 1992; Graham 1998, ch. 4). The Soviet-type compromise between Knowledge and Power was no longer acceptable to them. Having alienated a major portion of its own elite, the Soviet regime had lost its viability years before *perestroika*. In 1991 the pro-democracy opposition by white-collar, scientific, and technical workers in Moscow and Leningrad, assisted by nationalistic movements on the outskirts, finally brought down the rule of the communist party and the Soviet Union itself. At least for Soviet science, success immediately proved to be self-destructive: together with the regime and the party, its moderate privileges, financial support, and elite social status were also gone.

The Soviet cultural experience is of a more enduring character, as various aspects of it—by borrowing, disguise, or negation—have been transformed into features of present-day life. At the height of the Cold War, the Soviet and American metaphysics of Power and Knowledge had one thing in common: scientism, or the idea that Power depended upon Knowledge in ways that were most essential. They disagreed about the reciprocity of the relationship: in the United States, science was proclaimed independent of ideology and politics, while Soviets thought that the boundary was permeable both ways. Rejecting the ideology of pure science, Soviet Marxists proclaimed instead the ideology of practically oriented and politically laden research. A further aspect of the difference was the metaphysics of Power itself: for Soviets, power was indivisible and linked with ideology. Today, it is hardly possible to maintain either an unconditional belief in pure and apolitical science or an ideology of undivided totalitarian power. Having abandoned these elements of the metaphysics of the two Cold War rivals, societies today are trying to combine the rest, the idea of political pluralism and the view of science as entangled with society and its values.

Conclusion

In the midst of the Russian Civil War, the economist and philosopher Aleksandr Bogdanov (1873–1928) withdrew from revolutionary activities to devote himself to cultural work. Once a prominent Bolshevik, Bogdanov did not accept Lenin's revolution as proletarian and socialist—to him it looked more like an army revolt, dominated by a military mentality and producing a society of violent war communism rather than of emancipated labor. Seeing no prospect of an immediate truly socialist revolution anywhere in war-traumatized Europe, Bogdanov decided to work towards a future one, which he did not think could happen until the consciousness and cultural level of the working class rose to a qualitatively new level. While the Bolshevik regime struggled for its military survival and for control of the nation's economy, Bogdanov and his followers founded in 1918 the Proletkult, an independent movement for cultural and educational work among the proletariat.

In *Sotsializm nauki*, a seminal brochure published the same year, Bogdanov developed his views on the relationship between science and socialism. He argued there along a consistently Marxist line, but in one important respect went further than Marx. Despite understanding that science originated from human social and economic practices, Marx generally stopped short of extending his sociological analysis of ideologies and social and political theories to the natural sciences. Bogdanov took this logical next step, defining science as the "organized social and economic experience" and a way of organizing such experience that could be used as a tool for both class domination and class struggle. To him, it was no accident that the new European science emerged together with the development of modern capitalist production. Features of contemporary science that Bogdanov saw as bourgeois were its institutional organization—elitism and subordination to market relations—and the organization of scientific knowledge in a way that was too abstract and detached from the concerns and work practices of the masses. The creation of a new socialist society, in his view, would be accompanied by the development of a new "proletarian science," which would include all previously accumulated knowledge and experience, but would present it from the perspective of workers' values and labor practices—would change the point of view, like Marx had done in political economy, and Copernicus in astronomy. Bogdanov envisioned that proletarian science would also add new knowledge corresponding to the new kinds of social and economic experience, and would be taught and popularized in such a way as to be accessible to wide audiences, not just elites (Bogdanov [1918] 1990, 360–410).

As of today, Bogdanov's true socialist revolution has not happened, but the idea of extending sociological analysis to scientific knowledge has won wide popularity in the West since around 1970. It is almost commonplace nowadays to view and analyze science as grounded in social, economic, and cultural realities of human life. This idea that once constituted a definite sign of Marxists beliefs has developed further—usually without any reference to its historical roots in Marx and Marxist thought—and in generalized and varying formulations has become acceptable to people of different political and philosophical convictions. Personally, Bogdanov did not receive any credit for his suggestion; on the contrary, he took a lot of blame for a caricature of his views put forth by the Bolsheviks. Lenin, suspicious of the Proletkults' desire to remain independent from the party, made sure they were closed down, but the ideological justification for that decision was provided by accusations that by using the term "proletarian culture," the Proletkults were advocating radical and destructive policies towards the existing culture and science. After the demise of the Proletkult movement in 1921, Bogdanov's activities shifted again, this time to public health, where he directed an institute and organized a new system for blood transfusion. He died in a medical experiment that he performed on himself and was recognized by the Soviet government for his public health service, but with respect to Soviet ideology he remained a bogeyman, his name forever associated with cultural policies denounced as wrongfully radical. While later Soviet publications continued to blame the Proletkults for their alleged desire to abolish "bourgeois" science and create instead a completely different "proletarian" one, anticommunist propaganda adopted the same groundless charges and used them against the Bolsheviks themselves.

If, however, one abstains from propagandistic distortion, Bogdanov's theoretical views expressed in the 1918 essay have to be vindicated as an important precursor of modern ideas about science and society. They also provide a more appropriate framework for understanding and interpreting the history of Soviet science, than Popper's *Ansatz* about science and democracy mentioned in the introduction. Granted, neither a true socialist society in Bogdanov's sense nor a proletarian science in the same sense, has materialized. But to the extent that the revolutionary Russian breakthrough into modernity created a society with new and distinctive features—regardless of whether we call this society socialist, or not—it also developed its own, distinctive variety of modern science. The analysis in the preceding chapters of specific features of Soviet science substantiates this conclusion at several levels, albeit with varying degree of detail.

At the level of formal institutions, these features ranged from the overarching "big science" system of research institutes that influenced similar developments in other countries, to the more uniquely Stalinist institution of *sharashki*, or research prisons. With respect to the microsociology of the scientific community and academic culture, the early Soviet decades saw the development of new collectivist educational and research practices, while the later period favored more hierarchical "research schools." In the culture of scientific discourse there was a tradition of research colloquia with intensity and sincerity of critical engagements that can only be hoped for in other academic cultures, but which also lacked protection against, and could thus occasionally extend into, vicious attacks and ideological accusations. And finally, in the realm of scientific knowledge per se, Soviet science pioneered original approaches, ideas, and discoveries in various fields of research. Some of the specific knowledge claims that resulted remained overlooked and unappreciated and some failed, but others succeeded in changing modern science in fundamental ways. Contributions of the last kind constitute the usual criterion by which historians judge scientific developments in different cultures, and by this measure Soviet science stands among some of the most important cultural achievements of the 20th century.

In order to understand the reasons for Soviet success in science, which happened sometimes against the odds, one has to start with an acknowledgement of the extraordinary cultural value and importance that Soviet society attached to science. Contemporaries often took this feature for granted, sometimes lacking an appreciation for the fact, that no other country in the past century, neither Wilhelmian Germany nor Cold War America, equaled the Soviet Union in this respect. In absolute material terms, Soviet expenditures for the development of science and technology could not match those of more prosperous and stable nations, but they constituted a proportionally larger share of the country's wealth. As compensation for their relative poverty of resources and salaries, scientists in Soviet society enjoyed—though did not always fully appreciate—significantly more prestigious social status than their counterparts elsewhere.

The value attached to science and scholarship has fluctuated greatly in the course of human history, from one culture and time period to another. In Russia, the latest such cycle started around 1860, when science captured the imagination of the public and of the intelligentsia. The official embrace of the revolutionary Bolshevik regime further strengthened the cult of science in society, which reached its apogee around 1960 after the successes of the atomic and space projects. The century of rapid industrialization and modernization in Russia provided a background for this cult, and while science deserved much credit for its role as a modernizing force of the economy and the society, the level of expectations and adulation attached to it rose to an even higher degree, well into the realm of the unrealistic. One might interpret this phenomenon as an ironic transfer of a cultural function: while modernizers entertained high ideological hopes that the rational scientific worldview would replace religion, their trust in the power of science reached irrational heights and itself acquired some characteristics of quasi-religious belief. The fall of the communist regime brought with it a loss of societal attachment to the value of science and a plunge into an opposite ideological extreme—the illusions that science and scientists are not necessary for the country's economic prosperity and development.

However necessary, a high regard for science in general and adequate material support are not sufficient conditions for actual scientific accomplishment. To succeed in the Soviet context, a particular scientific approach or research field usually had to simultaneously satisfy two audiences: the country's political authorities and the international scientific community. The criteria and norms of these two equally crucial "reference groups" often diverged or even contradicted, making the task of Soviet scientists a non-trivial one. Solving only half of this life equation at the expense of the other half could lead to a major disaster, as was demonstrated by the examples of both early Soviet genetics and Lysenko's "Michurinist biology." Soviet physics, on the other hand, managed to satisfy both criteria marvelously, as was reflected in a series of Stalin and Nobel prizes awarded for its main achievements.

Of course, more interesting and meaningful than a certain number of prizes is that Soviet physics developed in an original way and had a unique and fundamental impact on science in the 20th century. Its recognition in the international arena, it is important to stress, should not be misinterpreted as a sign that physics was in any way less "Soviet": quite the contrary, factors specific to the Soviet situation contributed a great deal to some of its most ingenious accomplishments. As the preceding chapters have shown, advances made in Soviet physics were in no lesser measure products of Soviet conditions and society—for a period of time, Stalinist society—than were, for example, the stellar birth of Soviet genetics during the 1920s and early 1930s, or the subsequent rise of Lysenkoism. Seeing all these developments as parts of one fuller picture brings us closer to understanding the historical phenomenon of Soviet science, and if in the process some popular ideological prejudices have to be abandoned, so be it.

The case of Soviet science also illustrates the inadequacy of the ubiquitous dichotomy "Soviet (or Russian)" versus "Western," which in its usual meaning of opposition and contrast obviously does not work here. The tradition of discourse with such polar categories long precedes the Cold War, but decades of ideological and geopolitical struggle reinforced the habit of assuming the dichotomy without pausing to think whether it is actually applicable. Most grand oppositions suggested between "Russia" and the "West" break down once one starts paying attention to differences between individual countries and cultures in the supposedly uniform "West" and to historical changes. As an example, for several decades, it was proclaimed that a defining principle of "Western" science policy was attachment to the value of pure knowledge and its separation from applied research, while the Soviets were known for their refusal to recognize the boundary. Later, American science planners, as a lesson learned from the Sputnik shock, stopped making a strong distinction between pure and applied work, while Soviet scientists made the reverse move by using their increased political leverage to reestablish the official legitimacy of "fundamental" science (Ivanov 2002). The official rhetoric of ideological opposition between the two systems survived, but behind it one finds mutual learning and uncoordinated two-way borrowing.

Earlier chapters discussed other examples of such exchanges involving Soviet science and scientists. The transition to post-Cold War historiography requires a switch from the ideological discourse of grand divides and persistent contrasts to multicultural notions of flexible differences and interactions. The resulting comparative approach will integrate Soviet developments into the general history of science in the 20th century, without ignoring their cultural and political specificities. I hope this book will make a contribution towards achieving this major goal.

Bibliography

Aaserud, F. 1990. *Redirecting Science: Niels Bohr, Philanthropy, and the Rise of Nuclear Physics* (Cambridge: Cambridge University Press).

Abrikosov, A. A. 1965. *Akademik L. D. Landau* (M: Nauka).

Afiani, V. Yu. and S. S. Ilizarov. 1999. "'My razgonim k chertovoi materi akademiiu nauk,'—zaiavil 11 iiulia 1964 g. pervyi sekretar' TsK KPSS N. S. Khrushchev," *Voprosy Istorii Estestvoznaniia i Tekhniki* no. 1: 167–173.

[Akademiia]. 2000. *Akademiia Nauk v Resheniiakh Politbiuro TsK RKP(b)-VKP(b), 1922–1952*, edited by V. D. Esakov (M: ROSSPEN).

Akhiezer, A. I. 1993. "V starom UFTI," *Chteniia Pamiati A.F. Ioffe. 1990* (SPb.: Nauka), 71–79.

Akhundov, M. D. 1991. "Spasla li atomnaia bomba sovetskuiu fiziku?" *Priroda* no. 1: 90–97.

Aleksandrov, A. P., ed. 1988. *Vospominaniia ob Igore Vasil'eviche Kurchatove* (M.: Nauka).

Aleksandrov, D. A. 1996. "Pochemu sovetskie uchenye perestali pechatat'sia za rubezhom: stanovlenie samodostatochnosti i izolirovannosti otechestvennoi nauki, 1914–1940," *Voprosy Istorii Estestvoznaniia i Tekhniki* no. 3: 3–24.

Alexandrov, D. A. 1997. "The politics of scientific 'Kruzhok': study circles in Russian science and their transformation in the 1920s," in *Na Perelome: Sovetskaia Biologiia v 20-30-kh godakh*, edited by E. I. Kolchinsky (SPb.: SPb IIET RAN) Vyp. 1: 255–267.

Aleksandrov, A. F. and A. A. Rukhadze. 1997. "On the history of fundamental papers on the kinetic plasma theory," *Plasma Physics Reports* 23: 442–447.

Alekseev, P. V. 1987. *Revoliutsiia i Nauchnaia Intelligentsiia* (M.: Izdatel'stvo Politicheskoi Literatury).

Alikhanov, G. 1928. *Samokritika i Vnutripartiinaia Demokratiia* (L.).

Alkhazov, D. G., V. P. Shilov, and V. P. Eismont. 1982. *Pervyi v Evrope: Istoriia Sozdaniia Tsiklotrona Radievogo Instituta* (L.: Radievyi Institut).

Allen, J. F. and A. D. Misener. 1938. "Flow of liquid helium II," *Nature* 141: 75.

Alpatov, V. M. 1991. *Istoriia Odnogo Mifa. Marr i Marrizm* (M.: Nauka).

Andreev, A. V. 1993. "Ob ogranichennosti politizirovannogo podkhoda v sotsial'noi istorii fiziki," *Voprosy Istorii Estestvoznaniia i Tekhniki* no. 2: 116–118.

Andreev, A. V. 2000. *Fiziki ne Shutiat. Stranitsy Sotsial'noi Istorii Nauchno-Issledovatel'skogo Instituta Fiziki pri MGU (1922–1954)* (M.: Progress-Traditsiia).

Andreev, A. V., A. B. Kozhevnikov, and B. E. Yavelov, eds. 1994. "Kopengagenskaia operatsiia sovetskoi razvedki," *Voprosy Istorii Estestvoznaniia i Tekhniki* no. 2: 18–44.

[Atomnoe]. 1934. *Atomnoe Yadro. Sbornik Dokladov Pervoi Vsesoiuznoi Yadernoi Konferentsii* (M.-L.)

[Atomnyi]. 1998–2002. *Atomnyi Proekt SSSR*. 5 vols, continuing (M: Nauka).

Auerbakh, F. 1922. *Prostranstvo i Vremia. Materiia i Energiia. Elementarnoe Vvedenie v Teoriiu Otnositel'nosti. Perevod s dopolneniiami S. I. Vavilova* (M.: GIZ).

Babkin, B. P. 1949. *Pavlov: A Biography* (Chicago: University of Chicago Press).

Badash, L. 1985. *Kapitza, Rutherford and the Kremlin* (New Haven: Yale University Press).

Bailes, K. E. 1978. *Technology and Society under Lenin and Stalin: Origins of the Soviet Technical Intelligentsia* (Princeton: Princeton University Press).

Bailes, K. E. 1989. "Natural scientists and the Soviet system" in *Party, State, and Society in the Russian Civil War: Explorations in Social History*, edited by Diane P. Koenker, William G. Rosenberg, Ronald Grigor Suny (Bloomington: Indiana University Press), 267–295.

Bailes, K. E. 1990. *Science and Russian Culture in an Age of Revolutions: V. I. Vernadsky and His Scientific School, 1863–1945* (Bloomington: Indiana University Press).

Barsukov, E. Z. 1938. *Russkaia Artilleriia v Mirovuiu Voinu*. (M.)

Bastrakova, M. S. 1973. *Stanovlenie Sovetskoi Sistemy Organizatsii Nauki (1917–1922)* (M: Nauka).

Bastrakova, M. S. 1999. "Akademiia nauk i sozdanie issledovatel'skikh institutov (dve zapiski V. I. Vernadskogo)," *Voprosy Istorii Estestvoznaniia i Tekhniki* no. 1: 157–167.

Bastrakova, M. S., K. V. Ostrovitianov, *et al.*, eds. 1968. *Organizatsiia Nauki v Pervye Gody Sovetskoi Vlasti (1917–1925). Sbornik Dokumentov* (L: AN SSSR).

Bauman, Z. 1988. *Freedom* (Minneapolis: University of Minnesota Press).

Berezhkov, V. M. 1994. *At Stalin's Side: His Interpreter's Memoirs from the October Revolution to the Fall of the Dictator's Empire* (Secaucus, NJ: Carol).

Berg, A. I., ed. 1945. *Izobretenie Radio A. S. Popovym. Sbornik Dokumentov i Materialov* (M.-L.: AN SSSR).

Bernstein, J. 2002. "Heisenberg and the critical mass," *American Journal of Physics* 70: 911–916.

Bessarab, M. 1978. *Landau. Stranitsy Zhizni* (M: Moskovskii rabochii).

Bethe, H. A. 1980. "Recollections of solid state theory, 1926–1933" in (Mott 1980), 49–51.

Biew, A. M. 1954. *Kapitza: Der Atom-Zar* (München: Verlag Pohl).

Bloch, F. 1929. "Über die Quantenmechanik der Elektronen in Kristallgittern," *Zeitschrift für Physik* 52: 555–600.

Bloch, F. 1931. "Wellenmechanische Diskussion der Leitungs- und Photoeffekte," *Physikalische Zeitschrift* 32: 881–886.

Bloch, F. 1933. "Elektronentheorie der Metalle," *Handbuch der Radiologie* (Leipzig) 6: 1–226.

Bloch, F. 1980. "Memories of electrons in crystals" in (Mott 1980), 24–27.

Blok, G. 1921. "Izdatel'skaia deiatel'nost' K.E.P.S.," *Nauka i Ee Rabotniki* no. 1: 25–27.

Bogdanov, A. A. [1918] 1990. "Nauka i rabochii klass" in A. A. Bogdanov, *Voprosy Sotsializma* (M.: Izdatel'stvo politicheskoi literatury), 360–376.

Bogolubov, N. 1946a. "Expansions into a series of powers of a small parameter in the theory of statistical equilibrium," *Journal of Physics* 10: 257–264.

Bogolubov, N. 1946b. "Kinetic equations," *Journal of Physics* 10: 265–274.

Bogoliubov, N. N. 1946c. *Problemy Dinamicheskoi Teorii v Statisticheskoi Fizike* (M.-L.: GTTI).

Bogolubov, N. 1947. "On the theory of superfluidity," *Journal of Physics* 11: 23–32.

Bohm, D. 1949. "Note on a theorem of Bloch concerning possible causes of superconductivity," *Physical Review* 75: 502–504.

Bohm, D. and D. Pines. 1951. "A collective description of electron interactions. I. Magnetic interactions," *Physical Review* 82: 625–634.

Bohr, N. and L. Rosenfeld. 1933. "Zur Frage der Messbarkeit der elektromagnetischen Feldgrössen," *Kgl. Danske Vid. Selsk., Math.-Fys. Medd.* 12 (8): 3–65.

Bohr, N. and J. A. Wheeler. 1939. "The mechanism of nuclear fission," *Physical Review* 56: 426–450.

Boiko, V. V. and E. R. Vilensky. 1987. *Nikolai Ivanovich Vavilov, 1887–1987. Stranitsy Zhizni i Deiatel'nosti* (M.: Agropromizdat).

[Bol'shevistskaia]. 1950. "Bol'shevistskaia kritika i samokritika," *Bol'shaia Sovetskaia Entsiklopediia*, 2d ed., 5: 515–518.

Bonch-Bruevich, V. L. 1955. "Fizicheskie idei metoda elementarnykh vozbuzhdenii," *Uspekhi Fizicheskikh Nauk* 16, no. 1: 55–76.

Boudia, S. and X. Roqué, eds. 1997. *Science, Medicine, and Industry: The Curie and Joliot-Curie Laboratories*. Special issue, *History and Technology* 13: 241–343.

Brenchkevich, M. V. 1915. *Velikaia Evropeiskaia Voina, Ee Istoricheskie Osnovy i Smysl* (M.-Petrograd: A. S. Panafidin).

Bronshtein, M. P. 1931. "Novyi krizis teorii kvant," *Nauchnoe Slovo* no. 1: 38–55.

Brooks, N. M. 1997. "Chemistry in War, Revolution, and Upheaval: Russia and the Soviet Union, 1900–1929," *Centaurus* 39: 349–367.

Brown, L. M., A. Pais, and B. Pippard, eds. 1995. *Twentieth Century Physics*, 3 vols. (Bristol: Institute of Physics).

Brown, L. M. and H. Rechenberg. 1996. *The Origin of the Concept of Nuclear Forces* (Bristol: Institute of Physics).

Bukharin, N. I. [1931] 1971. "Theory and Practice from the Standpoint of Dialectical Materialism" in *Science at the Cross Roads. Papers Presented to the International Congress of the History of Science and Technology Held in London from June 29th to July 3rd, 1931 by the Delegates of the USSR*. 2nd ed. (London: F. Cass), 11–33.

Bukshpan, Ya. M. 1929. *Voenno-Khoziaistvennaia Politika. Formy i Organy Regulirovaniia Narodnogo Khoziaistva za Vremia Mirovoi Voiny 1914–1918 gg.* (M.-L.: Gosizdat; RANION).

Buryshkin, P. A. 1991. *Moskva Kupecheskaia* (M: Sovremennik).

Carr, E. H. 1947. *The Soviet Impact on the Western World* (New York: Macmillan).

Chanbarisov, Sh. Kh. 1988. *Formirovanie Sovetskoi Universitetskoi Sistemy* (M: Vysshaia Shkola).

Chenakal, V. L. 1947. "Optika v dorevoliutsionnoi Rossii." *Trudy Instituta Istorii Estestvoznaniia i Tekhniki* 1: 145–167.

Cherenkov, P. A. 1934. "Vidimoe svechenie chistykh zhidkostei pod deistviem γ-radiatsii," *Doklady AN SSSR* 2: 451–457.
[Cherenkov]. 1999. *Pavel Alekseevich Cherenkov: Chelovek i Otkrytie*, edited by A. N. Gorbunov and E. P. Cherenkova (M.: Nauka).
Chichibabin, A. E. 1914. "O medikamentakh," *Russkie Vedomosti* no. 188: 6; no. 189: 5.
Chichibabin, A. E. 1915. "Khimicheskaia promyshlennost' i nauka," *Russkie Vedomosti* no. 4: 6.
Chikobava, A. S. 1950. "O nekotorykh voprosakh sovetskogo yazykoznaniia," *Pravda*, 9 May.
Chikobava, A. S. 1985. "Kogda i kak eto bylo" in *Ezhegodnik Iberiisko-Kavkazskogo Yazykoznaniia* (Tbilisi), 9–52.
Coopersmith, J. 1992. *The Electrification of Russia, 1880–1926* (Ithaca: Cornell University Press).
David-Fox, M. 1997. *Revolution of the Mind: Higher Learning among the Bolsheviks, 1918–1929* (Ithaca: Cornell University Press).
David-Fox, M. 1999. "What Is Cultural Revolution?" *The Russian Review* 58: 181–201.
De Coppert, D., ed. 1992. *Understanding Rituals* (London: Routledge).
Dewey, P. E. 1988. "The New Warfare and Economic Mobilization" in *Britain and the First World War*, edited by John Turner (London: Unwin Hyman), 70–84.
Dirac, P. A. M. 1927. "The quantum theory of the emission and absorption of radiation," *Proceedings of the Royal Society of London* A 114: 243–265.
Dirac, P. A. M. 1928. "The quantum theory of the electron," *Proceedings of the Royal Society of London* A 117: 610–624.
Dirac, P. A. M. 1930. "A theory of electrons and protons," *Proceedings of the Royal Society of London* A 126: 360–365.
[Dirac]. 1990. *Pol' Dirak i Fizika XX Veka*, edited by B. V. Medvedev and A. B. Kozhevnikov (M: Nauka).
[Dirac and Tamm]. 1993–1996. *Paul Dirac and Igor Tamm. Correspondence*, edited by A. B. Kojevnikov, Part I: 1928–1933; Part II: 1933–1936 (München: Max-Planck-Institut für Physik), Preprint 93–80 (1993) and 96–40 (1996).
[Diskussiia]. 1947. "Diskussiia po knige G. F. Aleksandrova 'Istoriia zapadnoevropeiskoi filosofii' 16–25 iiunia 1947 g. Stenograficheskii otchet," *Voprosy Filosofii* no. 1.
[Diskussiia]. 1948a. "Diskussiia o prirode fizicheskogo znaniia. Obsuzhdenie stat'i M. A. Markova," *Voprosy Filosofii* no. 1: 203–224.

[Diskussiia]. 1948b. "Diskussiia o prirode fizicheskogo znaniia. Obsuzhdenie stat'i M. A. Markova," *Voprosy Filosofii* no. 3: 222–235 (the issue appeared in April 1949).

Dmitriev, I. S. 1996. " 'Osobaia missiia' Mendeleeva: fakty i argumenty," *Voprosy Istorii Estestvoznaniia i Tekhniki* no. 3: 126–141.

Dollezhal', N. A. 1999. *U Istokov Rukotvornogo Mira. Zapiski Konstruktora.* 2nd ed. (M.: NIKIET).

Drovenikov, I. S. and S. V. Romanov. 1998. "Trofeinyi uran, ili istoriia odnoi komandirovki" in (Vizgin 1998), 215–227.

Drude, P. 1900. *Lehrbuch der Optik* (Leipzig: Hirzel).

Edgerton, D. E. H. 1996. "British scientific intellectuals and the relations of science, technology, and war" in (Forman and Sánchez-Ron 1996), 1–35.

Ehrenfest, P. and A. F. Joffe. 1990. *Erenfest-Ioffe: Nauchnaia Perepiska* (L.: Nauka).

Esakov, V. D. 1991. "Mify i zhizn'," *Nauka i Zhizn'* no. 11: 110–118.

Esakov, V. D. 1993. "K istorii filosofskoi diskussii 1947 g.," *Voprosy Filosofii* no. 2: 83–106.

Esakov, V. D. and E. S. Levina. 1991. "Iz istorii bor'by s lysenkovshchinoi," *Izvestiia TsK KPSS* no. 4: 125–141; no. 6: 157–173; no. 7: 109–121.

Esakov, V. D. and E. S. Levina. 1994. "Delo KR (Iz istorii gonenii na sovetskuiu intelligentsiiu)," *Kentavr* no. 2: 54–69; no. 3: 96–118.

Evteeva, P. M. 1958. "A. E. Chichibabin," *Trudy Instituta Istorii Estestvoznaniia i Tekhniki* 18: 296–356.

Fabelinskii, I. L. 1978. "The discovery of combinatorial scattering of light (the Raman Effect)," *Soviet Physics Uspekhi* 21: 780–797.

Fainsod, M. 1963. *How Russia Is Ruled* (Cambridge: Harvard University Press).

Feffer, S. M. 1994. *Microscopes to Munitions: Ernst Abbe, Carl Zeiss, and the Transformation of Technical Optics, 1850–1914.* Ph.D. thesis (University of California, Berkeley).

Feinberg, E. L. 1995. "Rodoslovnaia rossiiskogo intelligenta," *Priroda* no. 7: 12–22.

Feinberg, E. L. 1999. *Epokha i Lichnost'. Fiziki. Ocherki i Vospominaniia* (M.: Nauka).

Fersman, A. E. 1921. "Puti nauchnogo tvorchestva," *Nauka i Ee Rabotniki* no. 1: 3–7.

Fersman, A. E. 1968. *Nash Apatit* (M: Nauka).

Filin, F. P. 1948. "O dvukh napravleniiakh v yazykovedenii," *Izvestiia AN SSSR. Otdelenie Literatury i Yazyka* 7, no. 6: 486–496.

[Filosofskie]. 1959. *Filosofskie Problemy Sovremennogo Estestvoznaniia. Trudy Vsesoiuznogo Soveshchaniia po Filosofskim Voprosam Estestvoznaniia* (M.: AN SSSR).
Fitzpatrick, S., ed. 1978. *Cultural Revolution in Russia, 1928–1931* (Bloomington: Indiana University Press).
Fitzpatrick, S. 1979. *Education and Social Mobility in the Soviet Union, 1921–1934* (Cambridge: Cambridge University Press).
Fitzpatrick, S. 1985. "Ordzhonikidze's takeover of Vesenkha: A case study in Soviet bureaucratic politics," *Soviet Studies* 37: 153–172.
Fitzpatrick, S. 1992. *The Cultural Front: Power and Culture in Revolutionary Russia* (Ithaca: Cornell University Press).
Fitzpatrick, S. 1999. "Cultural Revolution Revisited," *The Russian Review* 58: 202–209.
Flerov, G. N. and K. A. Petrzhak. 1940. "Spontaneous fission of uranium," *Journal of Physics* 3: 275–280.
Fock, V. 1926. "Zur Schrödingerschen Wellenmechanik," *Zeitschrift für Physik* 38: 242–260.
Forman, P. 1996. "Into quantum electronics: The maser as 'gadget' of Cold-War America," in (Forman and Sánchez-Ron 1996), 261–326.
Forman, P. and J. M. Sánchez-Ron, eds. 1996. *National Military Establishments and the Advancement of Science and Technology* (Dordrecht: Kluwer).
Fortescue, S. 1992. "The Russian Academy of Sciences and the Soviet Academy of Sciences: Continuity or disjunction," *Minerva* 30: 459–478.
Fosdick, R. B. 1952. *The Story of the Rockefeller Foundation* (New York: Harper).
Fowler, R. H. 1932. "Report on the theory of semiconductors," *Physikalische Zeitschrift der Sowjetunion* 2: 507–528; Russian translation: R. G. Fauler, "Teoriia poluprovodnikov," *Zhurnal Eksperimental'noi i Teoreticheskoi Fiziki* 3: 1–15 (1933).
Frank, I. M., ed. 1981. *Sergei Ivanovich Vavilov: Ocherki i Vospominaniia*, 2d ed. (M.: Nauka).
Frank, I. M. 1984. "A conceptual history of the Vavilov–Cherenkov radiation," *Soviet Physics Uspekhi* 27: 385–395.
Frank, I. M., ed. 1991a. *Sergei Ivanovich Vavilov: Ocherki i Vospominaniia*, 3d ed. (M.: Nauka).
Frank, I. M. 1991b. "Mysli o S. I. Vavilove," *Priroda* no. 3: 5–19.
Frenkel, J. 1924a. "Beitrag zur Theorie der Metalle," *Zeitschrift für Physik* 29: 214–240.

Frenkel, Ya. 1924b. "Teoriia elektroprovodnosti metallov," *Zhurnal Russkogo Fiziko-Khimicheskogo Obshchestva. Chast' Fizicheskaia* 56: 505–524; reprinted in (Frenkel 1958–1959), 2: 54–70.

Frenkel, J. 1926. "Über die Wärmebewegung in festen und flüssigen Körpern," *Zeitschrift für Physik* 35: 652–669; Russian translation in (Frenkel 1958–1959), 2: 254–268.

Frenkel, J. 1927. "Nouveaux développements de la théorie électronique des métaux" in *Atti Congresso Internazionale dei Fisici. Como-Pavia-Roma* (Bologna: Nicola Zanichelli), 2: 65–103.

Frenkel, J. 1931a. "On the transformation of light into heat in solids. I," *Physical Review* 37: 17–44.

Frenkel, J. 1931b. "On the transformation of light into heat in solids. II," *Physical Review* 37: 1276–1294.

Frenkel, J. 1931c. "Some remarks on the theory of the photoelectric effect," *Physical Review* 38: 309–320.

Frenkel, Ya. I. 1932a. "Teoriia metallov" in *Doklady, Predstavlennye k Torzhestvennoi Yubileinoi Sessii Akademii Nauk SSSR, Posviashchennoi XV-letiiu Oktiabr'skoi Revoliutsii* (L.: AN SSSR), 40–41; full version printed in (Frenkel 1970), 174–200.

Frenkel, J. 1932b. *Wave Mechanics: Elementary Theory* (Oxford: Clarendon Press).

Frenkel, J. 1932c. "On the elementary derivation of some relations in the electron theory of metals," *Physikalische Zeitschrift der Sowjetunion* 2: 247–253.

Frenkel, J. 1933. "On a possible explanation of superconductivity," *Physical Review* 43: 907–912.

Frenkel, J. 1936. "On the absorption of light and the trapping of electrons and positive holes in crystalline dielectrics," *Physikalische Zeitschrift der Sowjetunion* 9: 158–186; Russian version: Ya. I. Frenkel, "O pogloshchenii sveta i prilipanii elektronov i polozhitel'nykh dyrok v kristallicheskikh dielektrikakh," *Zhurnal Eksperimental'noi i Teoreticheskoi Fiziki* 6: 647–655.

Frenkel, J. 1939. "On the splitting of heavy nuclei by slow neutrons," *Physical Review* 55: 987; "Electro-capillary theory of the splitting of heavy elements by slow neutrons," *Journal of Physics* 1: 125–136.

Frenkel, Ya. I. [1945] 1959. *Kineticheskaia Teoriia Zhidkostei*, in (Frenkel 1958–59) vol. 3; English translation: (Frenkel 1946).

Frenkel, J. 1946. *Kinetic Theory of Liquids* (Oxford: Clarendon Press).

Frenkel, Ya. I. [1948] 1972. *Vvedenie v Teoriiu Metallov*. 4th ed. (L.: Nauka).

Frenkel, Ya. I. 1958–9. *Sobranie Izbrannykh Trudov*. 3 vols. (M.-L.: Nauka).

Frenkel, Ya. I. 1970. *Na Zare Novoi Fiziki. Sbornik Izbrannykh Nauchno-Populiarnykh Rabot* (L.: Nauka).

Frenkel, V. Ya. 1966. *Yakov Il'ich Frenkel'* (M.-L.: Nauka).
Frenkel, V. Ya. 1977. *Pol' Erenfest*, 2d ed. (M.: Atomizdat).
Frenkel, V. Ya. 1985. "Pervaia vsesoiuznaia yadernaia konferentsiia," *Chteniia Pamiati A. F. Ioffe. 1983* (L.: Nauka), 74–95.
Frenkel, V. Ya., ed. 1986. *Ya. I. Frenkel. Vospominaniia, Pis'ma, Dokumenty.* 2nd ed. (L.: Nauka).
Frenkel, V. Ya, ed. 1990. *Fiziki o Sebe* (L.: Nauka).
Frenkel, V. Ya. 1994. "George Gamow: World line 1904–1933 (On the ninetieth anniversary of G. A. Gamow's birth," *Physics-Uspekhi* 37: 767–789.
Frenkel, V. Ya. 1996. *Yakov Ilich Frenkel: His Work, Life, and Letters* (Basel: Birkhäuser).
Frenkel, V. Ya. 1997. "Yakov Ilich Frenkel: Sketches toward a civic portrait," *Historical Studies in the Physical and Biological Sciences* 27: 197–236.
Frenkel, V. 2002. "Einstein and Friedmann" in *Einstein Studies in Russia*, edited by Yuri Balashov and Vladimir Vizgin (Boston: Birkhäuser), 1–16.
Frenkel, V. Ya. and P. Josephson. 1990. "Sovetskie fiziki — stipendiaty Rokfellerovskogo fonda," *Uspekhi Fizicheskikh Nauk* 160: 103–143.
Friedman, H., L. B. Lokhart, and I. H. Blifford. 1996. "Detecting the Soviet bomb: Joe-1 in a rain barrel," *Physics Today* 49 (11): 38–41.
Friedman, A. 1922. "Über die Krümmung des Raumes," *Zeitschrift für Physik* 10: 377–387.
Friedmann, A. 1924. "Über die Möglichkeit einer Welt mit konstanter negativer Krümmung des Raumes," *Zeitschrift für Physik* 21: 326–333.
Frish, S. E. 1992. *Skvoz' Prizmu Vremeni: Vospominaniia* (M.: Izdatel'stvo Politicheskoi Literatury).
Frish, S. E. and A. I. Stozharov, eds. 1976. *Vospominaniia ob Akademike D. S. Rozhdestvenskom* (L.: Nauka).
Galison, P. 1994. "The ontology of the enemy: Norbert Wiener and the cybernetic vision," *Critical Inquiry* 21: 228–266.
Galison, P. and B. Hevly, eds. 1992. *Big Science: The Growth of Large-Scale Research* (Stanford: Stanford University Press).
Gamow, G. 1928. "Zur Quantentheorie des Atomkernes," *Zeitschrift für Physik* 51: 204–212.
Gamow, G. 1970. *My World Line: An Intellectual Autobiography* (New York: Viking).
Gamow, G. and D. Iwanenko. 1926. "Zur Wellentheorie der Materie," *Zeitschrift für Physik* 39: 865–868.
Gatrell, P. 1986. *The Tsarist Economy, 1850–1917* (New York: St. Martin's Press).

Gatrell, P. 1994. *Government, Industry and Rearmament in Russia, 1900–1914: The Last Argument of Tsarism* (Cambridge: Cambridge University Press).

Gavroglu, K. 1995. *Fritz London: A Scientific Biography* (Cambridge: Cambridge University Press).

Getty, J. A. 1985. *Origins of the Great Purges: The Soviet Communist Party Reconsidered, 1933–1938* (Cambridge: Cambridge University Press).

[GFTL] 1929. "Piatiletnii plan rabot GFTL" in *Piatiletnii Plan Nauchno-Eksperimental'noi Raboty v Sviazi s Rekonstruktsiei Promyshlennosti SSSR.* Vyp. 20 (M.).

Ginzburg, C. 1989. "The high and the low: the theme of forbidden knowledge in the sixteenth and seventeenth centuries," in Carlo Ginzburg, *Clues, Myths, and the Historical Method* (Baltimore: Johns Hopkins University Press), 60–76.

Ginzburg, V. L. 2001. *O Nauke, o Sebe i o Drugikh* (M.: Fizmatlit).

Ginzburg, V. and L. Landau. 1950. "K teorii sverkhprovodimosti," *Zhurnal Eksperimental'noi i Teoreticheskoi Fiziki* 20: 1064–1082.

Ginzburg, V., L. Landau, M. Leontovich, and V. Fok. 1946. "O nesostoiatel'nosti rabot A. A. Vlasova po obobshchennoi teorii plazmy i teorii tverdogo tela," *Zhurnal Eksperimental'noi i Teoreticheskoi Fiziki* 16: 246–252.

Goncharov, G. A. 1996. "Thermonuclear milestones," *Physics Today* 49 (11): 44–61.

Goncharov, G. A. 2002. "Termoiadernyi proekt SSSR: predystoriia i desiat' let puti k vodorodnoi bombe" in (Vizgin 2002a), 49–148.

Gorbunov, N. P. 1986. *Vospominaniia. Stat'i. Dokumenty* (M: Nauka).

Gordin, M. D. 2001. *The Ordered Society and Its Enemies: D. I. Mendeleev and the Russian Empire, 1861–1905.* Ph.D. thesis (Harvard University).

Gorelik, G. E. 1991. "Fizika universitetskaia i akademicheskaia," *Voprosy Istorii Estestvoznaniia i Tekhniki* no. 2: 31–46.

Gorelik, G. E. 1992. "Moskva, fizika, 1937 god," *Voprosy Istorii Estestvoznaniia i Tekhniki* no. 1: 15–32.

Gorelik, G. E. 1993. "Naturfilosofskie ustanovki v sovetskoi fizike (1933–1938 gg.), *Filosofskie Issledovaniia* no. 4: 313–334.

Gorelik, G. 1995a. *"Meine Antisowjetische Tätigkeit...": Rußische Physiker unter Stalin* (Braunschweig: Vieweg).

Gorelik, G. 1995b. "Lev Landau, prosocialist prisoner of the Soviet state," *Physics Today* 48 (5): 11–15.

Gorelik, G. 1998. "Andrei Sakharov: from Russian theoretical physics to international practical humanics" in *Physicists in the Postwar Political Arena: Comparative Perspectives. A Conference at the University of*

California, Berkeley, 22–24 January 1998. *Collected Conference Papers*, edited by Cathryn Carson et al. (unpublished), 44–55.

Gorelik, G. 2000. *Andrei Sakharov: Nauka i Svoboda* (M., Izhevsk: R&C Dynamics).

Gorelik, G. E. and V. Ya. Frenkel. 1994. *Matvei Petrovich Bronstein and Soviet Theoretical Physics in the Thirties* (Basel: Birkhäuser).

Gorelik, G. E. and A. B. Kozhevnikov. 1999. "Chto spaslo sovetskuiu fiziku ot lysenkovaniia. Dialog," *Priroda* no. 5: 95–104.

Gorelik, G. E. and G. A. Savina. 1993. "G. A. Gamov ... zamestitel' direktora FIANa," *Priroda* no. 8: 82–90.

Gorokhovsky, Yu. N., ed. 1968. *50 let Gosudarstvennogo Opticheskogo Instituta im. S. I. Vavilova (1918–1968)* (L.: Mashinostroenie).

Graham, L. R. 1967. *The Soviet Academy of Sciences and the Communist Party, 1927–1932* (Princeton: Princeton University Press).

Graham, L. R. 1975. "The formation of Soviet research institutes: a combination of revolutionary innovation and international borrowing," *Social Studies of Science* 5: 303–329.

Graham, L. R. 1985. "The sociopolitical roots of Boris Hessen: Soviet Marxism and the history of science," *Social Studies of Science* 15: 705–722.

Graham, L. R. 1987. *Science, Philosophy and Human Behavior in the Soviet Union* (New York: Columbia University Press).

Graham, L. R. 1992. "Big science in the last years of the big Soviet Union" in *Science after '40*, edited by Arnold Thackray, *Osiris* 7: 49–71.

Graham, L. R. 1993. *Science in Russia and the Soviet Union: A Short History* (Cambridge: Cambridge University Press).

Graham, L. R. 1998. *What Have We Learned About Science and Technology from the Russian Experience?* (Stanford: Stanford University Press).

Graham, L. R. 1999. "Is science a social construction? Some new thoughts based on the Russian experience" (lecture at the Mark M. Horblit Colloquium in the History of Science, Harvard University, April 1999).

Gray, G. W. 1941. *Education on an International Scale: A History of the International Education Board, 1923–1938* (New York: Harcourt, Brace).

Grebenshchikov, I. V. and D. S. Rozhdestvensky. [1918a] 1993. "Zapiska konsul'tantov Gosudarstvennykh Farforovogo i Stekliannogo Zavodov po voprosu o vydelenii Zavoda Opticheskogo Stekla," *Trudy Gosudarstvennogo Opticheskogo Instituta im. S. I. Vavilova (Transactions of the S. I. Vavilov State Optical Institute)* 83 (217), Supplement, 17–18.

Grebenshchikov, I. V. and D. S. Rozhdestvensky. [1918b] 1993. "O znachenii zavoda opticheskogo stekla dlia Rossii," *Trudy*

Gosudarstvennogo Opticheskogo Instituta im. S. I. Vavilova (Transactions of the S. I. Vavilov State Optical Institute) 83 (217), Supplement, 19–24.

Grebenshchikov, I. V. 1933. "Opticheskoe steklo i razvitie ego proizvodstva v SSSR" in *Trudy Noiabr'skoi Yubileinoi Sessii AN SSSR 1932 g.* (L.), 448–455.

Greene, J. M. 1997. *Ideology and Expertise: The Guomindang and Talent in China's Nanjing Decade, 1927–1937.* Ph.D. thesis (Washington University, St. Louis).

Grigor'ian, N. A., ed. 1997. *Akademik Leon Abgarovich Orbeli. Nauchnoe Nasledie* (M.: Nauka).

Grimm, E. D. 1915. "Bor'ba narodov" in *Voprosy Mirovoi Voiny. Sbornik Statei*, edited by M. I. Tugan-Baranovsky (Petrograd: Pravo), 1–19.

Grinevetsky, V. I. 1922. *Poslevoennye Perspektivy Russkoi Promyshlennosti* (Kharkov).

Gross, E. F. and N. A. Karyev. 1952. "Pogloshchenie sveta kristallom zakisi medi v infrakrasnoi i vidimoi chasti spektra," *Doklady AN SSSR* 84: 261–264.

Gross, E. P. 1987. "Collective variables in elementary quantum mechanics" in *Quantum Implications: Essays in Honour of David Bohm*, edited by B. J. Hiley and F. David Peat (London: Routledge), 46–65.

Gulo, D. D. and A. N. Osinovsky. 1980. *Dmitry Sergeevich Rozhdestvensky* (M: Nauka).

Haber, L. F. 1986. *The Poisonous Cloud: Chemical Warfare in the First World War* (Oxford: Clarendon Press).

Hacking, I. 1983. *Representing and Intervening: Introductory Topics in the Philosophy of Natural Sciences* (Cambridge: Cambridge University Press).

Hagen, A. 1996. "Export versus direct investment in the German optical industry: Carl Zeiss, Jena and Glaswerk Schott & Gen. in the UK, from their beginnings to 1933," *Business History* 38: 1–20.

Hahn, W. 1982. *The Fall of Zhdanov and the Defeat of Moderation, 1946–1953* (Ithaca: Cornell University Press).

Hall, K. P. 1999. *Purely Practical Revolutionaries: A History of Stalinist Theoretical Physics.* Ph.D. thesis (Harvard University).

Hall, K. Forthcoming. "'Think less about foundations': A short course on the *Course of Theoretical Physics* of Landau and Lifshitz" in *Pedagogy and Practice of Science: Historical and Contemporary Perspectives*, edited by David Kaiser (Cambridge: MIT Press).

Hardach, G. 1992. "Industrial mobilisation in 1914–1918: Production, planning and ideology" in *The French Home Front, 1914–1918*, edited by Patrick Friedenson (Providence: Berg), 57–88.

Hartcup, G. 1988. *The War of Invention: Scientific Developments, 1914–1918.* (London: Brassey's Defense Publishers).

Hayashi, M. and K. Katsuki. 1952. "Hydrogen-like absorption spectrum of cupruous oxide," *Journal of the Physical Society of Japan* 7: 599–603.

Heilbron, J. L. 1986. "The first European cyclotrons," *Rivista di Storia della Scienza* 3: 1–44.

Heisenberg, W. 1931. "Zum Paulischen Ausschließungsprinzip," *Annalen der Physik* 10: 888–904.

Heisenberg, W. and W. Pauli. 1929–1930. "Zur Quantentheorie der Wellenfelder," *Zeitschrift für Physik* 56: 30–88 (1929); 59: 168–190 (1930).

Hellbeck, J. 2000. "Writing the self in the time of terror" in *Self and Story in Russian History*, edited by Laura Engelstein and Stephanie Sandler (Ithaca: Cornell University Press).

Hessen, B. [1931] 1971. "The social and economic roots of Newton's 'Principia'" in *Science at the Cross Roads. Papers Presented to the International Congress of the History of Science and Technology Held in London from June 29th to July 3rd, 1931 by the Delegates of the USSR.* 2nd ed. (London: F. Cass), 151–212.

Hiley, B. J. 1997. "David Joseph Bohm, 20 December 1917–27 October 1992," Royal Society of London, *Biographical Memoirs*, 43: 107–131.

[HISAP]. 1997–1999. *Science and Society: History of the Soviet Atomic Project (1940s–1950s). Proceedings of the International Symposium*, 2 vols. (M: Izdat).

Hoddeson, L. H. and G. Baym. 1980. "The development of the quantum mechanical electron theory of metals: 1900-28" in (Mott 1980), 8–23.

Hoddeson, L., E. Braun, J. Teichmann, and S. Weart, eds. 1992. *Out of the Crystal Maze: Chapters from the History of Solid-State Physics* (New York: Oxford University Press).

Hoffmann, D. L. and P. Holquist. Forthcoming. *Cultivating the Masses: The Modern Social State in Russia, 1914–1941.*

Holloway, D. 1990. "The scientist and the tyrant," *The New York Review of Books* 1 March, 23–25.

Holloway, D. 1994. *Stalin and the Bomb: The Soviet Union and Atomic Energy, 1939–1956* (New Haven: Yale University Press).

Holquist, P. 1997. "Information is the alpha and omega of our work: Bolshevik surveillance in its pan-European context," *Journal of Modern History* 69: 415–450.

Hughes, T. P. 1989. *American Genesis: A Century of Invention and Technological Enthusiasm, 1870–1970* (New York: Viking).

Hull, A. 1999. "War of words: the public science of the British scientific community and the origins of the Department of Scientific and Industrial Research, 1914–16," *British Journal for the History of Science* 32: 461–81.

Igonin, V. V. 1975. *Atom v SSSR. Razvitie Sovetskoi Yadernoi Fiziki* (Saratov: Izdatel'stvo Saratovskogo Universiteta).

Ilizarov, S. S. and L. I. Pushkareva. 1994. "Beriia i teoriia otnositel'nosti," *Istoricheskii Arkhiv* no. 3: 215–223.

[Imperatorskaia]. 1917. *Materialy dlia Istorii Akademicheskikh Uchrezhdenii za 1889–1914 gg.* Chast' 1 (Petrograd: Imperatorskaia Akademiia nauk).

[Ioffe]. 1973. *Vospominaniia ob A. F. Ioffe* (L.: Nauka).

[Ioffe]. 1980. *Nauchno-Organizatsionnaia Deiatel'nost' Akademika A. F. Ioffe* (L.: Nauka).

Iovchuk, M. T. 1952. "Bor'ba mnenii i svoboda kritiki — vazhneishee uslovie razvitiia peredovoi nauki," *Voprosy Filosofii* no. 2: 14–31.

Ipatieff, V. N. 1946. *The Life of a Chemist: Memoirs of Vladimir N. Ipatieff.* (Stanford: Stanford University Press).

[Istoriia]. 1959. *Istoriia Kommunisticheskoi Partii Sovetskogo Soiuza* (M.: Gos. Izd. Polit. Lit-ry).

[Istoriia]. 1993. *Istoriia Akademii Nauk Ukraini, 1918–1923. Dokumenty i Materialy.* (Kyiv: Naukova Dumka).

Ivanenko, D. and I. Pomeranchuk. 1944. "O maksimal'noi energii, dostizhimoi v betatrone," *Doklady AN SSSR* 44: 343.

Ivanov, K. 2002. "Science after Stalin: forging a new image of Soviet science," *Science in Context* 15: 317–338.

Iwanenko, D. 1932. "The neutron hypothesis," *Nature* 129: 798.

Iwanenko, D. and L. Landau. 1926. "Zur Ableitung der Klein–Fockschen Gleichung," *Zeitschrift für Physik* 40: 161–162.

Iwanenko, D. and L. Landau. 1928. "Zur Theorie des magnetischen Elektrons," *Zeitschrift für Physik* 48: 340–348.

[Izluchenie]. 1984. "Izluchenie Vavilova–Cherenkova: 50 let otkrytiia," *Priroda* no. 10: 74–93.

Jammer, M. 1966. *The Conceptual Development of Quantum Mechanics* (New York: McGraw-Hill).

Joravsky, D. 1961. *Soviet Marxism and Natural Science, 1917–1932* (London: Routledge).

Joravsky, D. 1965. "The Vavilov brothers," *Slavic Review* 24: 381–394.

Joravsky, D. 1970. *The Lysenko Affair* (Cambridge: Harvard University Press).

Joravsky, D. 1983. "The Stalinist mentality and the higher learning," *Slavic Review* 42: 575–600.

Joravsky, D. 1985. "Return of the native," *The New York Review of Books* 5 December: 26.
Joravsky, D. 1989. *Russian Psychology: A Critical History* (Oxford: Basil Blackwell).
Josephson, P. R. 1988a. "Physics and Soviet–Western relations in the 1920s and 1930s," *Physics Today* 41 (9): 54–61.
Josephson, P. R. 1988b. "Physics, Stalinist politics of science and Cultural Revolution," *Soviet Studies* 40: 245–265.
Josephson, P. R. 1991. *Physics and Politics in Revolutionary Russia* (Berkeley: University of California Press).
Josephson, P. R. 1997. *New Atlantis Revisited: Akademgorodok, the Siberian City of Science* (Princeton: Princeton University Press).
Josephson, P. R. 2000. *Red Atom: Russia's Nuclear Power Program from Stalin to Today* (New York: W. H. Freeman).
Jungnickel, C. and R. McCormmach. 1986. *Intellectual Mastery of Nature: Theoretical Physics from Ohm to Einstein*, 2 vols. (Chicago: University of Chicago Press).
Kaftanov, S. V. 1985. "Po trevoge. Rasskaz upolnomochennogo gosudarstvennogo komiteta oborony S. V. Kaftanova," *Khimiia i Zhisn'* no. 6: 9–11.
Kaganov, M. I. 1998. *Shkola Landau. Chto Ya o Nei Dumaiu* (Troitsk: Trovant).
Kaganov, M. I. and V. Ya. Frenkel. 1981. *Vekhi Istorii Fiziki Tverdogo Tela* (M.: Znanie)
Kaganov, M. I. and I. M. Lifshitz. 1989. *Kvazichastitsy*, 2nd ed. (M.: Nauka).
Kapitza, P. L. 1923. "Some observations on α-particle tracks in a magnetic field," *Proceedings of the Cambridge Philosophical Society* 21: 511; reprinted in (Kapitza 1964), 1: 72–77.
Kapitza, P. 1929. "The change of electrical conductivity in strong magnetic fields," *Proceedings of the Royal Society of London* A 123: 292; reprinted in (Kapitza 1964), 1: 224–297.
Kapitza, P. 1934. "The liquefaction of helium by an adiabatic method," *Proceedings of the Royal Society of London* A 147: 189; reprinted in (Kapitza 1964), 1: 485–503.
Kapitza, P. L. [1937] 1998. "O stroitel'stve i razvertyvanii raboty Instituta fizicheskikh problem Akademii nauk SSSR" in (Kapitza 1988–1998), 4 (1998): 63–74.
Kapitza, P. 1938. "Viscosity in liquid helium below the λ-point," *Nature* 141: 74; reprinted in (Kapitza 1965), 2: 505–507.

Kapitza, P. 1939a. "Expansion turbine producing low temperatures applied to air liquefaction," *Journal of Physics* 1: 7–28; reprinted in (Kapitza 1965), 2: 521–550.

Kapitza, P. 1939b. "A new method of producing low temperatures for air liquefaction," in (Kapitza 1965), 3: 44–63.

Kapitza, P. L. 1941. "The study of heat transfer in helium II," *Journal of Physics* 4: 181; reprinted in (Kapitza 1965), 2: 581–624.

Kapitza, P. L. 1964–1965. *Collected Papers of P. L. Kapitza*, 2 vols, edited by D. ter Haar (Oxford: Pergamon Press).

Kapitza, P. L. [1969] 1998. "Khleb, maslo, no ne dzhem" in (Kapitza, 1988–1998), 4 (1998): 348–355.

Kapitza, P. 1986. "Pis'ma k materi," *Novyi Mir* no. 5: 192–216; no. 6: 194–218.

Kapitza, P. L. 1988–1998. *Nauchnye Trudy*. 4 vols. (M.: Nauka).

Kapitza, P. L. 1989a. *Pis'ma o Nauke, 1930–1980*, edited by P. E. Rubinin (M.: Moskovskii Rabochii).

Kapitsa, P. L. 1989b. *Letters to Mother: The Early Cambridge Period* (Ottawa).

Kapitza, P. L. 1989c. "Dvadtsat' dva otcheta akademika Kapitsy," in *Kratkii Mig Torzhestva* (M.: Nauka), 252–287.

[Kapitza]. 1990. *Kapitza in Cambridge and Moscow: Life and Letters of a Russian Physicist*, edited by J. W. Boag, P. E. Rubinin, and D. Shoenberg (Amsterdam: North Holland).

[Kapitza] 2003. *Kapitza, Kreml' i Nauka. T.1. Sozdanie Instituta Fizicheskikh Problem, 1934–1937 gg.*, edited by P. E. Rubinin and V. D. Esakov (M.: Nauka).

Kapitza, S. P. 1990. "[Predislovie k stat'e A. S. Sonina.]," *Priroda* no. 3: 90.

Kaptsov, N. A. 1970. "A. S. Popov" in *Razvitie Fiziki v Rossii (Ocherki)*, edited by A. S. Predvoditelev and B. I. Spasskii (M.: Vysshaia Shkola), 1: 264–269.

Kassow, S. D. 1989. *Students, Professors, and the State in Tsarist Russia*. (Berkeley: University of California Press).

Kedrov, B. M. 1948. "Znachenie kritiki i samokritiki v razvitii nauki (K voprosu o roli otritsaniia v dialektike i metafizike)," *Vestnik AN SSSR* no. 2: 68–100.

Kedrov, B. M. 1988. "Kak sozdavalsia nash zhurnal," *Voprosy Filosofii* no. 4: 92–104.

Kedrov, B. and G. Gurgenidze. 1955. "Za glubokuiu razrabotku leninskogo filosofskogo nasledstva," *Kommunist* no. 32: 45–56.

Kedrov, F. 1984. *Kapitza—Life and Discoveries* (M.: Mir).

Keldysh, L. V. 1985. "Tammovskie sostoianiia i fizika poverkhnosti tverdogo tela," *Priroda* no. 9: 17–33.

Keler, V. 1961. *Sergei Vavilov* (M.: Molodaia Gvardiia).
Kerber, L. L. 1996. *Stalin's Aviation Gulag. A Memoir of Andrei Tupolev and the Purge Era* (Washington, DC: Smithsonian Institution Press).
Khalatnikov, I. M., ed. 1989. *Landau: The Physicist and the Man. Recollections of L. D. Landau* (Oxford: Pergamon).
Khalatnikov, I. M. and V. P. Mineev, eds. 1996. *30 Years of the Landau Institute: Selected Papers* (Singapore: World Scientific).
Khariton, Yu. B. and Yu. N. Smirnov. 1994. *Mify i Real'nost' Sovetskogo Atomnogo Proekta* (Arzamas-16: VNIIEF).
Khlevniuk, O. V. 1995. *In Stalin's Shadow: The Career of "Sergo" Ordzhonikidze* (Armonk, N.Y.: M. E. Sharpe).
[Khrushchev] 1993. "Kak snimali N. S. Khrushcheva. Materialy plenuma TsK KPSS. Oktiabr' 1964 g.," *Istoricheskii Arkhiv* 1: 3–19.
Klein, M. J. 1970. *Paul Ehrenfest* (Amsterdam: North Holland).
Kleopov, I. L. 1964. *Geologicheskii Komitet, 1882–1929 gg. Istoriia Geologii v Rossii* (M.: Nauka).
Klimontovich, Yu. I. 1997. "Physics of collisionless plasma," *Physics-Uspekhi* 40: 21–51.
Kneen, P. 1989. "Soviet science policy under Gorbachev," *Soviet Studies* 41: 69–87.
Kojevnikov, A. 1996. "President of Stalin's Academy: the mask and responsibility of Sergei Vavilov," *Isis* 87: 18–50.
Kojevnikov, A. 1999. "Freedom, collectivism, and quasiparticles: social metaphors in quantum physics," *Historical Studies in the Physical and Biological Sciences* 29 (2): 295–331.
Kojevnikov, A. B. 2000. "Dialoge über Macht und Wissen" in *Im Dschungel der Macht: Intellektuelle Professionen unter Stalin und Hitler*, edited by Dietrich Beyrau (Göttingen: Vandenhoeck & Ruprecht), 45–64.
Kojevnikov, A. 2002a. "Introduction: A new history of Russian science," *Science in Context* 15: 177–182.
Kojevnikov, A. 2002b. "David Bohm and collective movement," *Historical Studies in the Physical and Biological Sciences* 33 (1): 161–192.
Kokin, L. 1981. *Yunost' Akademikov* (M.: Sovetskaia Rossiia).
Kol'tsov, A. V. 1982. *Razvitie Akademii Nauk kak Vysshego Nauchnogo Uchrezhdeniia SSSR, 1926–1932* (L.: Nauka).
Kol'tsov, A. V. 1999. *Sozdanie i Deiatel'nost' Komissii po Izucheniiu Estestvennykh Proizvoditel'nykh Sil Rossii, 1915–1930 gg.* (SPb.: Nauka).
Komlev, B. V., G. S. Sinitsina, and M. P. Koval'skaia. 1982. "V. G. Khlopin i uranovaia problema," *Voprosy Istorii Estestvoznaniia i Tekhniki* no. 1: 63–75.

Kononkov, A. F. and A. N. Osinovsky. 1968. "Sozdanie sovetskoi optiki i deiatel'nost' D. S. Rozhdestvenskogo" in *Ocherki po Istorii Sovetskoi Nauki i Kul'tury* (M: Izdatel'tvo Moskovskogo universiteta), 113–135.

Kopelev, L. 1981. *Utoli Moia Pechali* (Ann Arbor, MI: Ardis).

Kosarev, V. V. 1993. "Fiztekh, Gulag i obratno," *Chteniia pamiati A.F. Ioffe.* 1990 (SPb.: Nauka), 105–177.

Kovalevsky, V. I., ed. 1900. *Rossiia v Kontse XIX veka* (SPb.: Brokgauz-Efron).

Kozhevnikov, A. B. 1990. "Osnovnye etapy nauchnoi politiki v SSSR (1917–1941)," in *Tezisy Vtoroi Konferentsii po Sotsial'noi Istorii Sovetskoi Nauki* (M.: IIET AN SSSR), 26–27.

Kozhevnikov, A. B. 1993. *Filantropiia Rokfellera i Sovetskaia Nauka* (SPb.-M.: Mezhdunarodnyi Fond Istorii Nauki).

Kozhevnikov, A. B. 1994. "O nauke proletarskoi, partiinoi, marksistskoi..." in *Metafizika i Ideologiia v Istorii Estestvoznaniia*, edited by A. A. Pechenkin (M.: Nauka), 219–238.

Kozhevnikov, A. B. and O. I. Novik. 1989. "Analysis of information ties in early quantum mechanics," *Acta Historiae Rerum Naturalium nec non Technicarum* (Prague), 20: 115–159.

Kozlov, B. I. 1981. "Metrologicheskaia reforma v SSSR (1917–1927 gg.)," *Voprosy Istorii Estestvoznaniia i Tekhniki* no. 1: 24–34.

Kragh, H. 1990. *Dirac: A Scientific Biography* (Cambridge: Cambridge University Press).

Kragh, H. 1999. *Quantum Generations: A History of Physics in the Twentieth Century* (Princeton: Princeton University Press).

Krementsov, N. 1997. *Stalinist Science* (Princeton: Princeton University Press).

Krementsov, N. 2002. *The Cure: A Story of Cancer and Politics from the Annals of the Cold War* (Chicago: University of Chicago Press).

[Kriogenika]. 1978. *Razvitie Kriogeniki na Ukraine* (Kiev: Naukova Dumka).

Kruglov, A. 1994–1995. *Kak Sozdavalas' Atomnaia Promyshlennost' v SSSR* (M: TsNII AtomInform), 1st ed. (1994); 2nd ed. (1995).

Kruglov, A. K. 1998. "Samyi trudnyi moment v sozdanii atomnoi bomby: Zametki o pervom v SSSR promyshlennom reaktore dlia narabotki plutoniia" in (Vizgin 1998), 228–251.

Kurchatov, I. V. [1941] 1983. "Vsesoiuznoe soveshchanie po fizike atomnogo yadra (noiabr' 1940 g.)," *Uspekhi Khimii* 10: 350–358; reprinted in (Kurchatov 1983), 2: 336–347.

Kurchatov, I. V. 1983. *Izbrannye Trudy*. 3 vols. (M.: Nauka).

Landau, L. 1926. "Zur Theorie der Spektren der zweiatomiger Moleküle," *Zeitschrift für Physik* 40: 621–627, English translation: "On the theory of the spectra of diatomic molecules" in (Landau 1965), 1–7.

Landau, L. 1927. "Das Dämpfungsproblem in der Wellenmechanik," *Zeitschrift für Physik* 45: 430; English translation: "The damping problem in wave mechanics" in (Landau 1965), 8–18.

Landau, L. 1930. "Diamagnetismus der Metalle," *Zeitschrift für Physik* 64: 629; English translation: "Diamagnetism of metals" in (Landau 1965), 31–38.

Landau, L. 1932. "On the theory of stars," *Physikalische Zeitschrift der Sowjetunion* 1: 285–288; reprinted in (Landau 1965), 60–62.

Landau, L. 1933a. "Über die Bewegung der Elektronen im Kristallgitter," *Physikalische Zeitschrift der Sowjetunion* 3: 664–665; English translation: "Electron motion in crystal lattices" in (Landau 1965), 67–68.

Landau, L. 1933b. "Zur Theorie der Supraleitfähigkeit," *Physikalische Zeitschrift der Sowjetunion* 4: 43.

Landau, L. 1933c. "Eine mögliche Erklärung der Feldabhängigkeit der Suszeptibilität bei niedrigen Temperaturen," *Physikalische Zeitschrift der Sowjetunion* 4: 675; English translation: "A possible explanation of the field dependence of the susceptibility at low temperatures" in (Landau 1965), 73–76.

Landau, L. 1936. "Die kinetische Gleichung für den Fall Coulombischer Wechselwirkung," *Physikalische Zeitschrift der SowjetUnion* 10: 154–164; English translation: "The transport equation in the case of Coulomb interactions" in (Landau 1965), 163–170.

Landau, L. 1937a. "Zur Theorie der Phasenumwandlungen," *Physikalische Zeitschrift der Sowjetunion* 11: 26, 545; English translation: "On the theory of phase transitions" in (Landau 1965), 193–216.

Landau, L. 1937b. "Zur Theorie der Supraleitfähigkeit," *Physikalische Zeitschrift der Sowjetunion* 11: 129; English translation: "On the theory of superconductivity" in (Landau 1965), 217–225.

Landau, L. 1941. "Theory of superfluidity of helium II," *Journal of Physics* 5: 71–90; reprinted in (Landau 1965), 301–330.

Landau, L. 1946. "On the vibrations of the electronic plasma," *Journal of Physics* 10: 25–34; reprinted in (Landau 1965), 445–460; Russian version: "O kolebaniiakh elektronnoi plazmy," *Zhurnal Eksperimental'noi i Teoreticheskoi Fiziki* 16: 574–586.

Landau, L. 1949. "On the theory of superfluidity," *Physical Review* 75: 884–885; reprinted in (Landau 1965), 474–477.

Landau, L. D. 1956. "Teoriia Fermi-zhidkosti," *Zhurnal Eksperimental'noi i Teoreticheskoi Fiziki* 30: 1058; English translation: "The theory of a Fermi liquid," *Soviet Physics —JETP* 3: 920–925 (1957); reprinted in (Landau 1965), 723–730.

Landau, L. D. 1965. *Collected Papers of L. D. Landau*, edited by D. ter Haar (New York: Gordon and Breach, Science Publishers).

Landau, L. D. 1969. *Sobranie Trudov.* 2 vols. (M: Nauka).

[Landau]. 1991. "Lev Landau: god v tiur'me," *Izvestiia TsK KPSS* no. 3: 134–157.

Landau, L. and A. Kompaneets. [1935] 1965. "The electrical conductivity of metals" in (Landau 1965), 803–832.

Landau, L. and R. Peierls. 1930. "Quantenelektrodynamik im Konfigurationsraum," *Zeitschrift für Physik* 62: 188–200; English translation: "Quantum electrodynamics in configuration space" in (Landau 1965), 19–30.

Landau, L. and R. Peierls. 1931. "Erweiterung des Unbestimmheitsprinzips für die relativistische Quantentheorie," *Zeitschrift für Physik* 69: 56–69; English translation: "Extension of the uncertainty principle to relativistic quantum theory" in (Landau 1965), 40–51.

Landau, L. and I. Pomerantschuk. 1936. "Über die Eigenschaften der Metalle bei sehr niedrigen Temperaturen," *Physikalische Zeitschrift der Sowjetunion* 10: 649–665; English translation: "On the properties of metals at very low temperatures" in (Landau 1965), 171–183.

Landau-Drobantseva, K. 1999. *Akademik Landau. Kak My Zhili. Vospominaniia.* (M: Zakharov AST).

Larin, V. 2001. *Kombinat "Maiak"—Problema na Veka.* 2nd ed. (M: Ecopresscenter).

Lebedev, P. N. 1990. *Nauchnaia Perepiska P. N. Lebedeva*, edited by E. I. Pogrebysskaia (M.: Nauka).

Leipunsky, A. I. 1990. *Izbrannye Trudy. Vospominaniia.* (Kiev: Naukova Dumka).

Lenin, V. I. [1909] 1961. *Materializm i Empiriokrititsizm*, in Lenin, *Polnoe Sobranie Sochinenii* (M.: Gos. Izd. Polit. Lit-ry) 18: 7–384; English translation: *Materialism and Empirio-Criticism. Critical Comments on a Reactionary Philosophy*, in Lenin, *Collected Works* (M.: Foreign Languages Publishing House) 14: 17–361.

Lenin, V. I. [1918] 1965. "Ocherednye zadachi sovetskoi vlasti" in Lenin, *Polnoe Sobranie Sochinenii* (M.: Gos. Izd. Polit. Lit-ry) 36: 165–208; English translation: "The immediate tasks of the Soviet government" in Lenin, *Collected Works* (M.: Progress) 27: 235–277.

Levin, A. E. 1991. "Bitva bez izbieniia: soveshchanie po planetnoi kosmogonii 1951 g.," *Priroda* no. 9 : 99–107.
Levina, E. S. 1995. *Vavilov, Lysenko, Timofeev-Resovsky: Biologiia v SSSR—Istoriia i Istoriografiia* (M.: AIRO-XX).
Levshin, B. V., ed. 1986. *Dokumenty po Istorii Akademii Nauk SSSR, 1917–1925*. (L.: Nauka).
Levshin, V. 1954. "Sergei Ivanovich Vavilov (Ocherk zhizni i deiatel'nosti)" in (Vavilov 1954–1959), 1: 7–48.
Lifshitz, I. M. [1958] 1994. "Kvazichastitsy v sovremennoi fizike," *Priroda* no. 5: 11–20; reprinted in I. M. Lifshitz, *Izbrannye Trudy* (M.: Nauka), 2: 397–407.
Lindener, B. A. 1922. *Raboty Rossiiskoi Akademii Nauk v Oblasti Issledovaniia Prirodnykh Bogatstv Rossii; Obzor Deiatel'nosti KEPS za 1915–1921 gg.* (Petrograd: Rossiiskaia Akademiia nauk; KEPS).
Livanova, A. 1978. *L. D. Landau* (M.: Znanie).
Livanova, A. and V. Livanov. 1988. *"Vtoraia Stepen' Ponimaniia": Akademik L. I. Mandel'shtam* (M.: Znanie).
[Lomonosovsky] 1901. *Lomonosovsky Sbornik. Materialy dlia Istorii Razvitiia Khimii v Rossii* (M.: Mamontov).
London, F. 1947. "The present state of the theory of liquid helium," *International Conference on Fundamental Particles and Low Temperatures* (Cambridge: Taylor and Francis), 2: 1–19.
[Luzin]. 1999. *Delo Akademika Nikolaia Nikolaevicha Luzina*, edited by S. S. Demidov and B. V. Levshin (SPb.: RKhGI).
Maasen, S. 1995. "Who is afraid of metaphors," in *Biology as Society, Society as Biology: Metaphors*, edited by Sabine Maasen, Everett Mendelsohn, and Peter Weingart (Dordrecht: Kluwer), 11–35.
Machta, L. 1992. "Finding the site of the first Soviet nuclear test in 1949," *Bulletin of the American Meteorological Society* 73: 1797–1806.
MacLeod, K. and R. MacLeod. 1975. "War and economic development: Government and the optical industry in Britain, 1914–1918" in *War and Economic Development: Essays in Honour of David Joslin*, edited by Jay M. Winter (Cambridge: Cambridge University Press), 165–204.
MacLeod, R. 1993. "The chemists go to war: The mobilization of civilian chemists and the British war effort, 1914–1918," *Annals of Science* 50: 455–481.
MacLeod, R. 1998. "Chemistry for King and Kayser: Revisiting chemical enterprise and the European war" in *Determinants in the Evolution of the European Chemical Industry, 1900–1939*, edited by Anthony S. Travis et al. (Dordrecht: Kluwer), 25–49.

Maksimov, A. A., I. V. Kuznetsov, Ya. P. Terletsky, and N. F. Ovchinnikov, eds. 1952. *Filosofskie Voprosy Sovremennoi Fiziki* (M.: AN SSSR).

Mal'kov, V. L., ed. 1998. *Pervaia Mirovaia Voina: Prolog XX Veka* (M.: Nauka).

[Mandel'shtam]. 1979. *Akademik L. I. Mandel'shtam (k 100-letiiu so Dnia Rozhdeniia)* (M.: Nauka).

Mandelstam, L. and G. Landsberg. 1928. "Eine neue Erscheinung bei der Lichzerstreuung in Kristallen," *Die Naturwissenschaften* 16: 772.

Manevich, E. D. 1993. "Takie byli vremena," *Voprosy Istorii Estestvoznaniia i Tekhniki* no. 2: 119–132.

Manikovsky, A. A. 1920. *Boevoe Snabzhenie Russkoi Armii v Voinu 1914–1918 gg.* Chast' 1 (M: Voenno-Istoricheskaia Komissiia).

Man'kovsky, G. N. 1960. "K istorii izucheniia Kurskoi magnitnoi anomalii," *Trudy Instituta Istorii Estestvoznaniia i Tekhniki* 33: 3–16.

Markov, M. A. 1947. "O prirode fizicheskogo znaniia," *Voprosy Filosofii* no. 2: 140–176 (the issue appeared in February 1948).

Markov, M. A. 1990. "Glazami ochevidtsa," *Priroda* no. 5: 99–100.

Markovnikov, V. V. [1879] 1955. "Sovremennaia khimiia i russkaia khimicheskaia promyshlennost" in V. V. Markovnikov, *Izbrannye Trudy* (M.: AN SSSR), 637–676.

Martin, T. 2001. *The Affirmative Action Empire: Nations and Nationalism in the Soviet Union* (Ithaca: Cornell University Press).

Medvedev, R. 1972a. *Politicheskii Dnevnik, 1964–1970* (Amsterdam: Alexander Herzen Foundation).

Medvedev, R. 1972b. *Kniga o Sotsialisticheskoi Demokratii* (Amsterdam/Paris: Grasset & Fasquelle).

Medvedev, Zh. A. 2001. "Atomnyi GULAG," *Voprosy Istorii* no. 1: 44–59.

Mendelevich, G. A., ed. 1964. *Gor'kii i Nauka. Stat'i, Rechi, Pis'ma, Vospominaniia* (M: Nauka).

Meshchaninov, I. I. 1948. "O polozhenii v lingvisticheskoi nauke," *Izvestiia AN SSSR. Otdelenie Literatury i Yazyka* 7, no. 6: 473–485.

Mick, C. 2000. *Forschen für Stalin. Deutsche Fachleute in der Sowjetischen Rüstungsindustrie, 1945–1958* (München: R. Oldenburg Verlag).

Mikulinsky, S. R., ed. 1987. *Nikolai Ivanovich Vavilov. Ocherki. Vospominaniia. Materialy* (M.: Nauka).

Mochalov, I. I. 1982. *Vladimir Ivanovich Vernadsky, 1863–1945* (M.: Nauka).

Modin, Yu. 1994. *My Five Cambridge Friends: Burgess, Maclean, Philby, Blunt, and Cairncross by their KGB Controller* (New York: Farrar Straus Giroux).

[Moskovskaia]. 1948. "Moskovskaia konferentsiia po problemam darvinizma," *Priroda* no. 6: 85–87.

Mott, N. F. 1938. "Energy levels in real and ideal crystals," *Transactions of the Faraday Society* 34: 822–827, reprinted in *Sir Nevill Mott: 65 Years in Physics* (Singapore: World Scientific 1995), 145–151.
Mott, N., ed. 1980. *The Beginnings of Solid State Physics: A Symposium* (London: Royal Society).
Moyer, A. F. 1981. "Evaluations of Dirac's electron, 1928–1932," *American Journal of Physics* 49: 1055–1062.
Nametkin, S. S. 1954. "Nikolai Dmitrievich Zelinsky. Biografiia i ocherk nauchnoi deiatel'nosti" in N. D. Zelinsky, *Sobranie Trudov* (M.: AN SSSR), 1: 7–66.
[Nauchnaia]. 1950. *Nauchnaia Sessiia Posviashchennaia Problemam Fiziologicheskogo Ucheniia Akademika I. P. Pavlova. 28 iiunia-4 iiulia 1950 g. Stenograficheskii Otchet* (M.: AN SSSR).
[Nauchnye]. 1947. "Nauchnye diskussii," *Literaturnaia Gazeta* 29 November, 10 and 27 December.
[Nauka]. 1920–1922. *Nauka v Rossii. Spravochnyi Ezhegodnik (Dannye k 1 Yanvaria 1918 g.)* Vyp. 1. Petrograd (Petrograd 1920); Vyp. 2. Moskva (Petrograd 1922).
Negin, E. A., ed. 2000. *Sovetskii Atomnyi Proekt: Konets Atomnoi Monopolii. Kak eto Bylo..* 2nd ed. (Sarov: VNIIEF).
Nemenov, M. I. 1920. "Mediko-biologicheskii otdel Gosudarstvennogo rentgenologicheskogo i radiologicheskogo instituta i deiatel'nost' ego v 1919 godu," *Vestnik Rentgenologii* 1: 153–174.
Nemenov, M. I. 1928. "K istorii osnovaniia Gosudarstvennogo rentgenologicheskogo i radiologicheskogo instituta. Po lichnym vospominaniiam" in *Gosudarstvennyi Rentgenologicheskii, Radiologicheskii i Rakovyi Institut* (L.: GRRRI), 1–25.
Nettlau, M. 1996. *A Short History of Anarchism* (London: Freedom Press).
Newton, I. 1959. *The Correspondence of Isaac Newton*, Vol. 1, 1661–1675. Edited by H. W. Turnbull (Cambridge: Cambridge University Press).
[NKTP]. 1934. *Nauchno-Tekhnicheskoe Obsluzhivanie Tiazheloi Promyshlennosti* (M.-L.: NKTP).
[NKTP]. 1935. *Nauchno-Issledovatel'skie Instituty NKTP* (M.-L.).
Norris, R. S. 2002. *Racing for the Bomb: General Leslie R. Groves, the Manhattan Project's Indispensable Man* (South Royalton, VT: Steerforth Press).
Novik, O. I., unpublished. "Organizator fizicheskogo instituta AN SSSR."
Novikov, M. M. 1930. "Moskovskii universitet v pervyi period bol'shevistskogo rezhima" in *Moskovskii Universitet, 1755–1930* (Paris: Parizhskii komitet po oznamenovaniiu 175-letiia Moskovskogo universiteta), 156–192.

Novorussky, M. V. 1915. "Voina i novye otrasli russkoi promyshlennosti" in *Voprosy Mirovoi Voiny. Sbornik Statey*, edited by M. I. Tugan-Baranovsky (Petrograd: Pravo), 466–483.

Obreimov, I. V. 1973. "[Vospominaniia]" in (Ioffe 1973), 21–61.

[Obshchee 1939]. "Obshchee sobranie Akademii Nauk SSSR tovarishchu Stalinu," *Vestnik AN SSSR* no. 11–12: 2–3.

[Obshchee 1945]. "Obshchee sobranie Akademii Nauk SSSR 17 iiulia 1945 g.," *Vestnik AN SSSR* no. 7–8: 22–28.

[Obsuzhdenie]. 1948a. "Obsuzhdenie uchebnika prof. A. I. Denisova 'Sovetskoe Gosudarstvo i Pravo'," *Vestnik AN SSSR* no. 4: 103–105.

[Obsuzhdenie]. 1948b. "Obsuzhdenie knigi prof. B. M. Kedrova 'Engel's i Estestvoznanie'," *Vestnik AN SSSR* no. 3: 100–105.

[Obsuzhdenie]. 1948c. "Obsuzhdenie rabot po yazykoznaniiu," *Vestnik AN SSSR* no. 2: 113–118.

[Obsuzhdenie]. 1948d. "Obsuzhdenie raboty zhurnala 'Izvestiia AN SSSR, Otdelenie Literatury i Yazyka'," *Izvestiia AN SSSR, Otdelenie Literatury i Yazyka* 7, no. 5: 463–466.

[Ocherednye]. 1948. "Ocherednye zadachi sovetskogo yazykoznaniia," *Izvestiia AN SSSR, Otdelenie Literatury i Yazyka* 7, no. 5: 466–468.

Ol'denburg, S. F. 1927. "Vpechatleniia o nauchnoi zhizni v Germanii, Frantsii i Anglii," *Nauchnyi Rabotnik* no. 2: 88–101.

Olwell, R. B. 1990. "Princeton, David Bohm and the Cold War: A study in McCarthyism," junior paper submitted to the Department of History, Princeton University, May 1990.

Olwell, R. 1999. "Physical isolation and marginalization in physics: David Bohm's Cold War exile," *Isis* 90: 738–756.

Openkin, L. A. 1991. "I. V. Stalin: Poslednii prognoz budushchego (iz istorii napisaniia raboty 'Ekonomicheskie Problemy Sotsializma v SSSR')," *Voprosy Istorii KPSS* no. 7: 113–128.

[O povyshenii]. 1946. "O povyshenii okladov rabotnikov nauki i ob uluchshenii ikh material'no-bytovykh uslovii," Postanovlenie Soveta Ministrov no. 514, 6 March.

[O samokritike]. 1928. *O Samokritike* (Bibliotechka Srednego Povolzh'ia).

[O vnutrividovoi]. 1948. "O vnutrividovoi bor'be za sushchestvovanie sredi organizmov (Reshenie Biuro Otdeleniia biologicheskikh nauk AN SSSR)," *Vestnik AN SSSR* no. 3, 118.

Pavlenko, Yu. V., Yu. N. Ranyuk, and Yu. A. Khramov. 1998. *"Delo" UFTI, 1935–1938* (Kiev: Feniks).

Pauli, W. 1985. *Wissenschaftlicher Briefwechsel mit Bohr, Einstein, Heisenberg u. a.*, edited by Karl von Meyenn, Bd. 2, 1930–1939 (New York: Springer).

Peat, F. D. 1997. *Infinite Potential: The Life and Times of David Bohm* (Reading: Helix Books).
Pechenkin, A. A. 1994. "Antirezonansnaia kampaniia 1949–1951 gg." in *Metafizika i Ideologiia v Istorii Estestvoznaniia* (M.: Nauka), 184–219.
Peierls, R. E. 1929a. "Zur Theorie der galvanomagnetischen Effekte," *Zeitschrift für Physik* 53: 255–266.
Peierls, R. E. 1929b. "Zur Theorie des Hall-effekts," *Physikalische Zeitschrift* 30: 273–274.
Peierls, R. E. 1930. "Zur Theorie der elektrischen und thermischen Leitfähigkeit von Metallen," *Annalen der Physik* 4: 121–148.
Peierls, R. 1932a. "Elektronentheorie der Metalle," *Ergebnisse der Exakten Naturwissenschaften* 11: 264–322.
Peierls, R. 1932b. "Zur Theorie der Absorptionsspektren fester Körper (Züricher Habilitationsschrift)," *Annalen der Physik* 13: 905–952.
Peierls, R. 1932c. "Über die Absorptionsspektren fester Körper," *Physikalische Zeitschrift der Sowjetunion* 1: 297–298.
Peierls, R. E. 1980. "Recollections of early solid state physics" in (Mott 1980), 28–38.
Pekar, S. 1946a. "Local quantum states of an electron in an ideal ionic crystal," *Journal of Physics* 10: 341–346; Russian version: S. Pekar, "Lokal'nye kvantovye sostoianiia elektrona v ideal'nom ionnom kristalle," *Zhurnal Eksperimental'noi i Teoreticheskoi Fiziki* 16: 341–347.
Pekar, S. 1946b. "Autolocalization of an electron in a dielectric inertially polarizing medium," *Journal of Physics* 10: 347–350; Russian version: S. Pekar, "Avtolokalizatsiia elektrona v dielektricheskoi inertsionno poliarizuiushcheisia srede," *Zhurnal Eksperimental'noi i Teoreticheskoi Fiziki* 16: 335–340.
Perchenok, F. F. 1991. "Akademiia nauk na velikom perelome," *Zven'ia*, 163–238; reprinted: "'Delo Akademii nauk' i 'velikii perelom' v sovetskoi nauke" in *Tragicheskie Sud'by: Repressirovannye Uchenye Akademii Nauk SSSR* (M.: Nauka, 1995), 201–235.
[Pervye] 1948. "Pervye itogi tvorcheskikh diskussii," *Vestnik AN SSSR* no. 3: 5–15.
Pestre, D. and J. Krige. 1992. "Some thoughts on the early history of CERN" in (Galison and Hevly 1992), 78–99.
Pisarev, Yu. A. and V. L. Mal'kov, eds. 1994. *Pervaia Mirovaia Voina: Diskussionnye Problemy Istorii* (M.: Nauka).
[Polozhenie]. [1918] 1993. "Polozhenie o gosudarstvennom Opticheskom institute," *Trudy Gosudarstvennogo Opticheskogo Instituta im. S. I. Vavilova (Transactions of the S. I. Vavilov State Optical Institute)* 83 (217), Supplement: 49–52.

Pomeranchuk, I. 1941. "The thermal conductivity of the paramagnetic dielectrics at low temperatures," *Journal of Physics* 4: 357–379.
Popovsky, M. 1984. *The Vavilov Affair* (Hamden: Archon Books).
Popper, K. R. 1950. *The Open Society and Its Enemies* (Princeton: Princeton University Press).
Popper, K. R. 1957. *The Poverty of Historicism* (Boston: Beacon Press).
[Poslednii]. 1951. "Poslednii put' (Pokhorony S. I. Vavilova i rechi na traurnom mitinge)," *Vestnik AN SSSR* no. 2: 25–32.
[Postanovlenie]. 1950. "Prezidium Akademii Nauk. Postanovlenie," *Izvestiia AN SSSR. Otdelenie Literatury i Yazyka* 9, no. 1: 80–88.
Prostovolosova, L. N. 1991. "Neizvestnaia rukopis' S. I. Vavilova 'O dostoinstve sovetskogo uchenogo' (1947)," *Voprosy Istorii Estestvoznaniia i Tekhniki* no. 2: 102–104.
[Protiv]. 1948. "Protiv idealizatsii ucheniia A. Veselovskogo," *Izvestiia AN SSSR. Otdelenie Literatury i Yazyka* 7, no. 4: 362–364.
Rashba, É. I. 1985. "The prediction of excitons (On the 90th birthday of Ya. I. Frenkel')," *Soviet Physics Uspekhi* 27: 790–796.
[Rasshirennoe]. 1948a. "Rasshirennoe zasedanie Redaktsionno-Izdatel'skogo soveta AN SSSR," *Vestnik AN SSSR* no. 6: 73–80.
[Rasshirennoe]. 1948b "Rasshirennoe zasedanie Prezidiuma Akademii Nauk SSSR, 24–26 avgusta 1948 g.," *Vestnik AN SSSR* no. 9: 26.
Rhodes, R. 1986. *The Making of the Atomic Bomb* (New York: Touchstone).
Riehl, N. and F. Seitz. 1996. *Stalin's Captive: Nikolaus Riehl and the Soviet Race for the Bomb* (Washington, DC: American Chemical Society).
[Rockefeller]. 1953. *The Rockefeller Foundation. Directory of Fellowship Awards, 1917–1950*.
Roqué, X. Forthcoming. "Marie Curie's vision of the Institut du Radium" in *Radioactivity: Science and Culture*, edited by Christine Blondel.
Rosenbaum, T. E. 1989. "Rockefeller philanthropies in revolutionary Russia," *Newsletter of the Rockefeller Archive Center*, Summer 1989.
Rossianov, K. O. 1993a. "Editing nature: Joseph Stalin and the 'new' Soviet biology," *Isis* 84: 728–745.
Rossianov, K. O. 1993b. "Stalin kak redaktor Lysenko," *Voprosy Filosofii* no. 2: 56–69.
Rossiianov, K. 2002. "Beyond species: Il'ya Ivanov and his experiments on cross-breeding humans with anthropoid apes," *Science in Context* 15: 239–276.
Rozhdestvensky, D. S. [1918] 1993. "[Zapiska] v Komissariat narodnogo prosveshcheniia," *Trudy Gosudarstvennogo Opticheskogo Instituta im.*

S. I. Vavilova (Transactions of the S. I. Vavilov State Optical Institute) 83 (217), Supplement: 11–14.

Rozhdestvensky, D. S. 1920. "Spektral'nyi analiz i stroenie atoma," *Trudy Gosudarstvennogo Opticheskogo Instituta* 1 (6).

Rozhdestvensky, D. S. 1932. "Zapiska ob opticheskom stekle," *Trudy Gosudarstvennogo Opticheskogo Instituta* 8 (84).

Rozhdestvensky, D. S. 1951. *Raboty po Anomal'noi Dispersii v Parakh Metallov* (L.: AN SSSR).

Rubinin, P. E. 1994. "Svobodnyi chelovek v nesvobodnoi strane: K 100-letiiu so dnia rozhdeniia akademika P. L. Kapitsy," *Vestnik Rossiiskoi Akademii nauk* 64 (6): 497–523.

Sakharov, A. 1968. "Reflections on progress, peaceful coexistence, and intellectual freedom," *New York Times* 22 July.

Sakharov, A. 1990. *Memoirs* (New York: Knopf).

Sakharov, A. D. 1995. *Nauchnye Trudy* (M.: Tsentrkom).

Sakharov, A. D. [1964] 1996. "Stenogramma zasedaniia Obshchego sobraniia Akademii nauk SSSR, 22–26 iiunia 1964 goda" in *On Mezhdu Nami Zhil. ... Vospominaniia o Sakharove*, edited by B. L. Alt'shuler et al. (M.: Praktika), 857–866.

Samarin, A. M. 1950. "O sostoianii uchebnoi i nauchnoi raboty po yazykoznaniiu v vuzakh i mery po ee uluchsheniiu," *Vestnik Vysshei Shkoly* no. 9.

Shapin, S. and S. Schaffer. 1985. *Leviathan and the Air-Pump: Hobbes, Boyle, and the Experimental Life* (Princeton: Princeton University Press).

Shepilov, D. T. 1998. "Vospominaniia," *Voprosy Istorii* no. 3–6.

Shoenberg, D. 1942. "Physical research in the Soviet Union" in *Science in Soviet Russia by Seven British Scientists* (London: Watts), 6–17.

Shoenberg, D. 1985. "Piotr Leonidovich Kapitza, 9 July 1894–8 April 1984," Royal Society of London, *Biographical Memoirs* 31: 325–374.

Shreider, A. 1923. *Ocherki Filosofii Narodnichestva* (Berlin: Skify).

Shubnikov, A. V. 1991. "To, chto sokhranila pamiat' " in (Frank 1991a), 171.

Shubnikov, L. V. 1990. *Izbrannye Trudy. Vospominaniia* (Kiev: Naukova Dumka).

Sinelnikow, C., A. Leipunsky, A. Walther, and G. Latischev. 1932. "The disintegration of lithium by high velocity protons," *Physikalische Zeitschrift der Sowjetunion* 2: 285.

Singh, R. and F. Riess. 2001. "The 1930 Nobel Prize for physics: A close decision?" *Notes and Records of the Royal Society of London* 55: 267–283.

Singh, R. 2002. "C. V. Raman and the discovery of the Raman effect," *Physics in Perspective* 4: 399–420.
[Situation]. 1949. *The Situation in Biological Science. Proceedings of the Lenin Academy of Agricultural Sciences of the USSR. Verbatim Report* (M.: Foreign Languages Publishing House).
Slater, J. C. and W. Shockley. 1936. "Optical absorption by the alkali halides," *Physical Review* 50: 705–719.
Smirnov, Yu. N. 1996. "G. N. Flerov i stanovlenie sovetskogo atomnogo proekta," *Voprosy Istorii Estestvoznaniia i Tekhniki* no. 2: 100–125.
Smirnov, N. N., ed. 1999. *Rossiia i Pervaia Mirovaia Voina: Materialy Mezhdunarodnogo Nauchnogo Kollokviuma* (SPb.: DB).
Smirnova, T. M. 1979. "Istoriia razrabotki i provedeniia v zhizn' pervogo sovetskogo ustava vysshei shkoly" in *Gosudarstvennoe Rukovodstvo Vysshei Shkoloi v Dorevoliutsionnoi Rossii i v SSSR*, edited by N. P. Eroshkin (M: Moskovskii Gosudarstvennyi Arkhivnyi Institut), 6–38.
Smirnova T. M. 1984. *Organy Gosudarstvennogo Rukovodstva Vysshei Shkoloi RSFSR, 1917–1925 gg.* (M: Moskovskii Gosudarstvennyi Istoriko–Arkhivnyi Institut).
Sochor, Z. A. 1988. *Revolution and Culture: The Bogdanov–Lenin Controversy* (Ithaca: Cornell University Press).
Solomon, S. 2003. "Building bridges: Alan Gregg and Soviet Russia, 1925–1928," *Minerva* 41: 167–176.
Solov'ev, Yu. I. 1985. *Istoriia Khimii v Rossii. Nauchnye Tsentry i Osnovnye Napravleniia Issledovanii* (M.: Nauka).
Sominsky, M. S. 1964. *Abram Fedorovich Ioffe* (M.-L.: Nauka).
Sommerfeld, A. 1927a. "Zur Elektronentheorie der Metalle," *Die Naturwissenschaften* 15: 825–832.
Sommerfeld, A. 1927b. "Elekronentheorie der Metalle und des Volta-Effektes nach der Fermi'schen Statistik" in *Atti Congresso Internazionale dei Fisici. Como-Pavia-Roma* (Bologna: Nicola Zanichelli), 2: 449–473.
Sommerfeld, A. 1928. "Zur Elektronentheorie der Metalle," *Die Naturwissenschaften* 16: 374–381.
Sonin, A. S. 1994. *"Fizicheskii Idealizm." Istoriia Odnoi Ideologicheskoi Kampanii*. (M.: Fiziko-Matematicheskaia Literatura).
[Soviet]. 1951. *The Soviet Linguistic Controversy*. Translated from the Soviet Press by John V. Murra, Robert M. Hankin, and Fred Holling (Morningside Heights, NY: King's Crown Press).
Soyfer, V. 1994. *Lysenko and the Tragedy of Soviet Science* (New Brunswick: Rutgers University Press).
[Spravochnik]. 1930. *Spravochnik Partiinogo Rabotnika*, 7th ed. (M.-L.).

Spruch, G. M. 1979. "Pyotr Kapitza, octogenarian dissident," *Physics Today* 32 (9), 34–45.

Stalin, I. V. [1930] 1949. "Pis'mo A. M. Gor'komu, 17 yanvarya 1930 g.," in Stalin, *Sochineniia* (M.: Gos. Izd. Polit. Lit-ry) 12: 173–177; English translation in Stalin, *Works* (M.: Foreign Languages Publishing House) 12: 179–183 (1955).

Stalin, I. V. 1946. "New five-year plan for Russia. Election address," *Vital Speeches of the Day*, 12 (10) (1 March 1946), 300–304.

Stalin, I. V. 1950. *Marksizm i Voprosy Yazykoznaniia* (M.: Gos. Izd. Polit. Lit-ry); English translation: *Marxism and Linguistics* in (Stalin 1972), 407–444.

Stalin, I. V. 1952. *Ekonomicheskie Problemy Sotsializma v SSSR* (M.: Gos. Izd. Polit. Lit-ry); English translation: *Economic Problems of Socialism in the USSR* in (Stalin 1972), 445–481.

Stalin, J. V. 1972. *The Essential Stalin: Major Theoretical Writings, 1905–1952* (Garden City, NY: Anchor Books).

Stone, N. 1975. *The Eastern Front, 1914–1917* (New York: Charles Scribner's Sons).

Strekopytov, S. P. 1990. *Vysshii Sovet Narodnogo Khoziaistva i Sovetskaia Nauka, 1917–1932* (M.: Moskovskii Gosudarstvennyi Arkhivnyi Institut).

Strumilin, S. 1935. "Khimicheskaia promyshlennost' SSSR," *Problemy Ekonomiki* no. 2: 118–145.

Suris, R. A. and V. Ya. Frenkel. 1994. "Ya. I. Frenkel's studies of the theory of the electric conductivity of metals," *Physics-Uspekhi* 37: 357–373.

Tamm, I. 1926. "Novye printsipy statisticheskoi mekhaniki Boze-Einshteina v sviazi s voprosom o fizicheskoi prirode materii," *Uspekhi Fizicheskikh Nauk* 6: 112–141; reprinted in (Tamm 1975), 2: 254–286.

Tamm, I. E. 1928. "Elektronnaia teoriia metallov" in *Fizika* (M.-L.: Gosudarstvennoe Izdatel'stvo), 1: 62–77.

Tamm, Ig. 1930. "Über die Quantentheorie der molekularen Lichtzerstreuung in festen Körpern," *Zeitschrift für Physik* 60: 345–363.

Tamm, Ig. 1932a. "Some remarks on the theory of photoelectric effect in metals," *Physical Review* 39: 170–172.

Tamm, Ig. 1932b. "Über eine mögliche Art der Elektronenbindung an Kristalloberflächen," *Physikalische Zeitschrift der Sowjetunion* 1: 733–746; Russian version: I. Tamm, "O vozmozhnoi sviazi elektronov na poverkhnostiakh kristalla," *Zhurnal Eksperimental'noi i Teoreticheskoi Fiziki* 3: 34–35 (1933).

Tamm, I. E. 1962. "Yakov Il'ich Frenkel," *Soviet Physics Uspekhi* 5: 173–194.

Tamm, I. E. 1975. *Sobranie Nauchnykh Trudov*. 2 vols. (M.: Nauka).

Tamm, I. E. 1991. *Selected Papers*. Edited by B. M. Bolotovskii and V. Ya. Frenkel. (Berlin: Springer-Verlag).

[Tamm]. 1995. "I. E.Tamm v dnevnikakh i pis'makh," *Priroda* no. 7: 134–160.

Tamm, Ig. and I. Frank. 1937. "Coherent visible radiation of fast electrons passing through matter," *Comptes Rendus de l'Académie des Sciences de l'URSS* 14: 109–114, reprinted in (Tamm 1991), 30–35.

Tamm, Ig. and S. Schubin. 1931. "Zur Theorie des Photoeffektes an Metallen," *Zeitschrift für Physik* 68: 97–113.

Timasheff, N. 1946. *The Great Retreat* (New York: E. P. Dutton).

Timiriazev, K. A. 1911. "Novye potrebnosti nauki XX veka i ikh udovletvorenie na zapade i u nas" in (Timiriazev [1920] 1963), 56–66.

Timiriazev, K. A. 1915. "Nauka v sovremennoi zhizni" in (Timiriazev [1920] 1963), 342–357.

Timiriazev, K. A. [1920] 1963. *Nauka i Demokratiia. Sbornik Statei 1904–1919 gg.* (M: Izdatel'stvo Sotsial'no-Ekonomicheskoi Literatury).

Tisza, L. 1938. "Transport phenomena in helium II," *Nature* 141: 913.

Tobey, R. C. 1971. *The American Ideology of National Science, 1919–1930*. (Pittsburgh: University of Pittsburgh Press).

Todes, D. 1995. "Pavlov and the Bolsheviks," *History and Philosophy of the Life Sciences* 17: 379–418.

Todes, D. 2002. *Pavlov's Physiology Factory: Experiment, Interpretation, Laboratory Enterprise* (Baltimore: Johns Hopkins University Press).

Tolstoi, I. I. 1950. "Otrezvliaiushchii golos razuma," *Izvestiia AN SSSR. Otdelenie Literatury i Yazyka* 9, no. 1: 62–63.

Tolz, V. 1997. *Russian Academicians and the Revolution: Combining Professionalism and Politics* (New York: St. Martin's Press).

Tomilin, K. A. 1993. "Nesostoiavshiisia pogrom v teoreticheskoi fizike (1949)," *Filosofskie Issledovaniia* no. 4: 335–371.

Tonks, L. and I. Langmuir. 1929. "Oscillations in ionized gases," *Physical Review* 33: 195–210.

Topchiev, A. V. 1950. "I. V. Stalin o problemakh iazykoznaniia i zadachi Akademii nauk SSSR," *Vestnik AN SSSR* no. 7: 8–19.

Trebilcock, C. 1993. "Science, technology and the armaments industry in the UK and Europe, with special reference to the period 1880–1914." *Journal of European Economic History* 22: 565–580.

Trischler, H. 1990. "Historische Wurzeln der Grossforschung: Die Luftfahrtforschung vor 1945" in *Grossforschung in Deutschland*, edited by Margit Szöllösi-Janze and Helmut Trischler (Frankfurt am Main: Campus Verlag), 23–37.

Troitskaia, T. M. 1920. "K istorii Obshchestva Rossiiskikh Rentgenologov i Radiologov (nyne Rossiiskoi Assotsiatsii Rentgenologov i Radiologov)," *Vestnik Rentgenologii i Radiologii* 1 (1–2): 147–150.
Tropp, E. A., V. Ya. Frenkel, and A. D. Chernin. 1993. *Alexander A. Friedmann: The Man Who Made the Universe Expand* (Cambridge: Cambridge University Press).
Trubetskoi, E. N. 1915. *Voina i Mirovaia Zadacha Rossii* (M.: Sytin).
Turner, F. M. 1980. "Public science in Britain, 1880–1919," *Isis* 71: 589–608.
Unfried, B. 1994. "Rituale von Konfession und Selbstkritik: Bilder vom stalinistischen Kader," *Jahrbuch für Historische Kommunismusforschung*, 148–164.
Val'ter, A. K. 1933. *Ataka Atomnogo Yadra* (Kharkov).
Vavilov, S. I. 1925. "Novye opytnye podtverzhdeniia sledstvii obshchei teorii otnositel'nosti," *Uspekhi Fizicheskikh Nauk* 5: 457–460.
Vavilov, S. I. 1926. "Novye poiski efirnogo vetra", *Uspekhi Fizicheskikh Nauk* 6: 242–254.
Vavilov, S. I. 1928. *Eksperimental'nye Osnovaniia Teorii Otnositel'nosti* (M.-L.: GIZ).
Vavilov, S. I. 1932. "Do kontsa ispol'zovat' brigadno-laboratornyi," *Za Proletarskie Kadry* 3 March 1932.
Vavilov, S. I. 1934a. "O vozmozhnykh prichinakh sinego γ-svecheniia zhidkostei," *Doklady AN SSSR* 2: 457–461.
Vavilov, S. I. 1934b. "Dialektika svetovykh iavlenii," *Front Nauki i Tekhniki* no. 9: 38–45.
Vavilov, S. I. 1934c. "V. I. Lenin i fizika," *Priroda* no. 1: 35–38.
Vavilov, S. I. [1935] 1956. "Fizicheskaia optika Leonarda Eilera" in (Vavilov 1954–1959), 3: 138–147.
Vavilov, S. I. 1937. "Fizicheskii institut im. P. N. Lebedeva," *Vestnik AN SSSR* no. 10–11: 37–46.
Vavilov, S. I. 1939. "Novaia fizika i dialekticheskii materializm," *Pod Znamenem Marksizma* no. 12: 27–33.
Vavilov, S. I. 1944. "Lenin i sovremennaia fizika," *Uspekhi Fizicheskikh Nauk* 26: 113–132.
Vavilov, S. I. 1945. *Isaak N'iuton*. 2nd ed. (M.-L.: AN SSSR).
Vavilov, S. I. 1946. *Sovetskaia Nauka na Novom Etape* (M.: AN SSSR).
Vavilov, S. I. 1947. "Neskol'ko slov k stat'e M. A. Markova," *Voprosy Filosofii* no. 2: 138–139.
Vavilov, S. I. 1949a. "Zakon Lomonosova," *Pravda* 5 January.
Vavilov, S. I. 1949b. "Nauchnyi genii Stalina" in *Iosifu Vissarionovichu Stalinu Akademiia Nauk SSSR* (M.: AN SSSR), 5–18.

Vavilov, S. I. [1950a] 1956. "Lenin i filosofskie problemy sovremennoi fiziki" in *Velikaia Sila Idei Leninizma* (M.: Gospolitizdat), 171–186, reprinted in (Vavilov 1954–1959), 3: 85–96.
Vavilov, S. I. 1950b. "O dostoinstve i chesti sovetskogo uchenogo" in *O Sovetskom Patriotizme* (M.: Gospolitizdat), 438–462.
Vavilov, S. I. 1950c. *Nauka Stalinskoi Epokhi* (M.: AN SSSR).
Vavilov, S. I. 1954–1959. *Sobranie Sochinenii*. 4 vols. (M: AN SSSR).
Vavilov, S. I. 1981. "Nachalo avtobiografii" in (Frank 1981), 80–103.
[Vavilovy]. 1991. *Brat'ia Nikolai i Sergei Vavilovy* (M.: FIAN).
Veksler, V. 1944. "O novom metode uskoreniia reliativistskikh chastits," *Doklady AN SSSR* 44: 393; English translation: "A new method of acceleration of relativistic particles," *Journal of Physics* 9: 153–158 (1945).
Venkataraman, G. 1995. *Raman and His Effect* (London: Sangam).
Vernadsky, V. I. [1910] 1922. "Zadacha dnia v oblasti radiia" in (Vernadsky 1922), 31–44.
Vernadsky, V. I. [1911] 1922. "Radievye instituty" in (Vernadsky 1922), 44–52.
Vernadsky, V. I. [1915a] 1922. "Voina i progress nauki" in (Vernadsky 1922), 129–140.
Vernadsky, V. I. [1915b] 1922. "Ob izuchenii estestvennykh proizvoditel'nykh sil Rossii (Dolozheno v zasedanii Fiziko-Matematicheskogo otdeleniia 8 aprelia 1915 g.)" in (Vernadsky 1922), 5–25.
Vernadsky, V. I. [1916a] 1922. "Ob ispol'zovanii khimicheskikh elementov v Rossii" in (Vernadsky 1922), 52–69.
Vernadsky, V. I. [1916b] 1922. "O gosudarstvennoi seti issledovatel'skikh institutov" in (Vernadsky 1922), 25–31.
Vernadsky, V. I. 1921. "O nauchnoi rabote v Krymu," *Nauka i Ee Rabotniki* no. 4: 3–12.
Vernadsky, V. I. 1922. *Ocherki i Rechi*. T. 1. (Petrograd: Nauchnoe Khimiko-Tekhnicheskoe Izdatel'stvo NTO VSNKh RSFSR).
Vernadsky, V. 1993. *Zhizneopisanie. Izbrannye Trudy. Vospominaniia sovremennikov. Suzhdeniia Potomkov* (M.: Sovremennik).
Vernadsky, V. I. 1994. *Dnevniki 1917–1921* (Kiev: Naukova Dumka).
Vernsky, L. 1995. "V kabinete i vne ego (Iz razgovorov s dedom i iz semeinogo arkhiva)" in *Vospominaniia o Tamme* (M.: Nauka), 79–129.
Vizgin, V. P. 1990–1991. "Martovskaia (1936 g.) sessiia AN SSSR: sovetskaia fizika v fokuse," *Voprosy Istorii Estestvoznaniia i Tekhniki*, no. 1: 63–84 (1990); no. 3: 36–55 (1991).

Vizgin, V. P., ed. 1992. "U istokov sovetskogo atomnogo proekta: rol' razvedki, 1941–1946," *Voprosy Istorii Estestvoznaniia i Tekhniki* no. 3: 97–134.

Vizgin, V. P., ed. 1998–2002a. *Istoriia Sovetskogo Atomnogo Proekta: Dokumenty, Vospominaniia, Issledovaniia*. Vol. 1 (M.: Yanus-K), Vol. 2 (SPb.: RKhGU).

Vizgin, V. P. 2002b. "Fenomen kul'ta atoma v SSSR" in (Vizgin 2002a), 413–488.

Vizgin, V. P. and G. E. Gorelik. 1987. "The reception of the theory of relativity in Russia and the USSR" in *The Comparative Reception of Relativity*, edited by Thomas F. Glick (Dordrecht: Reidel), 265–326.

Vlasov, A. A. 1938. "O vibratsionnykh svoistvakh elektronogo gaza," *Zhurnal Eksperimental'noi i Teoreticheskoi Fiziki* 8: 291–318.

Vlasov, A. A. 1945a. "On the kinetic theory of an assembly of particles with collective interaction," *Journal of Physics* 9: 25–40.

Vlasov, A. A. 1945b. "On the theory of the solid state," *Journal of Physics* 9: 130–138.

Voinov, E. M. and A. G. Plotkina. 1995. "Razrabotka diffuzionnogo metoda razdeleniia izotopov urana," *Istoriia atomnogo proekta*. Vyp. 3 (M.: RNTs Kurchatovskii Institut).

Volkogonov, D. 1989. *Triumf i Tragediia: Politicheskii Portret I. V. Stalina*, T. 2 (2) (M.: APN).

Volkov, B. A. 1995. "Tammovskie sostoianiia pod tunnel'nym mikroskopom," *Priroda* no. 7: 45–47.

Vonsovsky, S. V. 1952. "Metod kvazichastits v kvantovoi teorii tverdogo tela," in *Pamiati Sergeia Ivanovicha Vavilova* (M.: AN SSSR), 363–374.

[Voprosy]. 1949. *Voprosy Istorii Otechestvennoi Nauki. Sessiia Akademii Nauk SSSR, 5–11 yanvaria 1949 g.* (M.-L.: AN SSSR).

Vucinich, A. 1984. *Empire of Knowledge: The Academy of Sciences of the USSR, 1917–1970* (Berkeley: University of California Press).

Vucinich, A. 2001. *Einstein and Soviet Ideology* (Stanford: Stanford University Press).

Walker, C. T. and G. A. Slack. 1970. "Who named the -ON's," *American Journal of Physics* 38: 1380–1389.

Wang, J. 1999. *American Science in an Age of Anxiety: Scientists, Anticommunism, and the Cold War* (Chapel Hill: University of North Carolina Press).

Wang, Z. 1995. "The First World War, academic science, and the 'Two Cultures': Educational reforms at the University of Cambridge," *Minerva* 33: 107–127.

Wannier, G. H. 1937. "The structure of electronic excitation levels in insulating crystals," *Physical Review* 52: 191–197.
Weart, S. 1979. *Scientists in Power* (Cambridge: Harvard University Press).
Wigner, E. and F. Seitz. 1933–1934. "On the constitution of metallic sodium," *Physical Review* 43: 804–810 (1933); 46: 509–524 (1934).
Williams, R. C. 1987. *Klaus Fuchs, Atom Spy* (Cambridge: Harvard University Press).
Wilson, A. H. 1931–1932. "The theory of electronic semi-conductors," *Proceeding of the Royal Society of London* A 133: 458–491 (1931), A 134: 277–287 (1932).
Windholz, G. and J. R. Kuppers. 1988. "Pavlov and the Rockefeller Foundation," *Pavlovian Journal of Biological Science* 23: 107.
Wittgenstein, L. 1958. *Philosophical Investigations* (London: Basil Blackwell & Mott).
Yavelov, B. E. 1985. *Ranniaia Istoriia Sverkhprovodimosti, 1911–1935*. Ph.D. thesis. (Institute for History of Science and Technology, Moscow).
Zakharchenya, B. P. and V. Ya. Frenkel. 1994. "History of the theoretical prediction and experimental discovery of the exciton," *Physics of the Solid State* 36: 469–474.
[Za bol'shevistkuiu]. 1948. "Za bol'shevistskuiu partiinost' v filosofii," *Voprosy Filosofii* no. 3: 3–12.
[Za svobodnuiu]. 1950. "Za svobodnuiu, tvorcheskuiu nauchnuiu kritiku," *Vestnik AN SSSR* no. 8: 10–20.
Zeldovich, Ya. B. 1993. *Selected Works of Yakov Borisovich Zeldovich*. 2 vols. (Princeton: Princeton University Press).
Zeldovich, Ya. B. and Yu. B. Khariton. [1941] 1983. "The mechanism of nuclear fission. II," *Soviet Physics Uspekhi* 26: 279–290.
Zelinsky, N. D. and V. S. Sadikov. 1918. *Ugol' kak Protivogaz* (Petrograd: Khimicheskii Komitet GAU).
Zelinsky, N. D. and V. S. Sadikov. [1941] 1960. "Ugol' kak sredstvo bor'by s udushaiushchimi i yadovitymi gazami. Eksperimental'noe issledovanie 1915–1916 gg." in N. D. Zelinskiy, *Sobranie trudov* (M: AN SSSR), 4: 59–145.
Zernov, M. S. 1912. *Obshchestvo Moskovskogo Nauchnogo Instituta* (M.).
Zhdanko, M. 1922. "O rabote mestnykh sil na severe Rossii," *Priroda* no. 3–5: 85–88.
Zhdanov, A. A. 1946. "Doklad o zhurnalakh *Zvezda* i *Leningrad*," *Pravda* 21 September.

Zhdanov, Yu. A. 1951. "O kritike i samokritike v nauchnoi rabote," *Bol'shevik* no. 21: 28–43.
Zhdanov, Yu. A. 1993. "Vo mgle protivorechii," *Voprosy Filosofii* no. 7: 65–92.
Zhezherun, I. F. 1978. *Stroitel'stvo i Pusk Pervogo v Sovetskom Soiuze Atomnogo Reaktora* (M.: Atomizdat).
Ziman, J. M. 1963. *Electrons in Metals. A Short Guide to the Fermi Surface* (London: Taulor and Francis).
Zinov'ev, G. 1925. *Nauka i Revoliutsiia*: Rech' 9 sentiabria 1925 g. po sluchaiu 200-letiia Akademii nauk (L.: GIZ).

Name Index

Abrikosov, Aleksei A. 164
Adrian, Edgar D. 107, 109
Akhiezer, Aleksandr I. 94, 98, 252, 263
Akulov, Nikolai S. 231
Alekseev, Igor S. xvi
Aleksandrov, Anatoly P. 145, 146
Aleksandrov, Georgy F. 189, 192–196, 200–202, 232
Aleksandrov, Pavel S. 297
Alexandrov, Daniel A. xvii
Alikhanov, Abram I. 129, 134, 293
Allen, John F. 117
Alpert, Yakov L. 166, 173
Andronov, Aleksandr A. 231
Aristotle 193
Arkadiev, Vladimir K. 163
Arkhipov, R. G. 164
Arsenieva, A. N. 79
Aston, Francis 179

Bacon, Francis 292
Bakh, Aleksei N. 107
Bakunin, Mikhail A. 50
Bardeen, John 273
Bardin, Ivan P. 124, 145, 171
Barlament, James xvii

Bauman, Karl Ya. 115
Bebel, August 176
Beletsky, Zinovy 194
Berezhkov, Valentin M. 295
Beria, Lavrenty P. 99, 119, 136, 137, 139, 141–144, 146, 148–150, 154, 217–219, 242, 243, 259, 291, 292
Bethe, Hans 60
Beyrau, Dietrich xviii
Bloch, Felix 52, 59, 61, 62, 63, 248, 249, 251, 255, 257, 263
Blokhintsev, Dmitry I. 150
Bogdanov, Aleksandr A. 281, 301–303
Bogolubov, Nikolai N. 271, 272
Bohm, David 52, 247, 265–269, 271–273
Bohr, Aage 272
Bohr, Niels 40, 54, 59, 73, 82–88, 94, 95, 108, 110, 111, 132, 149, 222, 223, 225, 241, 243
Boltzmann, Ludwig 34, 269, 270
Born, Max 66, 77, 83, 84
Bronstein, Matvei P. 87, 90, 91, 98, 250
Bruevich, Nikolai G. 170
Brush, Stephen xvii
Bukharin, Nikolai I. 105
Bursian, Viktor R. 40, 76, 98

Carson, Cathryn xvii
Charles I 178
Chel'tsov, Ivan M. 4
Cherenkov, Pavel A. 163, 164
Chernov, Dmitry K. 4
Chernyshev, Aleksandr A. 82
Chichibabin, Aleksei E. 10
Chikobava, Arnold S. 238
Cockroft, John D. 129
Curie, Marie 45

David-Fox, Michael 285
Davidovich, Semion A. 97
de Broglie, Louis 61, 78
Debye, Peter 66
Dietzgen, Joseph 176
Dirac, P. A. M. 57–59, 65–67, 78, 79, 94, 107–109, 130, 248, 264
Dollezhal', Nikolai A. 151
Dorfman, Yakov G. 101
Drude, Paul 54, 59, 60, 249
Dubinin, Nikolai P. 208
Dzyaloshinsky, Igor E. 164

Ehrenfest, Paul 34, 40, 55, 59, 67, 73, 76, 81–84, 94
Ehrenfest (Afanasieva), Tatiana A. 34
Einstein, Albert xiii, 41, 47, 66, 67, 91, 160, 164, 176, 217, 220–224, 241, 243
Elina, Olga Yu. xvii
Engelgardt, Vladimir A. 297
Engels, Friedrich 176, 206, 281
Epstein, Paul S. 73, 74, 82
Esakov, Vladimir D. 194
Euler, Leonhard 182
Ezhov, Nikolai I. 119

Fano, Ugo 52
Fedoseev, Piotr N. 194
Feinberg, Evgeny L. 173
Fermi, Enrico 60, 95, 130, 140, 264
Fersman, Aleksandr E. 28, 43, 133, 286
Figatner, Yu. I. 286
Filin, Fedot P. 236, 239
Finkel'shtein, Boris N. 97
Fischer, Emil 13
Fitzpatrick, Sheila xvii, 285, 288

Flerov, Georgy N. 132, 135, 136
Fock, Vladimir A. 39, 41, 42, 76, 77, 83–85, 91, 98, 119, 163, 169, 227, 233, 241–244, 250
Forman, Paul xvii
Fosdick, Raymond 80
Fowler, Ralph H. 250
Frank, Ilia M. 164, 168
Frenkel, Yakov I. xiii, 48, 49, 52–66, 68–72, 76–79, 83, 85, 91, 101, 106, 132, 163, 187, 231, 233, 247–258, 267, 269, 272, 273
Friedmann, Aleksandr A. 41, 47, 48
Frisch, Otto 135
Frish, Sergei E. 39, 168, 294
Frumkin, Aleksandr N. 115
Fuchs, Klaus 139, 140, 154, 155

Galkin, Ya. D. 134
Gamow, George (Gamov, Georgy A.) 76, 77, 79, 84, 87–91, 107, 108, 163, 166
Gershtein, S. S. 264
Gershun, Aleksandr L. 32
Getty, J. Arch 197, 200, 201
Ginzburg, Carlo 278
Ginzburg, Vitaly L. 96, 155, 227
Gorbachev, Mikhail S. xvi, 100, 159, 217, 300
Gorbunov, Nikolai P. 24, 25, 37, 108, 115
Gorky (Peshkov), Aleksei M. 197, 282, 283
Graham, Loren R. xvii, 44, 187, 224, 277
Grave, Dmitry A. 54
Grebenshchikov, Ilia V. 33, 35, 41
Gregg, Alan 83
Gross, Eugene 272
Gross, Evgeny F. 39, 257
Groves, Leslie 138, 144, 146–148
Guesde, Jules 64
Gurevich, Isai I. 134, 135, 138
Gurevich, L. E. 79

Hacking, Ian 274
Hahn, Otto 131

Name Index

Hale, George E. 45
Heilbron, J. L. xvii
Heisenberg, Werner 59, 61, 62, 77, 85–88, 248, 249, 272
Hessen, Boris M. 91, 169, 187, 220, 223, 224, 277
von Hevesy, George 57
Hill, Archibald V. 109
Hitler, Adolf 134, 149
Holloway, David 126, 137, 291
Hoover, Herbert 80
Hopfield, John J. 52
Houston, William 62
Humboldt, Wilhelm 236
Huxley, Thomas H. 14

Ignatovsky, Vladimir S. 163
Iovchuk, Mikhail T. 195, 204
Ipatieff, Vladimir N. 4, 5, 7, 9, 10, 26, 27, 109, 284
Ivanenko, Dmitry D. 76, 77, 79, 94, 129, 138

Joffe, Abram F. 34, 36, 41, 57, 59, 78, 82–84, 89–92, 94, 96, 100–103, 106, 115, 122, 129, 132, 133, 136, 145, 150, 161–163, 168, 184, 226, 228, 231, 233, 240
Joliot, Frédéric 30, 132
Joravsky, David xvii, 160, 188, 189, 281
Jordan, Pascual 77
Josephson, Paul R. xvii
Jost, Friedrich Wilhelm 59

Kachalov, Nikolai N. 33, 41, 42
Kaftanov, Sergei V. 136, 229, 233, 234
Kaganov, Moisei I. 245, 252, 272
Kalinin, Mikhail I. 158
Kamenev, Lev B. 105
Kapitza, Anna A. 106, 107, 111, 112, 120
Kapitza, Piotr L. xiv, 97–125, 136, 141–149, 163, 169, 171, 180, 185, 226, 233, 241, 245, 259–262, 264, 291–293
Karpinsky, Aleksandr P. 115
Kasso, Lev A. 12, 13, 15, 17, 220

Kasterin, Nikolai P. 221, 222
Kedrov, Bonifaty M. 195, 203, 204, 206, 210, 225, 231, 233
Keesom, Willem 116, 117
Keldysh, Mstislav V. 298
Kevles, Daniel J. xvii
Khalatnikov, Isaak M. 264
Khariton, Yuly B. 105, 132, 134, 135, 152–154, 156, 245
Khlopin, Vitaly G. 31, 129, 133, 138
Khrushchev, Nikita S. 146, 157, 158, 211, 214, 240, 297, 298
Klein, Oskar 77, 88
Kliueva, Nina G. 183
Kobzarev, Igor Yu. iv, xvi
Kolovrat-Chervinsky, Lev S. 36
Kol'tsov, Nikolai K. 16
Komarov, Vladimir L. 170, 171
Kompaneets, Aleksandr S. 94
Korets, Moisei A. 97
Korolev, Sergei P. 295
Korol'kov, A. L. 33
Kramers, Hendrik 88
Krementsov, Nikolai L. xvii
Krutkov, Yury A. 40, 76, 83, 98
Krylov, Aleksei N. 32, 41, 107, 163
Krzhizhanovsky, Gleb M. 167
Kummant, Eduard L. 10, 11
Kundt, August 15
Kurchatov, Igor V. 129, 130, 133–135, 137–141, 144, 147–151, 217–219, 241–243, 245, 293, 295, 296

Landau (Drobantseva), Kora 259
Landau, Lev D. xiii, 52, 56, 69, 72–79, 84–92, 94–98, 105, 117–120, 225, 227, 231, 233, 245, 247, 250–255, 257–266, 269–272
Landé, Alfred 58
Landsberg, Grigory S. 66, 163, 166, 226
Langmuir, Irving 270
Laurman, Emil Ya. 104
Lazarev, Piotr P. 16, 161–163, 166
Lebedev, Piotr N. 13, 15, 16, 34, 166, 220

Leipunsky, Aleksandr I. 92, 94, 96, 97, 105
Lenin, Vladimir I. 24, 25, 175, 176, 202, 222–224, 230, 281–284, 292
Leontovich, Mikhail A. 173, 227
Lifshitz, Evgeny M. 94, 98, 225, 263, 264
Lifshitz, Ilia M. 245, 272
Livingston, Katherine xvii
Lomonosov, Mikhail V. 16, 181, 182
London, Fritz 260, 262, 263
Lorentz, Hendrik Antoon 34, 40, 54, 59–61, 73, 85, 249
Lunacharsky, Anatoly V. 35, 36, 39, 176
Luzin, Nikolai N. 119
Lysenko, Trofim D. xi, 168, 171–174, 184–192, 207–214, 221, 225, 229, 235, 236, 244, 246, 247, 297, 298, 304

Maasen, Sabine 56
Mach, Ernst 65, 68, 176, 222
Machinsky, M. V. 79
Maksimov, Aleksandr A. 175, 194, 206, 222, 223, 225, 230, 233, 241–243
Malenkov, Georgy M. 112, 124, 142, 143, 146, 170, 194, 207, 209, 210, 229, 231, 234, 243
Malyshev, Viacheslav A. 123
Mandel, G. A. 79
Mandelstam, Leonid I. 66, 161, 163, 220, 221, 226, 231
Manikovsky, Aleksei A. 6, 7
Manuilov, Aleksandr A. 13
Markov, Moisei A. 206, 225, 231–234
Markovnikov, Vladimir V. 2, 3
Marr, Nikolai Ya. 189, 235–239
Marx, Karl 50, 176, 281, 292, 301, 302
Maxwell, James Clerk 15
McMillan, Edwin 139
Mendeleev, Dmitry I. 3, 4
Menzbir, Mikhail A. 13
Meshchaninov, Ivan I. 235–239
Mezhlauk, Valery I. 113–115, 119
Michurin, Ivan V. 208, 235
Migdal, Arkady B. 56
Miller, Dayton 220

Millikan, Robert 82, 83
Minakov, Piotr 13
Mints, Aleksandr L. 253
Misener, Donald 117
Mitin, Mark B. 194
Mitkevich, Vladimir F. 91, 162, 163, 222
Molotov, Viacheslav M. 107, 112, 113, 115, 119, 121, 123, 124, 129, 158, 288, 290
Mott, Neville F. 257
Mottelson, Ben 272
Mussolini, Benito 61
Mysovsky, Lev V. 129, 130

Nedelin, Mitrofan I. 295, 296
Nemenov, Mikhail I. 35, 36, 41
Newton, Isaac 159, 165, 170, 178–181, 185, 224
Nikiforov, Pavel M. 286
Novikov, Mikhail M. 26
Nuzhdin, Nikolai I. 213, 297, 298

Obreimov, Ivan V. 67, 82, 94, 105, 120
Ol'denburg, Sergei F. 25–27, 282, 283, 286
Omel'ianosvky, Mikhail E. 68
Oparin, Aleksandr I. 213
Oppenheimer, J. Robert 266
Orbeli, Leon A. 201, 213, 240,
Ordzhonikidze, Grigory (Sergo) K. 129, 130

Papaleksi, Nikolai D. 163, 166, 226
Pauli, Wolfgang 60, 71, 85–88, 95, 250
Pavlov, Ivan P. 83, 107, 111, 112, 114, 189, 235, 240, 289, 290
Peierls, Rudolf 62, 63, 69, 71, 86, 87, 135, 248, 249, 255, 257
Pekar, Solomon I. 52, 253, 254
Perepelkin, Yakov 32
Perkin, William 3
Peter I (the Great) xi, 292
Petrzhak, Konstantin A. 132, 136
Pines, David 52, 272, 273
Pitaevskii, Lev P. 264
Planck, Max 160

Pohl, Robert Wichard 256
Pomeranchuk, Isaak Ya. 52, 94, 138, 263
Popov, Aleksandr S. 289
Popper, Karl R. x–xii, 303
Predvoditelev, Aleksandr S. 226–228
Pringsheim, Peter 160
Prozorova, L. A. 264
Pushkin, Aleksandr S. 191

Raman, Chandrasekhara Venkata 66
Rapoport, Iosif A. 212
Rasetti, Franco 130
Rechenberg, Helmut xvii
Riabinin, Yury N. 97
Riehl, Nikolaus 150
Röntgen, Wilhelm 34
Rose, Wickliffe 81, 83
Rosenfeld, Léon 88
Roskin, Grigory I. 183
Rossiianov, Kirill O. xvii
Rozenkevich, Lev V. 79, 94, 117
Rozhdestvensky, Dmitry S. 33–37, 39–42, 91, 161–163, 168
Rubinin, Pavel E. 100
Rutherford, Ernest 102–107, 109–111, 114, 115, 169

Saburov, Maksim Z. 145, 146
Sakharov, Andrei D. 155, 156, 245, 295–299
Schottky, Walter 59
Schrödinger, Erwin 61, 77
Semenov, Nikolai N. 82, 101, 106, 132, 163
Serdiuchenko, Georgy P. 236, 237, 239
Seitz, Frederick 249
Serov, V. A. 282
Shchodro, Nikolai K. 163
Shepilov, Dmitry T. 204, 208, 209, 219, 233, 234
Shockley, William 256
Shoenberg, David 100, 104, 111
Shubin, Semion P. 250, 251, 272
Shubnikov, Aleksei V. 241
Shubnikov, Lev V. 39, 92, 96, 97, 105, 106, 117

Shuleikin, Vasily V. 163
Sinel'nikov, Kirill D. 84, 92, 105
Sirotkina, Irina E. xvii
Skobel'tsyn, Dmitry V. 84, 169
Slater, John C. 256
Sommerfeld, Arnold 40, 59, 60–62, 73, 86, 251, 257
Soyfer, Valery N. 209
Stalin (Dzhugashvili), Iosif V. xii, xiv, 76, 100, 105, 107, 112, 115, 119, 121, 124, 125, 129, 130, 136, 137, 140–144, 146, 148, 149, 157, 172, 173, 180, 183, 184, 189, 190, 194, 195, 198, 200, 202, 203, 209–211, 215–217, 224, 237–239, 246, 259, 286, 288, 290–293
Steklov, Vladimir A. 282
Stoner, Edmund C. 272
Sukov, M. K. 142
Suslov, Mikhail A. 194, 204, 209, 234, 237, 243
Strassmann, Friedrich 131

Tamm, Igor E. 52, 64–69, 83–85, 98, 107, 155, 156, 164, 166, 226, 227, 233, 245, 247, 250, 251, 254, 255, 257, 261, 262, 272, 297
Terenin, Aleksandr N. 39, 42, 163
Terletsky, Yakov P. 149, 150
Timiriazev, Arkady K. 220, 221, 243
Timiriazev, Kliment A. 12–17, 24, 44, 220
Tisza, Laszlo 260, 262
Todes, Oskar M. 63
Tonkov, V. N. 282
Tolstoy, Lev N. 12
Tonks, Lewi 270
Topchiev, Aleksandr V. 218, 229, 231, 232, 234
Trapeznikova, Olga M. 67, 97
Trotsky, Lev D. 76, 98, 105
Trowbridge, Augustus 84
Truman, Harry S. 140, 155

Unfried, Berthold 199

Vainshtein, L. A. 264
Val'ter, Anton K. 92

Varga, Evgeny S. 204
Vavilov, Nikolai I. 158–160, 168, 171, 184, 185, 207
Vavilov, Sergei I. xiv, 32, 115, 131, 157–185, 187, 198, 213, 218–220, 223–227, 229–234, 240, 244, 288
Veksler, Vladimir I. 139
Vernadsky, George (Georgy V.) 31, 132
Vernadsky, Vladimir I. xiii, 16–22, 27, 28, 30, 31, 35, 36, 43, 44, 127–129, 132, 133, 290
Vinogradov, Ivan M. 171
Vinogradov, Viktor V. 236, 238, 239
Vizgin, Vladimir P. xvi
Vlasov, Anatoly A. 227, 270, 271
Volta, Alessandro 61
Vonsovsky, Sergei V. 250, 272
Voznesensky, Nikolai A. 143
Vyshinsky, Andrei Ya. 173

Wagner, Carl 59
Walton, E. T. S. 129
Wang, Jessica A. xvii
Wannier, Gregory 256, 257

Weart, Spencer xvii
Wentzel, Gregor 251
Wheeler, John A. 132
Wigner, Eugene P. 249
Williams, Vasily R. 235
Wilson, Alan H. 248–251
Wittgenstein, Ludwig 201, 215

Yudin, Pavel F. 194

Zabudsky, Grigory A. 7, 8
Zavadovsky, Boris M. 212
Zaveniagin, Avraamy P. 241
Zeldovich, Yakov B. 132, 134, 135, 153, 155, 156, 245
Zelinsky, Nikolai D. 10, 11
Zhdanov, Andrei A. 183, 192–196, 198, 201–207, 209, 210, 236
Zhdanov, Yury A. 192, 204, 208–210, 240, 243
Zhebrak, Anton R. 207, 208
Zhigulev, A. V. 134
Zhukovsky, Piotr M. 212
Zinin, Nikolai N. 3
Zinoviev, Grigory E. 279, 280, 285

Subject Index

Academic degrees 15, 34, 75, 78, 95, 164, 289, *see also* Professionalization of science
Academic societies 10, 20, *see also* Moscow Society for Scientific Institute; Royal Society of London; Russian Association of Physicists; Russian Physico-Chemical Society; Russian Society of Roentgenologists and Radiologists; United Council of Learned Institutions
Academy of Sciences
    elections to 25, 28, 90, 91, 118, 122, 131, 146, 160–163, 169–173, 181, 184, 193, 226–229, 259, 286, 288, 297–299
    Imperial 16–22, 30, 181
    President xiv, 157, 158, 168, 170–173, 178–184, 198, 213, 219, 229, 282, 298
    Presidium 114, 115, 166, 167, 169–171, 213
    Russian 26–28, 30, 31, 41, 279, 282
    Secretary 25, 27, 108, 170, 218, 283
    Soviet 21, 22, 96, 107, 108, 114, 161–163, 165–171, 174, 180–184, 189, 190, 198, 204, 208, 213, 221, 226–231, 234, 237, 286, 288, 294, 298, 300
    see also KEPS, Radium Commission, Uranium Commission
Academy (other), *see* Agriculture; Artillery; Communist; Georgian; Medicine; Pedagogical; Ukrainian
Affirmative action 75
Agitprop (Department of Propaganda and Agitation) 183, 193, 194, 196, 204, 206–210, 213, 219, 225, 233–237
Agriculture
    Academy of Agricultural Sciences (VASKhNiL) 168, 171, 188, 207–212, 229
    Agricultural Scientific Committee 283, 284
    Department (of the Central Committee) 210
    see also Commissariat; Ministry
Akademgorodok, *see* Scientific cities

American Relief Administration (ARA) 80, 81
Anarchists 50, 51
Anti-communism xi, 80, 83, 102, 188, 189, 216, 219, 267, 277, 281, 290, 303, *see also* Cold War ideology; Propaganda; Sovietology
Anti-cosmopolitan campaign 188, 189, 231–233, 240
Anti-Semitism 158, 166, 231, 233, 241, 258, 267
Apparat, apparatchik 105, 158, 186, 287, 294, 295, *see also* Bureaucracy
Artillery
  Artillery Academy 4, 5, 32
  Chief Artillery Administration 7, 8, 33, 37
Arzamas-16 (Sarov; Volga branch) 152, 153, 156
Astronomy 174, 190, 225, *see also* Cosmology; Pulkovo Observatory
Aspirant, see Students-Graduate
Atomic bomb xiv, 99, 125, 132, 134–137, 140–144, 148–157, 217–219, 228, 242, 243, 268, 291, 292, 294, 296, 298, *see also* Atomic project; Critical mass; Espionage; Hiroshima; Hydrogen bomb; Manhattan Project; Nuclear physics; Uranium
Atomic Commission 41
Atomic project
  Anglo-American (*see* Manhattan Project)
  British (Tube Alloys) 135–140
  German 135, 137, 138, 291
  Soviet xiv, 80, 99, 100, 122, 126, 127, 134–138, 140–156, 217, 231, 234, 242, 245, 264, 279, 291, 293, 295, 304 (*see also* Arzamas-16; Laboratory # 2; Special Committee; Uranium)
August Session, *see* Lysenko Affair

Band theory 63, 72, 247–251, 254–258, 261, 269, *see also* Bloch electron
Big science xiii, xiv, 23, 24, 37, 38, 42–46, 127, 147, 148, 156, 303

Biology xiii, 12, 14, 16, 19, 29, 83, 84, 184, 218, 221
  ideological discussion 188, 189, 204, 207–214, 217–219
  see also Evolution; Genetics; Lysenko affair; Michurinist biology
Bloch electron 59, 61, 62, 248, 249, 255
Bolsheviks 21, 24–29, 31, 36, 37, 45, 64, 65, 68, 80, 83, 158, 178, 220, 279–284, 287, 290, 301–304, *see also* Government
Bourgeois specialists 26, 89, 161, 172, 173, 178, 284–285, 287, 289, *see also* Red specialists
Bourgeois science 184, 187, 188, 302, 303, *see also Proletkult*; Socialist science; Sociology of knowledge
Britain (contacts and comparisons) xii, 1, 3, 4, 11, 14, 17, 18, 33, 35, 41, 45, 77, 80, 84, 102–106, 109, 115, 116, 133, 135, 174, 181, 183, 247, 249, 250
Bureaucracy 13, 16, 29, 31, 111, 169–171, 185, 196, 197, 203, 209, 215, 226, 234, 283, *see also Apparat*; Career pattern

Career pattern 1, 3, 75, 78, 159, 169, 170, 171, 177, 180, 287–289, 294, 295, *see also* Professionalization of Science
Carnegie Institution 14
"Catch up and Surpass!" 126, 144, 245, 246, 293–295
Central Committee (of the Communist Party) 112, 170, 175, 183, 190–194, 202, 207, 209, 210, 212, 214, 233, 237, 242, 243, 288, 298, 299, *see also Agitprop*; Agricultural Department; Politburo; Secretariat; Science Department
Chemistry xiii, 2–5, 8–10, 16, 18, 19, 21, 27, 38, 93, 135, 150, 190, 284
Chemical warfare 7–11, 284
  War Chemical Committee 8–10, 27, 29, *see also* Military

Subject Index

Cherenkov, or Vavilov–Cherenkov radiation 163, 164, 167, 169
China 46
Civil War
  British 178, 179
  Russian xii, xiii, 9, 23, 28–32, 37, 40–43, 45, 47, 53, 54, 65, 80, 81, 101, 283, 301
Class 12, 17, 165, 175, 177, 184, 239, 277, 281, 284–288, 301, 302, *see also* Elite; Intelligentsia; Privileges and hierarchy; Proletariat
Cold War xi, xii, xv, 46, 80, 141, 148, 149, 158, 175, 182, 183, 192, 246, 258, 278, 294, 300, 304, 305
  ideology xi, xii, 183, 189, 277, 278, 300, 305, *see also* Propaganda
Collective excitations, *see* Quasiparticles
Collectivism 38, 43, 44, 46, 50, 51, 64, 182, 184, 199, 211, 265, 267–269, *see also* Freedom
Collectivist approach in physics xiv, 48, 49, 55, 56, 59–63, 71, 72, 187, 247, 250, 254–258, 261–265, 267–273, *see also* Electrons; Many-Body; Quasipartiles
Collectivization xii, 50, 51, 56, 63, 70, 79, 89, 168, 258, 285, *see also* Kolkhoz; Mir
Commissariat(s) 29, 89, 283, 288
  of Agriculture (*Narkomzem*) 29, 284
  of Army and Navy (*Narkomvoenmor*) 29, 284
  of Arms Production 168
  of Enlightenment *see Narkompros*
  of Heavy Industry *see* NKTP
  of Posts and Telegraphs (*Narkompochtel*) 29
  of Public Health (*Narkomzdrav*) 29, 81, 84, 284
  see also Government; VSNKh
Communist Academy 175
Communist Party, communists xi, xiii, 24, 25, 89, 105, 158, 168, 177, 184, 188, 190, 197, 200–202, 205, 215, 218, 266–268, 279–286, 289, 290, 293, 300
Post-communism 74, 178, 300, 304
  see also Agitprop; Bolsheviks; Central Committee; Politburo
Condensed matter physics xiv, 48, 49, 51, 58, 69, 96, 227, 265, 269, 273, 274, *see also* Collectivist approach; Electrons; Low-temperature; Many-body; Plasma physics, Quasiparticles; Solid state physics; Transport phenomena
Conservatism, conservatives 12, 47, 95, 163, 165, 184, 185, 220, 221, 287, 289, 299, 300, *see also* Monarchism; White Russians
Cosmology, Big Bang 41, 47, 48, 108, 190, *see also* Astronomy
Council of Ministers, *see* Government
Court of honor 183, *see also* Patriotism
Crisis in science 86–88, 95, *see also* Revolution in science
Critical mass 127, 134, 135, 153, 154, *see also* Uranium
Criticism
  in politics 13, 20, 26, 28, 110–114, 124, 142, 143, 180, 183, 191–196, 200, 201, 209, 231, 233
  in science xiv, 86, 95, 150, 151, 186–189, 205, 206, 214, 225, 227, 242
  and self-criticism (*kritika i samokritika*) 177, 191, 195–206, 208, 209, 211, 213, 230, 233, 235–238
Cult of science, scientism 46, 156, 157, 243, 266, 300, 304
Cultural Revolution xiii, 73–79, 88–92, 95–98, 159, 161, 165, 168, 172, 178, 184, 241, 279, 285, 286, 287, 290, *see also* Education; Great Break; Great Retreat; Pedagogical experiments; Red specialists; Revolution and science; Students
Cyclotrons, particle accelerators 92, 128–131, 133, 134, 138, 139, 164

Democracy
and science x–xii, 13, 14, 24, 177, 181, 182, 184, 278, 303
bourgeois 203
intraparty 191, 197, 199, 202
Soviet xi, 112, 181, 182, 186, 197, 203, 205, 211, 293
Democratic centralism 197, 211
Denmark 82–88, 149
Department of Scientific and Industrial Research (DSIR) 45, 105, 108
Dialectics, see Materialism-dialectical
Dictatorship x, xi, 27, 58, 43, 44, 85, 110, 143, 148, 158, 191, 253, 259, 276, 278, 290
totalitarian model 184, 215, 216, 276, 281, 300
see also Democracy-Soviet; Democratic centralism; Gulag; Security police
Dirac equation 79, 87
hole theory 57–59, 248, 259
see also Quantum electrodynamics
Diskussiia (disputation) 197–202, 204–208, 210–213, 225, 229, 230, 234–239
Dissidents, see Resistance

Education
in Russian empire 75, 100, 101
Soviet 26, 29, 30, 75, 78, 89, 95, 161, 283, 285–289, 301
see also Commissariat; *Gymnasium*; Ministry; Pedagogical; Universities
Electrons
arrested 252, 253
bound 49, 248, 249, 251, 252, 254, 256–258, 273
collectivized 56, 59–63, 72, 250, 255–258, 269, 272
free 248–252, 254–256, 258, 273
organized 267–269, 271, 272
trapped 248, 251–254, 256
see also Bloch electron; Quasiparticles

Elementary excitations, see Quasiparticles
Elite groups 157, 186, 276, 288, 292–295, 299, 300, 302, see also Class; Intelligentsia; Privileges and hierarchy
Emigration x, 26, 31, 73, 74, 108, 166, 177
Engineers 4, 10, 19, 25, 36, 38, 41, 46, 89, 100, 101, 104, 151, 156, 186, 253, 284–288, 294
Espionage, intelligence 127, 136, 137, 139, 140, 141, 149, 150, 154, 155
paranoia 80, 117, 119, 155, 146, 267
Evolution, Darwinism 12, 14, 204, 208, 212
Exciton 51, 70, 252, 254–258
Exploration 20, 25, 28, 30, 148, see also Geology

Federation of Architects, Engineers, Chemists and Technicians (FAECT) 266, 267
Fermi-liquid, see Quantum liquid
FIAN (Physical Institute of the Academy of Sciences) 69, 91, 98, 131, 134, 155, 163, 164, 166–169, 174, 181, 225, 226, 227
Fission 127, 131, 132 133, 136
spontaneous 132, 136
France (contacts and comparisons) xii, 4, 11, 17, 18, 33, 45, 50, 77, 133, 260
Freedom 5, 15, 49–51, 250, 253, 259, 261, 272
collectivist xiv, 49, 55, 62, 72
in physics 48, 49, 55, 60, 62, 63, 67, 249, 253, 261
of criticism x, xii, 186, 215
sexual 259
see also Collectivism; Criticism; Electrons; Ideal gas
Frenkel defect 59

Games 185, 186, 191, 197, 201, 202, 204–208, 211, 212, 215, 244, see also Rituals; Rules

Subject Index

Gender 290, 291, 296, 297
Genetics xi, xiv, 44, 168, 171, 185, 187, 188, 207, 208, 214, 221, 236, 244, 297, 304, 305, *see also* Biology-ideological discussion; Lysenko affair; Michirunist biology
Geology xiii, 16–21, 28, 30, 38, 93, 133, 138, 140, 147, 148
    Geological Committee (of the Mining Department) 283, 284
    see also Exploration; KEPS; Uranium
Georgian Academy of Sciences 238
German Specialists in the Soviet Union 147, 150, 151
Germany (contacts and comparisons) x, 2–8, 10–17, 18, 33, 41, 54, 60, 73, 74, 76, 77, 83, 84, 85, 133, 135, 145, 146, 160, 249, 284, 304, *see also* Atomic project; German specialists; Kaiser Wilhelm Gesellschaft
Gosplan (State Planning Commission) 284
Government
    Council of Ministers (Imperial) 10, 13, 16, 27, 29, 283, 284
    Council of Ministers (Soviet after 1946) 144–146, 186, 214, 295
    Provisional 26, 27, 30, 35
    Sovnarkom (Bolshevik) 6, 9, 14, 22, 23, 25–29, 31, 35, 37, 41, 53, 81, 90, 165, 166, 284, 285
    Stalinist xiv, 98, 113–115, 120, 122, 133, 161, 287, 288, 292
    State Committee of Defense, GKO (emergency wartime) 123–125, 136, 137, 141, 172
    see also Commissariats; *Gosplan*; Ministry; State support for science; VSNKh
Great Break 63, 79, 285, *see also* Cultural revolution
Great Retreat 165, 184, 287, *see also* Conservatism; Cultural revolution
Gulag 28, 141, 148, 156, *see also* Sharashka

Gymnasium (advanced high school) 75, 100, 101, 175

Hiroshima 125, 140, 141, *see also* Atomic bomb
Historiography of Soviet Science xi–xvii, 21, 126, 127, 179–181, 218–221, 230, 277, 278, 285, 303–306
Hole 51, 56–59, 248, 249, 252, 254–256, *see also* Dirac-hole theory; Quasiparticles
HUAC (House Un-American Activities Committee) 267, 268
Hydrogen bomb 155, 156, 246, 295

Ideal gas 48, 49, 261, 263, 269
Idealism 187–189, 212, 222, 223, 225, 229–231, 241, 258, *see also* Materialism
Ideology xi, 60, 158, 183, 184, 205, 206, 215, 216, 239, 280, 281
    and science xiv, 26, 36, 87, 157, 181, 187–190, 210, 213–225, 230, 232, 235, 240, 242, 243, 247, 258, 273, 279
Ideological discussions 175, 185, 187–191, 205, 206, 214, 216, 217, *see also* Biology; Criticism and self-criticism; *Diskussiia*; Linguistics; Philosophy; Physics; Physiology; Political economy
IEB (International Education Board) 81, 83, 84, *see also* Rockefeller philanthropy
IFP (Institute of Physical Problems, Moscow) 98, 106, 107, 110, 116, 118, 121, 122, 144–146, 264
Individuality 67, 68, 179, 263, 271, *see also* Quantum statistics
Industrialization xii, 3, 8, 79, 89, 90, 93, 120, 285, 304, *see also* Modernization
Industry and science xiii, 1–3, 5–11, 22, 24, 27, 32, 33, 35, 37–44, 92–94, 107, 120–125, 130, 131, 147, 156, 172, 180, 284, *see also* Atomic project; Optical glass; Oxygen; Science-Applied

Institute (*see also* Research institutes)
  of Applied Chemistry 29
  of Applied Mineralogy
    (*Lithogaea*) 29
  Automotive (NAMI) 29
  of Brain Research (Bekhterev) 29
  California Institute of
    Technology 74, 82, 265
  Central Aero-Hydrodynamic
    (TsAGI) 29
  of Chemical Physics 132, 155
  of Crystallography 241
  of Experimental Biology 16, 29
  of Experimental Medicine 162
  of Genetics 171, 207, 208
  Lomonosov 16
  of Natural Sciences (Lesgaft) 29
  State Optical Institute (GOI)
    35–42, 122, 162, 163, 166, 168,
    180, 181
  of Philosophy 194, 196, 206, 231
  Physics (Leningrad
    University) 33, 34, 39, 168
  Physics Research (Moscow
    University), *see* NIIF
  Physical (Academy of Sciences),
    see FIAN
  of Physical Problems, *see* IFP
  Physico-Mathematical (Academy
    of Sciences) 90, 162, 163, 166
  Physico-Technical, Leningrad
    (GFTI, LFTI), *see* LFTI
  of Physics and Biophysics 16, 29,
    160, 162, 163, 166
  Polytechnical, Petrograd-
    Leningrad 34, 36, 101–106,
    257, 258
  Radium (RI, RIAN) 31, 36, 128–134
  State Roentgenological and
    Radiological (GRRI) 35, 36,
    38, 101
  State Scientific-Technical
    (GONTI) 27
  Technological, Petrograd 25
  of Theoretical Geophysics 162
  of Theoretical Physics (Academy
    of Sciences) 90, 91, 265
  of Theoretical Physics,
    Copenhagen 85, 87, 88,
    130, 149
  Ukrainian Physico-Technical, *see*
    UFTI
  Women's Medical 35, 36
Intelligentsia 27, 186, 279, 283,
  286–288, 290, 300, 304, *see also*
  Class
International contacts, travel, and
  comparisons xi, 3, 4, 9, 10, 12, 15, 17,
  18, 23, 32, 40, 41–45, 54, 77, 79–85, 88,
  94, 96, 97, 102, 105, 108, 109, 160, 174,
  175, 181, 299 (by country, *see* Britain,
  China, Denmark, France, Germany,
  Netherlands, USA), *see also*
  Emigration
Internationalism 17, 24, 43, 85, 111,
  165, 231, 239, 289, *see also*
  Nationalism; Patriotism

Kaiser Wilhelm Gesellschaft 12, 13
Kapitza-club 104, *see also Kruzhok*
Kasso affair 12, 13, 15, 17
KEPS (Commission for the Study of
  Natural Productive Forces) 18–22,
  28, 30, 31, 35, 43, 129
KGB, *see* Security police
Kolkhoz (collective farm) 50, 63, 258,
  see also Collectivization
Kritika i samokritika, see Criticism and
  self-criticism
Kruzhok (Russian-style seminar) 34,
  104, 177

Laboratory
  # 2 137, 147
  Cavendish 102, 104, 105
  Central Scientific-Technical, of the
    War Ministry 4, 7, 8, 27
  Los Alamos 139, 140, 153–155,
    266
  Metallurgical (Chicago) 140
  Mond 105, 109, 110, 115, 116
  Navy Scientific-Technical 4
  Physico-Techical, Leningrad
    (GFTL) *see* LFTI

Subject Index

Radio, Nizhnii Novgorod 29
see also Research imperative
Landau diamagnetism 86
  phase transition 96
  school 94, 95, 98, 258, 264, 265
LFTI (Leningrad Physico-Technical Institute) 36, 54, 78, 79, 82, 90–94, 96, 101, 122, 129–134, 163, 240, 253, 254
Liberalism, constitutional democrats x, 17, 26, 27
Light quantum, *see* Photon
Linguistics xiii, 204, 218, 224
  ideological discussion xiv, 188, 189, 190, 216, 219, 235–240, 244
Liquid helium 67, 92, 104–105, 116–118, 260–265, *see also* Oxygen; Superconductivity and Superfluidity
Literature, literary criticism 12, 190, 205, 282, 283, *see also* Socialist realism
Low-temperature physics 67, 92, 97, 104–106, 120–121, 241, *see also* Liquid helium; Oxygen liquefaction; Superconductivity and Superfluidity
Luminescence 160, 163, 164, 166, 169, 180, *see also* Cherenkov radiation
Lysenko affair xi, xiv, 157, 168, 171, 185, 187–191, 207–214, 216, 221, 225, 229, 230, 232, 235, 236, 240, 244, 246, 297, 298, 305, *see also* Biology-ideological discussion; Genetics; Michurinist biology

Magnetism 86, 96, 103, 104, 258, 263, 272
Magnon 263
Manhattan project xiv, 127, 136–140, 143, 144, 146–149, 156, 266, 291, *see also* Laboratory-Los Alamos; Laboratory-Metallurgical
Many-body interactions 48, 268–270, 274
Marxism, Marxist philosophy 50, 51, 53, 64, 176, 184, 191, 193–195, 202, 203, 212, 266, 280

view of science 28, 31, 45, 169, 180, 182, 188, 217–220, 219–224, 235, 238- 244, 277, 286, 297, 300
see also Ideology; Materialism Philosophy; Social democrats
Materialism 95, 176, 219, 220, 222, 225, 235, 258
  Dialectical xiv, 68, 91, 175, 176, 182, 187–189, 219, 222–224, 231, 238, 243, 244, 266
  Historical 91
  Materialism and Empiriocriticism 175, 176, 223, 224
  Scientific realism 223
  see also Idealism
Mathematics 3, 34, 47, 48, 73, 83, 84
Medicine, public health 10, 19, 29, 35, 36, 38, 81, 83, 84, 101, 102, 183, 296, 302
  Academy of Medical Sciences 189
  Military Medical Academy 282
  see also Commissariat; Ministry
Mensheviks 64, 65, 68, 212
Metaphors 187, 276, 292, 296
  in science 49, 55, 56, 60, 63, 66, 253, 267–273
  see also Political language
Michurinist biology 188, 208, 210, 211, 213, 304, *see also* Lysenko Affair
Military and science xiii, 1, 2, 4–11, 17, 18, 22, 32, 42, 46, 96, 97, 123, 133, 135, 136, 147, 172, 180, 284, 296, *see also* Artillery; Atomic project; Chemical warfare; Medicine; Mobilization; Laboratory (Navy, War)
Ministry
  Navy 4
  of Agriculture 209–211, 283
  of Enlightenment 13, 33
  of Higher Education 213, 227, 229, 233, 237
  of Public Health 183
  War 4
Mir (village commune) 52, 55
Mobilization 6, 10, 19, 22, 33, 37, 123, 135, 172

Modernism in/of science xiii, 23, 31, 42, 43, 100, 217, 221, 224, 225, 278, 279, 303
Modernization (economic) xiii, 2, 6, 30, 32, 37, 42, 43, 46, 184, 304
Monarchism, monarchists 12, 17, 26, 27, 284
Monopolism 189, 207
Moral dilemmas, compromises xii, 17, 18, 120, 159, 160, 169, 174, 178, 179, 180, 184, 185, 245, 295, 296, 297
Moscow Society for Scientific Institute 16, 29

Narkompros (Commissariat of Enlightenment) 19, 26–29, 35, 37, 39, 41, 42, 53, 78, 84, 85, 89, 161, 285
Nationalism 10, 17, 31, 43, 158, 165, 184, 188, 23, 289, 290, 300, *see also* Internationalism; Patriotism
NEP (New Economic Policy) 32, 202, 285
Netherlands (contacts) 34, 40, 67, 73, 82–84
Neutron 129, 130, 132, 133, 134, 138
NIIF (Physics Research Institute, Moscow University) 15, 91, 163, 220, 221
NKTP (Commissariat of Heavy Industry) 92–94, 96, 122, 129–131, 165
Nobel Prize 99, 163, 164, 265, 305
Nuclear physics 69, 84, 86, 87, 90, 92–94, 108, 127–133, 135, 142, 156, 163, 166, 186, 234, 241, 243, 272, *see also* Cyclotron; Radioactivity
Nuclear tests 140, 146, 150, 154, 155, 156, 267, 295–297, *see also* Atomic bomb

Optical glass and industry 7, 32–35, 37–42, 44, 172, 180
Optics 15, 32, 37, 41, 162, 164, 166, 167, *see also* Luminescence; Institute-Optical
Oxygen liquefaction and industry 120, 121, 123, 124, 142–145, 291

Oxygen Trust 123–125, 142–145, 291, 292

Pact between science and politics, *see* Power and Knowledge
Paradox of Soviet science xi–xiii, 303–305
Partiinost' (party-mindedness) 182, 184, 224, 280, 281, *see also* Sociology of knowledge
Patriotism, unpatriotic behavior 183, 184, 189, 245, *see also* Anti-cosmopolitan campaign; Court of honor; Nationalism
Patronage xiii, xiv, 2, 16, 20, 24, 31, 83, 92, 114, 118, 125, 130, 283, *see also* Philanthropy; State support
Pedagogical
  Academy of Pedagogical Sciences 237
  experiments 75, 76, 89, 161
Perestroika xvi, 74, 159, 217, 300
Philanthropy 2, 16, 80, *see also* American Relief Administration; Carnegie Institution; Patronage; Rask-Ørsted Foundation; Rockefeller philanthropy; State support
Philosophy, philosophers 91, 95, 219–223, 243, 244
  ideological discussion 188, 189, 191–196, 200–206
Philosophy and/of science x, 87, 95, 176, 180, 218, 219, 222–225, 243, 277–279, *see also* Idealism; Marxism; Materialism; Positivism; Sociology of knowledge
Phonon 64, 68–71, 254, 255, 261–263, *see also* Quasiparticles
Photon 66, 68–71, 86, 160, 164, 256
Physical Review 131, 132, 269
Physics xiii, 16, 19, 38, 41, 78, 83, 90, 184, 244
  classical 219–221, 224
  experimental 15, 73, 84, 91, 93, 96, 162, 164, 166, 167, 254, 257, 262, 263

Subject Index

ideological discussion xiv, 190, 216–221, 225–234, 241–243, 246
theoretical xiii, 41, 72–80, 88, 90, 91, 94, 98, 162, 166, 174, 226, 227, 247, 254
see also Condensed matter; Low-temperature; Magnetism; Nuclear; Optics; Plasma; Quantum; Relativity; X-rays
Physikalische Zeitschrift der Sowjetunion 94
Physiology 3, 289, 290
ideological discussion 188, 189, 200, 201, 240
Planning 6, 20, 21, 37, 44, 45, 111, 122, 284, see also Gosplan; Mobilization
Plasma physics 49, 227, 247, 263, 266, 269–272
Plasmon 265, 272
Plutonium 127, 138, 140, 144, 151, 153, 154, 242
Polaron 252–254
Politburo 76, 107, 109, 112, 124, 130, 141, 144, 158, 171, 172, 192, 209–211, 279, 288, 291
Political economy 2, 3, 204, 239
ideological discussion 188, 190, 206, 207
Political language, rhetoric xiii, xiv, 49, 112–114, 159, 160, 173–186, 189–191, 197–199, 206, 214, 215, 218, 223, 258, 272–276, 280, 281, 289, 292–299, see also Metaphors; Propaganda
Positivism 65, 68, 69, 176, 219, 222, 261
Positron 95, 129
Power and Knowledge, and Pact between them xii–xv, 14, 22, 24, 276–287, 290–296, 298–300, see also Scientists and politicians
Privileges and hierarchy 77, 87, 91, 95, 283, 288, 289, 290, 292–295, 298, 299, see also Elite
Professionalization of science 3, 12, 13, 16, 19, 32, 44, 45, 78, 89, see also Career; Research Institutes

Proletarian science, *see Proletkult*; Socialist science
Proletariat 281, 286
Proletkult 301–303
Propaganda xi, 17, 46, 181–183, 189, 278, 303
Public image of science xiv, 46, 93, 133, 156, 157, 159, 172, 173, 180–184, 243, 287, 290, 294, 295, 298, 304
Pulkovo Observatory 174
Purges, political persecution xii, 25, 96–99, 106, 130, 155, 165, 197, 199, 207, 223, 253
of scientists 26, 89, 100, 117–120, 155, 162, 168, 171, 187, 223, 253, 259–261, 267, 268, 285–290
ideological purges see *Zhdanovshchina*

Quantum electrodynamics, theory of radiation xvi, 66, 69, 86–88, 95, 96, 129
Quantum liquid 261–264, see also Superconductivity and Superfluidity
Quantum mechanics, quantum theory xiii, 34, 40–42, 48, 54, 68, 73, 76–79, 82–87, 96, 187, 217–225, 231, 234, 241, 243, 250, 263, 267
Quantum statistics (Bose–Einstein; Fermi–Dirac) 67, 68, 261, 263, 264, see also Individuality
Quasiparticles (Collective or Elementary Excitations) xiv, 50, 51, 69, 247, 262–264, 272–275, see also Electrons; Exciton; Hole; Magnon; Phonon; Plasmon; Polaron

Racism 236, 238
Radio 7, 44, 289
Radioactivity 44, 84, 127–130, 151, 155, 164, 169, 296, 297, see also Nuclear physics; Nuclear tests; Radium
Radium Commission 129, see also Institute-Radium
Raman-effect (Combinatorial scattering) 66

Rask-Ørsted Foundation 84
Red specialists 89, 285, 286, *see also* Bourgeois specialists; Cultural revolution
Relativity theory xiii, 41, 47, 48, 73, 91, 164, 176, 187, 217–224, 231, 234, 242, 243
Religion xii, xiv, 12, 17, 47, 152, 178, 179, 185, 200, 219, 278, 279, 281, 292, 304
Research imperative (research versus teaching) 2–5, 12–16, 20, 23–26, 34, 39, 44, 73, *see also* Research institutes; Scientific schools
Research institutes xiii, 2, 12–16, 19–25, 28–31, 35, 42–45, 89, 90, 156, 160–166, 227, 293, 298, 303, *see also* listing under Institutes
Resistance and opposition 26, 118, 119, 142–144, 159, 290, 295–300
Revolution xii, 12, 16, 21–23, 46, 47, 77, 279, 287, 292, 303
    of 1905: 18, 177, 178 (*see also Vekhi*)
    February 26, 35, 53, 64
    October (Bolshevik) 19, 26, 27, 28, 65, 301
    see also Cultural Revolution
Revolution and science xiii, 26, 31, 35, 40, 43–48, 76, 77, 87, 90–93, 129, 159, 161, 165, 172, 187, 221, 304, *see also* Soviet science
Revolution in science xiii, 76, 77, 85–87, 187, 219, 221, 222, *see also* Crisis; Cultural revolution
Rituals 112, 159, 169–173, 184, 190, 191, 195, 197–202, 214, 299, *see also* Games
Rockefeller philanthropy 74, 79–81, 83–85, *see also* IEB
Royal Society of London 104, 170
Rules, rule following 15, 159, 170, 180, 187, 191, 198, 200, 208, 210, *see also* Games
Russian Association of Physicists 39, 57, 82
Russian Physico-Chemical Society 29, 54, 101, 289
Russian Society of Roentgenologists and Radiologists 35

Science
    Applied 1–5, 18, 20, 21, 22, 28, 31, 32, 42–45, 93, 94, 122, 133, 145, 162, 168, 180, 207, 245, 305 (*see also* Industry, *Vnedrenie*)
    Fundamental 42, 94, 122, 162, 166, 168, 245, 278, 305
    Popularization 181
    Pure 1–3, 5, 9, 18–20, 22, 28, 32, 34, 37, 89, 180, 278, 300
    see also Bourgeois science; Military; Socialist science; Soviet science; and by disciplines; Value in society
Science Department (of the Central Committee) 115, 204, 208, 243
Science policy 24–26, 28–31, 44–46, 78, 84, 85, 88–98, 102, 109–111, 122, 130, 131, 133, 137, 156, 165, 166, 171, 186, 187, 210, 214, 277–280, 283–295, 298–300, 305, *see also* Atomic project-Soviet
Scientific cities 156, 157, 298
Scientific schools 15, 73, 112, 265, 303, *see also* Landau school
Scientism, *see* Cult of science
Scientists
    as elite 156, 186, 298–300
    party membership 158, 168, 169, 171, 220, 266, 280, 289
    political views 13–15, 24–27, 37, 48, 53, 64, 65, 91, 112, 116, 167, 175–180, 184, 185, 259, 260, 266, 267, 280, 284–295
Scientists and Politicians xii, xiv, 89, 93, 111–114, 142, 143, 150, 157, 161, 162, 168–171, 178, 182, 185, 186, 190, 191, 205–210, 214–216, 238, 242, 276, 277, 279, 280–291, 296, 297, *see also* Ideological discussions; Moral dilemmas; Power and knowledge
Secrecy 4, 97, 132, 133, 135, 136, 139, 140, 149–153, 156, 159, 183, 218
Secretariat (of the Central Committee) 194, 196, 202, 209, 213, 229, 230, 234, 237, 243

Subject Index

Security police (NKVD, MGB, KGB) 89, 136, 141, 149, 162, 171, 259, *see also* Gulag
Sharashka (research prison) 156, 253, 303
Slavery 48, 50, 51
Social Democrats 64, 68, *see also* Bolsheviks; Mensheviks; Tzotskyism
Socialism, socialists 37, 50, 56, 64, 68, 119, 177, 202, 247, 269, 273, 280, 295, 301–303
   and science 110, 111, 301
   see also Soviet science
Socialist realism 281–283, *see also* Literature
Socialist-revolutionaries (SR) 52, 53
Socialist science xiv, 23, 24, 31, 46, 48, 49, 184, 247, 273–275, 302, 303, *see also* Bourgeois science; Sociology of knowledge
Sociology of knowledge, Social construction 47–49, 223, 281, 300, 302, 303, *see also* Theory and practice; *Partiinost'*; Power and Knowledge
Solid state physics 48, 49, 51, 54–57, 59–64, 66–72, 86, 92, 96, 248–251, 255–261, 263, 264, 267, 268, *see also* Band theory; Condensed matter; Electrons
Soviet science xii, 1, 277, 278, 300
   Soviet system of research xiii, 2, 12, 21–23, 31, 32, 42–46, 127, 305
   specificities 181, 183
   successes and failures x–xiii, 41, 188, 189, 214, 247, 305
   see also Research institutes
Sovietology 99, 197, *see also* Anti-communism
Special Committee on Atomic Bomb 141, 143, 144, 148, 291, 293
Stalin Prize 99, 124, 150, 153, 164, 171, 305
Stalinism xiii, 22, 76, 95–96, 99, 112, 125, 144, 157–159, 165, 172, 173, 179–189, 197, 198, 214–216, 219, 244, 253, 259, 260, 279, 285, 287, 288, 294, 295
   Post-Stalin changes 156, 243, 295–298
   see also Dictatorship; Purges
State support for science xiii, 16, 22, 25, 28, 29, 41, 44, 45, 80, 89, 90, 110, 111, 115, 129, 131, 133, 141, 165, 166, 283, 284, 290, 293–295, 300, 304
Students, student culture 12, 39, 76, 78, 89, 90, 91, 94, 95, 285
   Graduate student (*aspirant*) 15, 75, 78, 92, 95, 163, 166
   Postdoctoral students and fellowships 79, 82–85
Superconductivity and Superfluidity 96, 116–118, 120, 260–263, 267–269, *see also* Liquid helium

Technology 3, 4, 6, 7, 10, 12, 24, 29, 38, 41–45, 104–105, 120–124, 144, 145, 147, 148, 151–154, 245, 284, *see also* Atomic project; Engineers; Industry; Optical glass; Oxygen; *Vnedrenie*
Theory and practice 182, 191, 203, 277, 279–281
Transport phenomena (Boltzmann and Vlasov kinetic equations) 252, 269–271
Trotskyism 76, 98
Tube Alloys, *see* Atomic project (British)

UFTI (Ukrainian Physico-Technical Institute) 67, 90, 92–94, 96–98, 105, 117, 120, 129–131
Ukrainian Academy of Sciences 31, 131
United Council of Learned Institutions 282
Universities
   in revolutionary Russia 23, 26, 30, 44, 81, 283
   in Russian Empire 1, 3, 12, 13, 15–17, 20, 29, 30
   Soviet 165, 226, 233
   see also Cultural revolution; Education; Students

University
- Baku 75
- Berkeley 130, 139, 265–267
- Cambridge 103–107, 109, 110, 115–117, 130, 169, 170
- Higher Technological School, Moscow 10
- Illinois 273
- Kharkov 97, 98
- Leiden 34, 73, 82, 116
- Lyceum 36
- Minnesota 69
- Moscow 2, 10, 12, 13, 15–17, 26, 64, 66, 73, 91, 150, 160, 161, 166, 175, 190, 194, 204, 208, 220, 226–230, 234, 236, 241–243, 270 (*see also* Kasso affair; NIIF)
- Princeton 267, 268
- St. Petersburg–Petrograd–Leningrad 4, 5, 33, 34, 39, 40, 53, 75, 78, 98, 168
- Tauride 31, 53, 65
- Tbilisi 238
- Yale 132

Uranium 127, 133, 136–138, 140, 146, 147, 148, 150
  chain reaction 131–134, 151, 153
Uranium Commission 133, 138
  isotopes and isotope separation 132, 134–136, 140, 144, 146, 151, 153, 266
  reactor 138, 140, 144, 147, 150, 151, 242, 295
  see also Fission; Plutonium
USA (contacts and comparisons) xii, 4, 7, 9, 14, 45, 46, 71, 72, 74, 77, 80–83, 131, 133, 174, 183, 221, 245, 249, 265–267, 273, 274, 292, 294, 296, 300, 304, 305, *see also* Manhattan project

Value of science in society 13, 74, 186, 214, 218, 283, 294, 295, 298, 303, 304
VASKhNIL, *see* Agriculture-Academy
Vekhi, Smena Vekh 178
Vlasov equation, *see* Transport phenomena
Vnedrenie (industrial application of research) 122, 123, 145
VSNKh (All-Union Soviet of People's Economy; economic commissariat) 27, 29, 37, 42, 78, 89, 92, 105

West 13, 15, 18, 187, 188, 277, 305
  exchanges with xiv, 4, 6, 13, 15, 17, 35, 45, 46, 77, 81, 83, 105, 111, 132, 136–141, 149, 155, 207, 223, 231, 235, 245, 246, 257, 258, 263, 264, 273–275, 300
  obsequiousness before the West 189
White Russians 30, 31, 53, 65
Wrecking 89, 162, 168, 177, 285
WWI (The Great War) xii, xiii, 1, 4–12, 16–24, 27, 29, 31, 32, 35, 44, 45, 73, 160, 289, 291, 293
WWII (The Great Patriotic War) x, xii, 46, 51, 96, 123–125, 127, 134–137, 145–149, 156, 170, 173, 174, 181, 186, 188, 227, 253
  Post-war changes 186, 187, 207, 291

X-rays 35, 36, 44, 92, 257

Zeitschrift für Physik 47, 77, 94
Zhdanovshchina (ideological campaigns in culture) 158, 183, 188, 192, 205, 213